T0213610

Although topology was recognized by Gauss and Maxwell to play a pivotal role in the formulation of electromagnetic boundary value problems, it is a largely unexploited tool for field computation. The development of algebraic topology since Maxwell provides a framework for linking data structures, algorithms, and computation to topological aspects of three-dimensional electromagnetic boundary value problems. This book attempts to expose the link between Maxwell and a modern approach to algorithms.

The first chapters lay out the relevant facts about homology and cohomology, stressing their interpretations in electromagnetism. These topological structures are subsequently tied to variational formulations in electromagnetics, the finite element method, algorithms, and certain aspects of numerical linear algebra. A recurring theme is the formulation of and algorithms for the problem of making branch cuts for computing magnetic scalar potentials and eddy currents. An appendix bridges the gap between the material presented and standard expositions of differential forms, Hodge decompositions, and tools for realizing representatives of homology classes as embedded manifolds.

Mathematical Sciences Research Institute
Publications

48

Electromagnetic Theory and Computation
A Topological Approach

Mathematical Sciences Research Institute Publications

1 Freed/Uhlenbeck: *Instantons and Four-Manifolds*, second edition
2 Chern (ed.): *Seminar on Nonlinear Partial Differential Equations*
3 Lepowsky/Mandelstam/Singer (eds.): *Vertex Operators in Mathematics and Physics*
4 Kac (ed.): *Infinite Dimensional Groups with Applications*
5 Blackadar: *K-Theory for Operator Algebras*, second edition
6 Moore (ed.): *Group Representations, Ergodic Theory, Operator Algebras, and Mathematical Physics*
7 Chorin/Majda (eds.): *Wave Motion: Theory, Modelling, and Computation*
8 Gersten (ed.): *Essays in Group Theory*
9 Moore/Schochet: *Global Analysis on Foliated Spaces*
10–11 Drasin/Earle/Gehring/Kra/Marden (eds.): *Holomorphic Functions and Moduli*
12–13 Ni/Peletier/Serrin (eds.): *Nonlinear Diffusion Equations and Their Equilibrium States*
14 Goodman/de la Harpe/Jones: *Coxeter Graphs and Towers of Algebras*
15 Hochster/Huneke/Sally (eds.): *Commutative Algebra*
16 Ihara/Ribet/Serre (eds.): *Galois Groups over* $\mathbb{Q}$
17 Concus/Finn/Hoffman (eds.): *Geometric Analysis and Computer Graphics*
18 Bryant/Chern/Gardner/Goldschmidt/Griffiths: *Exterior Differential Systems*
19 Alperin (ed.): *Arboreal Group Theory*
20 Dazord/Weinstein (eds.): *Symplectic Geometry, Groupoids, and Integrable Systems*
21 Moschovakis (ed.): *Logic from Computer Science*
22 Ratiu (ed.): *The Geometry of Hamiltonian Systems*
23 Baumslag/Miller (eds.): *Algorithms and Classification in Combinatorial Group Theory*
24 Montgomery/Small (eds.): *Noncommutative Rings*
25 Akbulut/King: *Topology of Real Algebraic Sets*
26 Judah/Just/Woodin (eds.): *Set Theory of the Continuum*
27 Carlsson/Cohen/Hsiang/Jones (eds.): *Algebraic Topology and Its Applications*
28 Clemens/Kollár (eds.): *Current Topics in Complex Algebraic Geometry*
29 Nowakowski (ed.): *Games of No Chance*
30 Grove/Petersen (eds.): *Comparison Geometry*
31 Levy (ed.): *Flavors of Geometry*
32 Cecil/Chern (eds.): *Tight and Taut Submanifolds*
33 Axler/McCarthy/Sarason (eds.): *Holomorphic Spaces*
34 Ball/Milman (eds.): *Convex Geometric Analysis*
35 Levy (ed.): *The Eightfold Way*
36 Gavosto/Krantz/McCallum (eds.): *Contemporary Issues in Mathematics Education*
37 Schneider/Siu (eds.): *Several Complex Variables*
38 Billera/Björner/Green/Simion/Stanley (eds.): *New Perspectives in Geometric Combinatorics*
39 Haskell/Pillay/Steinhorn (eds.): *Model Theory, Algebra, and Geometry*
40 Bleher/Its (eds.): *Random Matrix Models and Their Applications*
41 Schneps (ed.): *Galois Groups and Fundamental Groups*
42 Nowakowski (ed.): *More Games of No Chance*
43 Montgomery/Schneider (eds.): *New Directions in Hopf Algebras*
44 Buhler/Stevenhagen (eds.): *Algorithmic Number Theory*
45 Jensen/Ledet/Yui: *Generic Polynomials: Constructive Aspects of the Inverse Galois Problem*
46 Rockmore/Healy (eds.): *Modern Signal Processing*
47 Uhlmann (ed.): *Inside Out: Inverse Problems and Applications*
48 Gross/Kotiuga: *Electromagnetic Theory and Computation: A Topological Approach*

Volumes 1–4 and 6–27 are published by Springer-Verlag

Electromagnetic Theory and Computation: A Topological Approach

Paul W. Gross
MSRI and HP/Agilent

P. Robert Kotiuga
Boston University

CAMBRIDGE
UNIVERSITY PRESS

Paul Gross
pwgross@alum.bu.edu

P. Robert Kotiuga
Department of Electrical
and Computer Engineering
Boston University
8 Saint Mary's Street
Boston, MA 02215
United States
prk@bu.edu

Series Editor
Silvio Levy
Mathematical Sciences
Research Institute
17 Gauss Way
Berkeley, CA 94720
United States

MSRI Editorial Committee
Hugo Rossi (chair)
Alexandre Chorin
Silvio Levy
Jill Mesirov
Robert Osserman
Peter Sarnak

The Mathematical Sciences Research Institute wishes to acknowledge support by the National Science Foundation. This material is based upon work supported by NSF Grant 9810361.

CAMBRIDGE UNIVERSITY PRESS
Cambridge, New York, Melbourne, Madrid, Cape Town,
Singapore, São Paulo, Delhi, Tokyo, Mexico City

Cambridge University Press
The Edinburgh Building, Cambridge CB2 8RU, UK

Published in the United States of America by Cambridge University Press, New York

www.cambridge.org
Information on this title: www.cambridge.org/9780521175234

© Mathematical Sciences Research Institute 2004

This publication is in copyright. Subject to statutory exception
and to the provisions of relevant collective licensing agreements,
no reproduction of any part may take place without the written
permission of Cambridge University Press.

First published 2004
First paperback edition 2011

A catalogue record for this publication is available from the British Library

ISBN 978-0-521-80160-7 Hardback
ISBN 978-0-521-17523-4 Paperback

Cambridge University Press has no responsibility for the persistence or
accuracy of URLs for external or third-party internet websites referred to in
this publication, and does not guarantee that any content on such websites is,
or will remain, accurate or appropriate.

Table of Contents

Preface

The authors are long-time fans of MSRI programs and monographs, and are thrilled to be able to contribute to this series. Our relationship with MSRI started when Paul Gross was an MSRI/Hewlett-Packard postdoctoral fellow and had the good fortune of being encouraged by Silvio Levy to coauthor a monograph. Silvio was there when we needed him, and it is in no way an understatement to say that the project would never have been completed without his support.

The material of this monograph is easily traced back to our Ph.D. theses, papers we wrote, and courses taught at Boston University over the years. Our apologies to anyone who feels slighted by a minimally updated bibliography. Reflecting on how the material of this monograph evolved, we would like to thank colleagues who have played a supporting role over the decades. Among them are Alain Bossavit, Peter Caines, Roscoe Giles, Robert Hermann, Lauri Kettunen, Isaak Mayergoyz, Peter Silvester, and Gilbert Strang. The authors are also indebted to numerous people who read through all or part of the manuscript, produced numerous comments, and provided all sorts of support. In particular, Andre Nicolet, Jonathan Polimeni, and Saku Suuriniemi made an unusually thorough effort to review the draft.

Paul Gross would like to acknowledge Nick Tufillaro at Hewlett-Packard and Agilent Technologies for mentoring him throughout his post-doc at MSRI. Tim Dere graciously provided his time and expertise for illustrations. This book could not have happened without help and encouragement from Tanya.

Robert Kotiuga is grateful to the students taking his courses at Boston University, to Nevine, Michele, Madeleine, Peter and Helen for their support, and to Boston University for granting him a leave while the book was in its final stages.

Paul Gross and Robert Kotiuga
August 2003

We are here led to considerations belonging to the Geometry of Position, a subject which, though its importance was pointed out by Leibnitz and illustrated by Gauss, has been little studied.

James Clerk Maxwell, *A Treatise on Electricity and Magnetism*, 1891

Introduction

The title of this book makes clear that we are after connections between electromagnetics, computation and topology. However, connections between these three fields can mean different things to different people. For a modern engineer, computational electromagnetics is a well-defined term and topology seems to be a novel aspect. To this modern engineer, discretization methods for Maxwell's equations, finite element methods, numerical linear algebra and data structures are all part of the modern toolkit for effective design and topology seems to have taken a back seat. On the other hand, to an engineer from a half-century ago, the connection between electromagnetic theory and topology would be considered "obvious" by considering Kirchhoff's laws and circuit theory in the light of Maxwell's electromagnetic theory. To this older electrical engineer, topology would be considered part of the engineer's art with little connection to computation beyond what Maxwell and Kirchhoff would have regarded as computation. A mathematician could snicker at the two engineers and proclaim that all is trivial once one gets to the bottom of algebraic topology. Indeed the present book can be regarded as a logical consequence for computational electromagnetism of Eilenberg and Steenrod's *Foundations of Algebraic Topology* [ES52], Whitney's *Geometric Integration Theory* [Whi57] and some differential topology. Of course, this would not daunt the older engineer who accomplished his task before mathematicians and philosophers came in to lay the foundations.

The three points of view described above expose connections between pairs of each of the three fields, so it is natural to ask why it is important to put all three together in one book. The answer is stated quite simply in the context of the three characters mentioned above. In a modern "design automation" environment, it is necessary to take the art of the old engineer, reduce it to a science as much as possible, and then turn that into a numerical computation. For the purposes of computation, we need to feed a geometric model of a device such as a motor or circuit board, along with material properties, to a program which exploits algebraic topology in order to extract a simple circuit model from a horrifically complicated description in terms of partial differential equations and boundary value problems. Cohomology and Hodge theory on manifolds with boundary are the bridge between Maxwell's equations and the lumped parameters of circuit theory, but engineers need software that can reliably make this connection in an accurate manner. This book exploits developments in algebraic topology since the time of Maxwell to provide a framework for linking data structures, algorithms, and computation to topological aspects of 3-dimensional electromagnetic boundary value problems. More simply, we develop the link between Maxwell and a modern topological approach to algorithms for the analysis of electromagnetic devices.

To see why this is a natural evolution, we should review some facts from recent history. First, there is Moore's law, which is not a physical law but the observation that computer processing power has been doubling every eighteen months. In practical terms this means that in the year 2003 the video game played by five-year-old playing had the same floating-point capability as the largest supercomputer 15 years earlier. Although the current use of the term "computer" did not exist in the English language before 1950, Moore's law can be extrapolated back in time to vacuum tube computers, relay computers, and mechanical computing machines of the 1920's. Moving forward in time, the economics of building computers will bring this exponential increase to a halt before physics predicts the demise of Moore's law, but we are confident this trend will continue for at least another decade. Hence we should consider scientific computing and computational electromagnetics in this light.

The second set of facts we need to review concern the evolution of the tools used to solve elliptic boundary value problems. This story starts with Dirichlet's principle, asserting the existence of a minimizer for a quadratic functional whose Euler–Lagrange equation is Laplace's equation. Riemann used it effectively in his theory of analytic functions, but Weierstrass later put it into disrepute with his counterexamples. Hilbert rescued it with the concept of a minimizing sequence, and in the process modern functional analysis took a great step forward. From the point of view of finite element analysis, the story really starts with Courant, who in the 1920's suggested triangulating the underlying domain, using piecewise-polynomial trial functions for Ritz's method and producing a minimizing sequence by subdividing the triangulation. Courant had a constructive proof in mind, but three decades later his idea was the basis of the finite element method. Issues of adaptive mesh refinement can be interpreted as an attempt to produce a best approximation for a fixed number of degrees of freedom, as the number of degrees of freedom increase. In the electrical engineering of the 1960's, the finite element method started making an impact in the area of two-dimensional static problems that could be formulated in terms of a scalar potential or stream function. With the development of computer graphics in the 1970's, electrical engineers were beginning to turn their attention to the representation of vector fields, three-dimensional problems, and the adaptive generation of finite element meshes. At the same time, it is somewhat unfortunate that the essence of electromagnetic theory seen in Faraday and Maxwell's admirable qualitative spatial reasoning was lost under the vast amounts of numerical data generated by computer. In the 1980's came the realization that differential form methods could be translated to the discrete setting and that the hard work had already been done by André Weil and Hassler Whitney in the 1950's, but this point of view was a little slow to catch on. Technology transfer from mathematics to engineering eventually happened, since all of this mathematics from the 1950's was set in terms of simplicial complexes which fit hand in glove with the data structures of finite element analysis.

Before outlining the book in detail, there is one more observation to make about the process of automating the topological aspects that were once considered to be the engineer's art. Not only has the exponential increase in computing power given us the means to tackle larger and higher dimensional problems, but it has fundamentally changed the way we interact with computers. It took less than twenty years from "submitting a job" with a stack of punched cards at the university computing center to simulating an electromagnetic field in a personal virtual reality environment. With the continuing evolution of three-dimensional, real-time video games, we are assured of improved environments for having computers deal with the topological aspects of electromagnetic design. The task at hand is to identify the interactions between electromagnetics and algebraic topology that can have the greatest impact on formalizing the design engineer's intuition so that computers can be integrated more effectively into the design process.

Outline of Book. Chapter 1 develops homology and cohomology in the context of vector calculus, while suppressing the formalism of exterior algebra and differential forms. This enables practicing engineers to appreciate the relevance of the material with minimal effort. Although Gauss, Helmholtz, Kirchhoff and Maxwell recognized that topology plays a pivotal role in the formulation of electromagnetic boundary value problems, it is still a largely unexploited tool in problem formulation and computational methods for electromagnetic fields. Most historians agree that Poincaré and Betti wrote the seminal papers on what is now known as algebraic topology. However, it is also clear that they stood on the shoulders of Riemann and Listing. A glimpse into the first chapter of [Max91] shows that these same giants were under the feet of Maxwell. Correspondence between Maxwell and Tait reveals that Maxwell consciously avoided both Grassmann's exterior algebra and Hamilton's quaternions as a formalism for electromagnetism in order to avoid ideological debates. Credit is usually given to Oliver Heaviside for fitting Maxwell's equations into a notation accessible to engineers. Hence it is fair to say that the wonderful insights into three-dimensional topology found in Maxwell's treatise have never been exploited effectively by engineers. Thus our first chapter is a tunnel from some of the heuristic topological instincts of engineers to the commutative algebraic structures that can be extracted from the data structures found in electromagnetic field analysis software. A mathematician would make this all rigorous by appealing to the formalism of differentiable manifolds and differential forms. We leave the reader the luxury of seeing how this happens in a Mathematical Appendix.

Chapter 2 underlines the notion of a quasistatic electromagnetic field in the context Maxwell's equations. Quasistatics is an engineer's ticket to elliptic boundary value problems, variational principles leading to numerical algorithms, and the finite element method. We make certain physical assumptions in order to formulate the quasistatic problem, and the reader gets to see how circuit theory in the sense of Kirchhoff arises in the context of quasistatic boundary value problems. Besides promoting the boundary value problem point of view, the variational principles discussed in chapter 2 tie duality theorems for manifolds with boundary to the lumped parameters of circuit theory.

Having had a intuitive glimpse into the uses of duality theorems for manifolds with boundary in the first two chapters, Chapter 3 goes on to formalize some of the underlying ideas. After presenting the traditional Poincaré and Lefschetz duality theorems in the context of electromagnetics and circuit theory, we move to Alexander duality and present it in the context of linking numbers and magnetic scalar potentials. This approach is closest to Gauss' understanding of the matter and is completely natural in the context of magnetoquasistatics. Finally, for subsets of three-dimensional Euclidean space that have a continuous retraction into their interiors, we show that the absolute and relative (modulo boundary) homology and cohomology groups, as commutative groups, are torsion-free. This is significant for two reasons. First, it tells us why coming up with simple examples of torsion phenomena in three dimensions is a bit tricky, and second, it paves the way to using integer arithmetic in algorithms which would otherwise be susceptible to rounding error if implemented with floating point operations. With this result we are ready to return to the primary concerns of the engineer.

In Chapter 4 we finally arrive at the finite element method. It is introduced in the context of Laplace's equation and a simplicial mesh. The simplicial techniques used in topology are shown to translate into effective numerical algorithms that are naturally phrased in terms of the data structures encountered in finite element analysis. Although this opens the door to many relatively recent developments in computational electromagnetics, we focus on how the structures of homology and cohomology arise in the context of finite element algorithms for computing 3-dimensional electric and magnetic fields. In this way, the effectiveness of algebraic topology can be appreciated in a well-studied computational setting. Along the way we also get to see how the Euler characteristic is an effective tool in the analysis of algorithms.

One of the main strengths of the book comes to center stage in Chapter 5. This chapter addresses the problem of coupling magnetic scalar potentials in multiply-connected regions to stream functions which describe currents confined to conducting surfaces. This problem is considered in detail and the topological aspects are followed from the problem formulation stage through to the matrix equations arising from the finite element discretization. In practice this problem arises in non-destructive evaluation of aircraft wings, pipes, and other places. This problem is unique in that it is a three-dimensional magnetoquasistatic problem which admits a formulation solely in terms of scalar potentials, yet the topological aspects can be formulated in full generality while the overall formulation is sufficiently simple that it can be presented concisely. This chapter builds on all of the concepts developed in previous chapters, and is an ideal playground for illustrating how the tools of homological algebra (long exact sequences, duality theorems, etc.) are essential from problem formulation to interpretation of the resulting matrix equations.

Chapter 5 is self-contained except that one fundamental issue is acknowledged but sidestepped up to this point in the book. That issue is computation of cuts for magnetic scalar potentials. This is a deep issue since the simplest general definition of a cut is a realization, as an embedded orientable manifold with boundary, of an element of the second homology group of a region modulo its boundary. Poincaré and Maxwell took the existence of cuts for granted, and it was Pontryagin and Thom who, in different levels of generality, pointed out the need for an existence proof and gave a general framework for realizing homology classes as manifolds in the case that there is such a realization. It is ironic that historically, this question was avoided until the tools for its resolution were developed. For our purposes, an existence proof is given in the Mathematical Appendix, and the actual algorithm for computing a set of cuts realizing a basis for the second homology group is given in chapter six.

Chapter 6 bridges the gap between the existence of cuts and their realization as piecewise-linear manifolds which are sub-complexes of a finite element mesh (considered as a simplicial complex). Any algorithm to perform this task is useful only if some stringent complexity requirements are met. Typically, on a given mesh, a magnetic scalar potential requires about an order of magnitude less work to compute than computing the magnetic field directly. Hence if the computation of cuts is not comparable to the computation of a static solution of a scalar potential subject to linear constitutive laws, the use of scalar potentials in multiply-connected regions is not feasible for time-varying and/or nonlinear problems. We present an algorithm that involves the formulation and finite element solution of a Poisson-like equation, and additional algorithms that involve only integer arithmetic. We then have a favorable expression of the overall complexity in terms of a familiar finite element solution and the reordering and solution of a large sparse integer matrix equation arising for homology computation. This fills in the difficult gap left over from Chapter 5.

Chapter 7, the final chapter, steps back and considers the techniques of homological algebra in the context of the variational principles used in the finite element analysis of quasistatic electromagnetic fields. The message of this chapter is that the formalism of homology, and cohomology theory via differential form methods, are essential for revealing the conceptual elegance of variational methods in electromagnetism as well as providing a framework for software development. In order to get this across, a paradigm variational problem is formulated which includes as special cases all of the variational principles considered in earlier chapters. All the topological aspects considered in earlier chapters are then seen in the light of the homology and cohomology groups arising in the analysis of this paradigm problem. Because the paradigm problem is n-dimensional, this chapter no longer emphasizes the more visual and intuitive aspects, but exploits the formalism of differential forms in order to make connections to Hodge theory on manifolds with boundary, and variational methods for quasilinear elliptic partial differential equations. The engineer's topological intuition has now been obscured, but we gain a paradigm variational problem for which topological aspects which lead to circuit models are reduced by Whitney form discretization to computations involving well-understood algorithms.

The Mathematical Appendix serves several purposes. First, it contains results that make the book more mathematically self-contained. These results make the algebraic aspects accessible to the uninitiated, tie differential forms to cohomology, make clear what aspects of cohomology theory depend on the metric or constitutive law, and which do not. Second, certain results, such as the proof of the existence of cuts, are presented. This existence proof points to an algorithm for finding cuts, but involves tools from algebraic topology not found in introductory treatments. Having this material in an appendix makes the chapters of the book more independent.

Having stated the purpose of the book and outlined its contents, it is useful to list several problems not treated in this book. They represent future work which may be fruitful:

(1) Whitney forms and Whitney form discretizations of helicity functionals, their functional determinants, and applications to impedance tomography. There is already a nice exposition on Whitney forms accessible to engineers [Bos98].
(2) Lower central series of the fundamental group and, in three dimensions, the equivalent data given by Massey products in the cohomology ring. This algebraic structure contains more information than homology groups but, unlike the fundamental group, the computation of the lower central series can be done in polynomial time and gives insight into computational complexity of certain sparse matrix techniques associated with homology calculations.
(3) Additional constraints on cuts. Although we present a robust algorithm for computing cuts for magnetic scalar potentials, one may consider whether, topologically speaking, these cuts are the simplest possible. Engineers should not have to care about this, but the problem is very interesting as it relates to the computation of the Thurston norm on homology. Furthermore, if one introduces force constraints into the magnetoquasistatic problems considered in this book, the problem is related to the physics of "force-free magnetic fields" and has applications from practical magnet design to understanding the solar corona.
(4) Common historical roots between electromagnetism, computation and topology. Electromagnetic theory developed alongside topology in the works of Gauss, Weber, Möbius and Riemann. These pioneers also had a great influence on each other which is not well documented. In addition, Courant's paper, which lead to the finite element method, was written when triangulations of manifolds were the order of the day, and about the time when simplicial techniques in topology were undergoing rapid development in Göttingen.

We hope that the connections made in this book will inspire the reader to take this material beyond the stated purpose of developing the connection between algebraic structures in topology and methods for 3-dimensional electric and magnetic field computation.

Any problem which is nonlinear in character... or whose structure is initially defined in the large, is likely to require considerations of topology and group theory in order to arrive at its meaning and its solution.

Marston Morse, The Calculus of Variations in the Large, 1934

1

From Vector Calculus to Algebraic Topology

1A. Chains, Cochains and Integration

Homology theory reduces topological problems that arise in the use of the classical integral theorems of vector analysis to more easily resolved algebraic problems. Stokes' theorem on manifolds, which may be considered the fundamental theorem of multivariable calculus, is the generalization of these classical integral theorems. To appreciate how these topological problems arise, the process of integration must be reinterpreted algebraically.

Given an n-dimensional region Ω, we will consider the set $C_p(\Omega)$ of all possible p-dimensional objects over which a p-fold integration can be performed. Here it is understood that $0 \leq p \leq n$ and that a 0-fold integration is the sum of values of a function evaluated on a finite set of points. The elements of $C_p(\Omega)$, called *p-chains*, start out conceptually as p-dimensional surfaces, but in order to serve their intended function they must be more than that, for in evaluating integrals it is essential to associate an orientation to a chain. Likewise the idea of an orientation is essential for defining the oriented boundary of a chain (Figure 1.1).

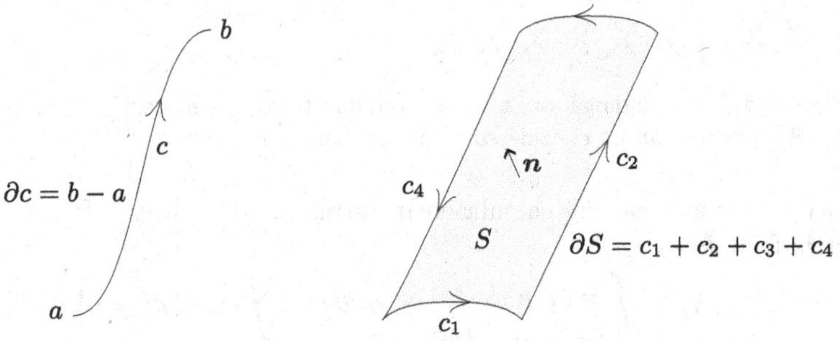

Figure 1.1. Left: a 1-chain. Right: a 2-chain.

At the very least, then, we wish to ensure that our set of chains $C_p(\omega)$ is closed under orientation reversal: for each $c \in C_p(\Omega)$ there is also $-c \in C_p(\Omega)$.

The set of integrands of p-fold integrals is called the set of *p-cochains* (or *p*-forms) and is denoted by $C^p(\Omega)$. For a chain $c \in C_p(\Omega)$ and a cochain $\omega \in C^p(\Omega)$, the integral of ω over c is denoted by $\int_c \omega$, and integration can be regarded as a mapping

$$\int : C_p(\Omega) \times C^p(\Omega) \to \mathbb{R}, \quad \text{for } 0 \leq p \leq n,$$

where $\mathbb{R}$ is the set of real numbers. Integration with respect to p-forms is a linear operation: given $a_1, a_2 \in \mathbb{R}$, $\omega_1, \omega_2 \in C^p(\Omega)$ and $c \in C_p(\Omega)$, we have

$$\int_c a_1\omega_1 + a_2\omega_2 = a_1 \int_c \omega_1 + a_2 \int_c \omega_2.$$

Thus $C^p(\Omega)$ may be regarded as a vector space, which we denote by $C^p(\Omega, \mathbb{R})$. Reversing the orientation of a chain means that integrals over that chain acquire the opposite sign:

$$\int_{-c} \omega = -\int_c \omega.$$

More generally, it is convenient to regard $C_p(\Omega)$ as having some algebraic structure — for example, an abelian group structure, as follows:

Example 1.1 Chains on a transformer. This example is inspired by electrical transformers, though understanding of a transformer is not essential for understanding the example. A current-carrying coil with n turns is wound around a toroidal piece of magnetic core material. The coil can be considered as a 1-chain, and it behaves in some ways as a multiple of another 1-chain c', a single loop going around the core once (see Figure 1.2). For instance, the voltage V_c

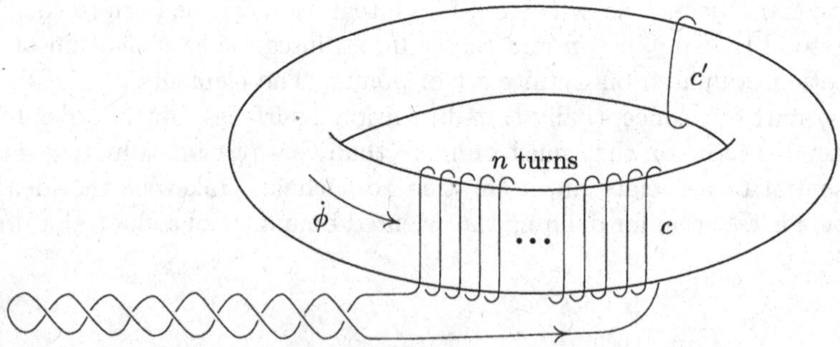

Figure 1.2. Windings on a solid toroidal transformer core. A 1-chain c in $C_1(\mathbb{R}^3 - \text{core})$ can be considered as a multiple of the 1-chain c'.

induced in loop c can be calculated in terms of the voltage of loop c' from the electric field $\boldsymbol{E}$ as

$$V_c = \int_c \boldsymbol{E} \cdot \boldsymbol{t}\, dl = \int_{nc'} \boldsymbol{E} \cdot \boldsymbol{t}\, dl = n \int_{c'} \boldsymbol{E} \cdot \boldsymbol{t}\, dl = nV_{c'},$$

where $\boldsymbol{t}$ is the unit vector tangential to c (or c'). □

For this reason it is convenient to regard as a 1-chain any integer multiple of a 1-chain, or even any linear combination of 1-chains. That is, we insist that our set of 1-chains be closed under chain addition (we had already made it closed under inversion or reversal). Moreover we insist that the properties of an abelian group (written additively) should be satisfied: for 1-chains c, c', c'', we have

$$c + (-c) = 0, \quad c + 0 = c, \quad c + c' = c' + c, \quad c + (c' + c'') = (c + c') + c''.$$

Given any n-dimensional region Ω, the set of "naive" p-chains $C_p(\Omega)$ can be extended to an abelian group by this process, the result being the set of all linear combinations of elements of $C_p(\Omega)$ with coefficients in $\mathbb{Z}$ (the integers). This group is denoted by $C_p(\Omega, \mathbb{Z})$ and called the *group of p-chains with coefficients in* $\mathbb{Z}$.

If linear combinations of p-chains with coefficients in the field $\mathbb{R}$ are used in the construction above, the set of p-chains can be regarded as a vector space. This vector space, denoted by $C_p(\Omega, \mathbb{R})$ and called the *p-chains with coefficients in* $\mathbb{R}$, will be used extensively. In this case, for $a_1, a_2 \in \mathbb{R}$, $c_1, c_2 \in C_p(\Omega, \mathbb{R})$, $\omega \in C^p(\Omega, \mathbb{R})$,

$$\int_{a_1c_1+a_2c_2} \omega = a_1 \int_{c_1} \omega + a_2 \int_{c_2} \omega.$$

In a similar fashion, taking a ring R and forming linear combinations of p-chains with coefficients in R, we have an R-module $C_p(\Omega, R)$, called the *p-chains with coefficients in R*. This construction has the previous two as special cases. It is possible to construct analogous groups for p-cochains, but we need not do so at the moment. Knowledge of rings and modules is not crucial at this point; rather the construction of $C_p(\Omega, R)$ is intended to illustrate how the notation is developed.

For coefficients in $\mathbb{R}$, the operation of integration can be regarded as a bilinear pairing between p-chains and p-forms. Furthermore, for reasonable p-chains and p-forms this bilinear pairing for integration is nondegenerate. That is,

$$\text{if } \int_c \omega = 0 \text{ for all } c \in C_p(\Omega), \text{ then } \omega = 0$$

and

$$\text{if } \int_c \omega = 0 \text{ for all } \omega \in C^p(\Omega), \text{ then } c = 0.$$

Although this statement requires a sophisticated discretization procedure and limiting argument for its justification [Whi57, dR73], it is plausible and simple to understand.

In conclusion, it is important to regard $C_p(\Omega)$ and $C^p(\Omega)$ as vector spaces and to consider integration as a bilinear pairing between them. In order to reinforce this point of view, the process of integration will be written using the linear space notation

$$\int_c \omega = [c, \omega];$$

that is, $C^p(\Omega)$ is to be considered the dual space of $C_p(\Omega)$.

1B. Integral Laws and Homology

Consider the fundamental theorem of calculus,

$$\int_c \frac{\partial f}{\partial x}\, dx = f(b) - f(a), \quad \text{where } c = [a, b] \in C_1(\mathbb{R}^1).$$

Its analogs for two-dimensional surfaces Ω are:

$$\int_c \operatorname{grad}\phi \cdot \boldsymbol{t}\, dl = \phi(p_2) - \phi(p_1) \quad \text{and} \quad \int_S \operatorname{curl}\boldsymbol{F} \cdot \boldsymbol{n}\, dS = \int_{\partial S} \boldsymbol{F} \cdot \boldsymbol{t}\, dl,$$

where $c \in C_1(\Omega)$, $\partial c = p_2 - p_1$, and $S \in C_2(\Omega)$. In three-dimensional vector analysis ($\Omega \subset \mathbb{R}^3$) we have

$$\int_c \operatorname{grad}\phi \cdot \boldsymbol{t}\, dl = \phi(p_2) - \phi(p_1), \qquad \int_S \operatorname{curl}\boldsymbol{F} \cdot \boldsymbol{n}\, dS = \int_{\partial S} \boldsymbol{F} \cdot \boldsymbol{t}\, dl,$$

$$\int_V \operatorname{div}\boldsymbol{F}\, dV = \int_{\partial V} \boldsymbol{F} \cdot \boldsymbol{n}\, dS,$$

where $c \in C_1(\Omega)$, $\partial c = p_2 - p_1$, $S \in C_2(\Omega)$, and $V \in C_3(\Omega)$. Note that here we are regarding p-chains as point sets but retaining information about their orientation.

These integral theorems, along with four-dimensional versions that arise in covariant formulations of electromagnetics, are special instances of the general result called Stokes' theorem on manifolds. This result, discussed at length in Section MA-H (page 232), takes the form

$$\int_c d\omega = \int_{\partial c} \omega,$$

where the linear operators for boundary (∂) and exterior derivative (d) are defined in terms of direct sums:

$$\partial : \textstyle\bigoplus_p C_p(\Omega) \to \bigoplus_p C_{p-1}(\Omega), \qquad d : \bigoplus_p C^{p-1}(\Omega) \to \bigoplus_p C^p(\Omega).$$

When p-forms are called p-cochains, d is called the coboundary operator. For an n-dimensional region Ω the following definition is made:

$$C^p(\Omega) = 0 \ \text{ for } p < 0, \qquad C_p(\Omega) = 0 \text{ for } p > n.$$

In this way, the boundary operator on p-chains has an intuitive meaning which carries over from vector analysis. On the other hand, the exterior derivative must be regarded as the operator which makes Stokes' theorem true. When a formal definition of the exterior derivative is given in a later chapter, it will be a simple computation to verify the special cases listed above.

For the time being, let the restriction of the boundary operator to p-chains be denoted by ∂_p and the restriction of the exterior derivative to p-forms be denoted by d^p. Thus

$$\partial_p : C_p(\Omega) \to C_{p-1}(\Omega) \quad \text{and} \quad d^p : C^p(\Omega) \to C^{p+1}(\Omega).$$

Considering various n-dimensional regions Ω and p-chains for various values of p, it is apparent that the boundary of a boundary is zero

$$(\partial_p \partial_{p+1})c = 0 \ \text{ for all } c \in C_{p+1}(\Omega).$$

An interesting question which arises regards the converse. If the boundary of a p-chain is zero, then is this chain the boundary of some chain in $C_{p+1}(\Omega)$? In general this is false, however more formalism is required in order to give a detailed answer to this question and to see its implications for vector analysis.

Rewriting the equation above as

$$\text{im}\,\partial_{p+1} \subset \ker \partial_p \tag{1–1}$$

the question above reduces to asking if the inclusion is an equality. In order to regain the geometric flavor of the question, define

$$B_p(\Omega) = \text{im}\,\partial_{p+1} \quad \text{and} \quad Z_p(\Omega) = \ker \partial_p$$

where elements of $B_p(\Omega)$ are called p-boundaries and elements of $Z_p(\Omega)$ are called p-cycles. The inclusion (1–1) can be rewritten as $B_p(\Omega) \subset Z_p(\Omega)$, and the question at hand is an inquiry into the size of the quotient group

$$H_p(\Omega) = Z_p(\Omega)/B_p(\Omega),$$

is called the pth (absolute) homology group of Ω. This construction can be made with any coefficient group, and in the present case $Z_p(\Omega)$ and $B_p(\Omega)$ are vector spaces and $H_p(\Omega)$ is a quotient space.

The following equivalence relation can be used to refer to the cosets of $H_p(\Omega)$. Given $z_1, z_2 \in Z_p(\Omega)$, we write $z_1 \sim z_2$ and say that z_1 is homologous to z_2 if $z_1 - z_2 = b$ for some $b \in B_p(\Omega)$. Hence z_1 is homologous to z_2 if z_1 and z_2 lie in the same coset of $H_p(\Omega)$. In the present case $H_p(\Omega)$ is a vector space and the dimension of the pth homology "group" is called the pth Betti number,

$$\beta_p(\Omega) = \dim\left(H_p(\Omega)\right).$$

The following examples are intended to give a geometric sense for the meaning of the cosets of $H_p(\Omega)$.

Example 1.2 Concentric spheres: $\Omega \subset \mathbb{R}^3$, $\beta_2 \neq 0$. Consider three concentric spheres and let Ω be the three-dimensional spherical shell whose boundary is formed by the innermost and outermost spheres. Next, let $z \in Z_2(\Omega)$ be the sphere between the innermost and outermost spheres, oriented by the unit outward normal. Since z is a closed surface, $\partial_2 z = 0$ however z is not the boundary of any three-dimensional chain in Ω, that is $z \neq \partial_3 c$ for any $c \in C_3(\Omega)$. Hence z represents a nonzero coset in $H_2(\Omega)$. In this case $\beta_1(\Omega) = 1$ and $H_2(\Omega)$ is generated by cosets of the form $az + B_2(\Omega)$, where $a \in \mathbb{R}$. □

Example 1.3 Curves on a knotted tube: $\Omega \subset \mathbb{R}^3$, $\beta_1 \neq 0$. Suppose $\Omega \in \mathbb{R}^3$ is the region occupied by the knotted solid tube in Figure 1.3. Let $z \in Z_1(\Omega)$ be a closed curve on the surface of the knot while $z' \in Z_1(\mathbb{R}^3 - \Omega)$ is a closed curve which links the tube. In the figure, $z \notin B_1(\Omega)$ and $\beta_1(\Omega) = 1$. The cosets of $H_1(\Omega)$ can be expressed as $az + B_1(\Omega)$ where $a \in \mathbb{R}$. Dually, $z' \in B_1(\mathbb{R}^3 - \Omega)$ and $\beta_1(\mathbb{R}^3 - \Omega) = 1$, hence the cosets of $H_1(\mathbb{R}^3 - \Omega)$ are $a'z' + B_1(\mathbb{R}^3 - \Omega)$ where $a' \in \mathbb{R}$. □

Example 1.4 3-d solid with internal cavities: $\Omega \subset \mathbb{R}^3$, $H_2(\Omega)$, $H_0(\mathbb{R}^3 - \Omega)$ of interest. Suppose Ω is a compact connected subset of $\mathbb{R}^3$. In this case we

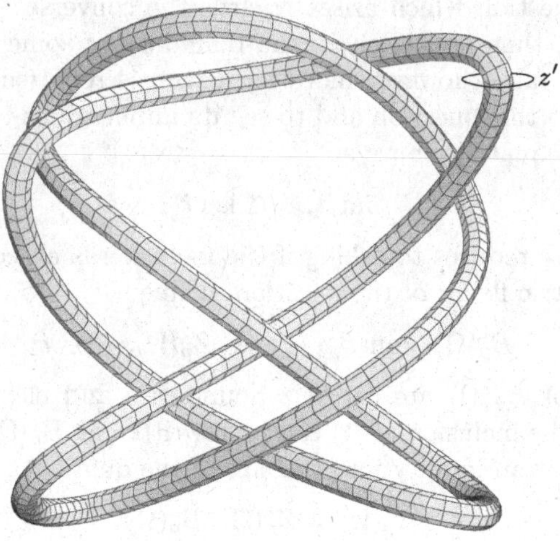

Figure 1.3. A (5,2) torus knot, illustration for Example 1.3.

take a compact set to mean a closed and bounded set. By an abuse of language, we assume $\Omega \in C_3(\Omega)$ where, when considered as a chain, $\partial\Omega$ has the usual orientation so that Ω is considered as both a chain and set. The boundary $\partial_3\Omega = S_0 \cup S_1 \cup S_2 \cup \cdots \cup S_n$ where $S_i \in Z_2(\Omega)$, for $0 \leq i \leq n$, are the connected components of $\partial_3\Omega$ (think of Ω as a piece of Swiss cheese). Furthermore, let S_0 be the connected component of $\partial_3\Omega$ which, when taken with the opposite orientation, becomes the boundary of the unbounded component of $\mathbb{R}^3 - \Omega$. Given that Ω is connected, it is possible to find $n+1$ components Ω'_i of $\mathbb{R}^3 - \Omega$ such that

$$\partial_3\Omega'_i = -S_i \quad \text{for } 0 \leq i \leq n.$$

It is obvious that surfaces S_i cannot possibly represent independent generators of $H_2(\Omega)$ since their sum (as chains) is homologous to zero, that is,

$$\sum_{i=0}^{n} S_i = \partial_3\Omega \quad \text{or} \quad \sum_{i=0}^{n} S_i \sim 0.$$

However, $H_2(\Omega)$ is generated by cosets of the form

$$\sum_{i=1}^{n} a_i S_i + B_2(\Omega).$$

This can be rigorously shown through duality theorems for manifolds which are the topic of Chapter 3, but a heuristic justification of the statement is the following. Choose 0-cycles p_i (points), $0 \leq i \leq n$, such that $p_i \in Z_0(\Omega'_i)$ and define 1-chains (curves) $c_i \in C_1(\mathbb{R}^3)$, $1 \leq i \leq n$, by the following:

$$\partial c_i = p_i - p_0.$$

The points p_i are $n+1$ generators of $H_0(\mathbb{R}^3 - \Omega)$ while the c_i connect the components of $\mathbb{R}^3 - \Omega$. It is apparent that for $1 \leq i, j \leq n$, the curves c_i can be arranged to intersect S_i once and not intersect S_j if $i \neq j$. If the curves c_i are regarded as point sets,

$$\beta_2\left(\Omega - \left(\bigcup_{i=1}^{n} c_i\right)\right) = 0$$

and

$$\beta_0\left(\left(\overline{\mathbb{R}^3 - \Omega}\right) \cup \left(\bigcup_{i=1}^{n} c_i\right)\right) = 1,$$

where, in the latter case, multiples of the 0-cycle p_0 can be taken to generate the zeroth homology group. This property cannot be achieved by taking fewer than n such c_i. That is, for every curve c_i which goes through Ω there corresponds one and only one generator of $H_2(\Omega)$. In summary

$$\beta_2(\Omega) = n = \beta_0(\overline{\mathbb{R}^3 - \Omega}) - 1$$

where the n independent cosets of the form

$$\sum_{i=1}^{n} a_i S_i + B_2(\Omega) \quad \text{for} \quad a_i \in \mathbb{R}$$

and

$$\sum_{i=0}^{n} a_i' P_i + B_0(\overline{\mathbb{R}^3 - \Omega}) \quad \text{for} \quad a_i' \in \mathbb{R}$$

can be used to generate $H_2(\Omega)$ and $H_0(\overline{\mathbb{R}^3 - \Omega})$, respectively. These arguments are essentially those of Maxwell [Max91, Art. 22]. In Maxwell's terminology the periphractic number of a region Ω is $\beta_2(\Omega)$. The general case where Ω consists of a number of connected components is handled by applying the same argument to each connected component of Ω and choosing the same p_0 for every component. In this case it will also be true that

$$\beta_0(\overline{\mathbb{R}^3 - \Omega}) = \beta_2(\Omega) + 1.$$

□

Example 1.5 Curves on an orientable surface: Ω an orientable surface, $H_1(\Omega)$ of interest. It is a fact that any bounded orientable 2-dimensional surface is homeomorphic to a disc with n handles and k holes. For some integers n and k, any orientable 2-dimensional surface with boundary can be mapped in a 1-1 continuous fashion into some surface like the one shown in Figure 1.4 (see [Mas67, Chapter 1] or [Cai61, Chapter 2] for more pictures and explanations).

Let Ω be the surface described above and let $\beta_1 = 2n + k - 1$. Consider 1-cycles $z_i \in Z_1(\Omega)$ for $1 \leq i \leq \beta_1$ where z_{2j-1} and z_{2j} $(1 \leq j \leq n)$ is a pair of cycles which correspond to the jth handle, while z_{2n+j}, where $1 \leq j \leq k-1$, corresponds to the jth hole as shown in Figure 1.5. The kth hole is ignored as far as the z_i are concerned. It is clear that for $1 \leq i \leq \beta_1$, the z_i are nonbounding cycles. What is less obvious is that $H_1(\Omega)$ can be generated by β_1 linearly independent cosets of the form

$$\sum_{i=1}^{\beta_1} a_i z_i + B_1(\Omega) \quad \text{for} \quad a_i \in \mathbb{R}.$$

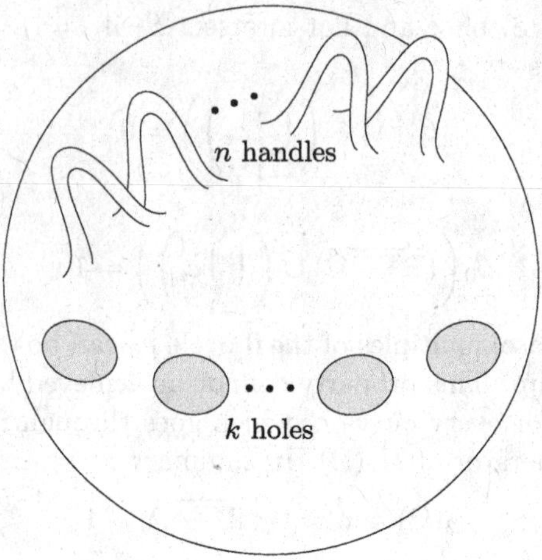

Figure 1.4. Disc with handles and holes.

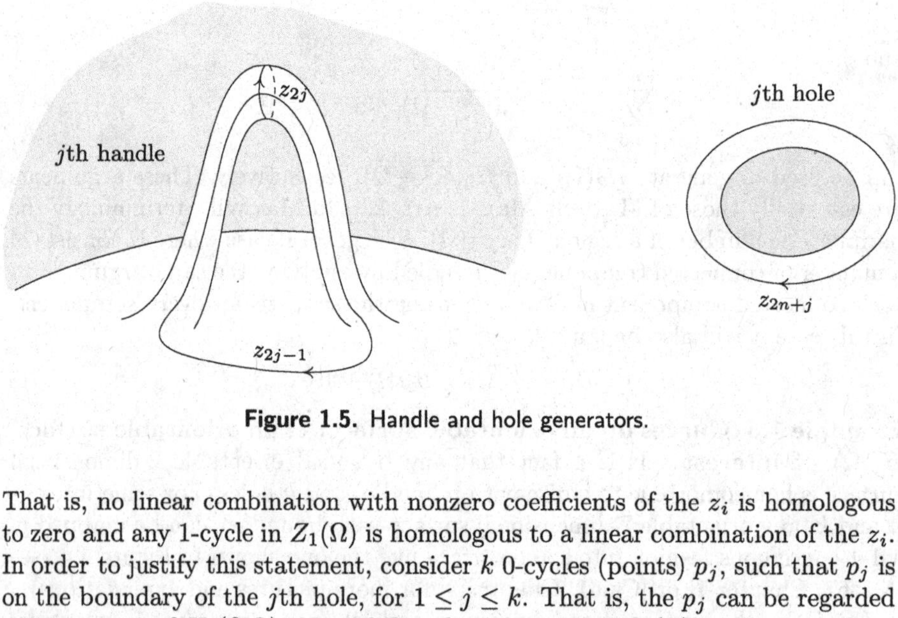

Figure 1.5. Handle and hole generators.

That is, no linear combination with nonzero coefficients of the z_i is homologous to zero and any 1-cycle in $Z_1(\Omega)$ is homologous to a linear combination of the z_i. In order to justify this statement, consider k 0-cycles (points) p_j, such that p_j is on the boundary of the jth hole, for $1 \le j \le k$. That is, the p_j can be regarded as generators of $H_0(\partial_1 \Omega)$. Next define β_1 1-chains $c_i \in C_1(\Omega)$ such that

$$z_{2j} = c_{2j-1} \text{ and } z_{2j-1} = c_{2j} \text{ for } 1 \le j \le n,$$

and

$$\partial c_{2n+j} = p_j - p_k \text{ for } 1 \le j \le k-1.$$

Note c_i intersects z_i once for $1 \le i \le \beta_1$ and does not intersect z_l if $i \neq l$.

If the surface is cut along the c_i, it would become simply connected while remaining connected. Furthermore it is not possible to make Ω simply connected with fewer than β_1 cuts. Hence, regarding the c_i as sets, one can write

$$H_1\left(\Omega - \left(\bigcup_{i=1}^{\beta_1} c_i\right)\right) = 0$$

and the c_i can be said to act like branch cuts in complex analysis. Removing the cuts along c_i successively introduces a new generator for $H_1(\Omega)$ at each step, so that

$$\beta_1(\Omega) = 2n + k - 1$$

and the z_i, are indeed generators of $H_1(\Omega)$.

Throughout this construction the reader may have wondered about the special status of the kth hole. It should be clear that

$$0 \sim \partial_2 \Omega \sim \sum_{i=1}^{k-1} z_{2n+i} + \partial_2(k\text{th hole})$$

hence associating z_{2n+k} with the kth hole as z_{2n+j} is associated with the jth hole does not introduce an independent new generator to $H_1(\Omega)$. Finally, if Ω is not connected, then the above considerations can be applied to each connected component of Ω. □

In examples 1.2, 1.3, 1.4, and 1.5 the ranks of $H_p(\Omega)$ were $1, 1, n$, and $2n+k-1$, respectively. In order to prove this fact, it is necessary to have a way of computing homology, but from the definition $H_p(\Omega) = Z_p(\Omega)/B_p(\Omega)$ involving the quotient of two infinite groups (vector spaces) it is not apparent that the homology groups should even have finite rank. In general, compact manifolds have homology groups of finite rank, but it is not worthwhile to pursue this point since no method of computing homology has been introduced yet. Instead the relation between homology and vector analysis will now be explored in order to show the importance of homology theory in the context of electromagnetics.

1C. Cohomology and Vector Analysis

To relate homology groups to vector analysis, consider Stokes' theorem

$$\int_c d\omega = \int_{\partial c} \omega$$

rewritten for the case of p-chains on Ω:

$$[c, d^{p-1}\omega] = [\partial_p c, \omega].$$

Stokes' theorem shows that d^{p-1} and ∂_p act as adjoint operators. Since $\partial_p \partial_{p+1} = 0$, we have

$$[c, d^p d^{p-1}\omega] = [\partial_{p+1} c, d^{p-1}\omega] = [\partial_p \partial_{p+1} c, \omega] = 0$$

for all $c \in C_p(\Omega)$ and $\omega \in C^p(\Omega)$. This results in the operator equation

$$d^p d^{p-1} = 0 \quad \text{for all } p,$$

when integration is assumed to be a nondegenerate bilinear pairing. Hence, surveying the classical versions of Stokes' theorem, we immediately see that the vector identities

$$\text{div curl} = 0$$

and

$$\text{curl grad} = 0$$

follow as special cases.

As in the case of the boundary operator, the identity $d^p d^{p-1} = 0$ does not imply that $\omega = d^{p-1}\eta$ for some $\eta \in C^{p-1}(\Omega)$ whenever $d^p\omega = 0$ and it is useful to define subgroups of $C^p(\Omega)$ as follows. The group of *p-cocycles* (or closed forms) on Ω is denoted by

$$Z^p(\Omega) = \ker d^p,$$

and the group of *p-coboundaries* (or exact forms) on Ω is denoted by

$$B^p(\Omega) = \operatorname{im} d^{p-1}.$$

The equation

$$d^p d^{p-1} = 0$$

can thus be rewritten as

$$B^p(\Omega) \subset Z^p(\Omega)$$

and, in analogy to the case of homology groups, we can define

$$H^p(\Omega) = Z^p(\Omega)/B^p(\Omega),$$

the *pth cohomology group* of Ω. This is a measure of the extent by which the inclusion misses being an equality. The groups $B^p(\Omega)$, $Z^p(\Omega)$ are vector spaces, while $H^p(\Omega)$ is a quotient space in the present case since the coefficient group is $\mathbb{R}$. Cosets of $H^p(\Omega)$ come about from the following equivalence relation. Given $z^1, z^2 \in Z^p(\Omega)$, $z^1 \sim z^2$ (read z^1 is cohomologous to z^2) if $z^1 - z^2 \in b$ for some $b \in B^p(\Omega)$. That is, z^1 is cohomologous to z^2 if z^1 and z^2 lie in the same coset of $H^p(\Omega)$.

The topological problems of vector analysis can now be reformulated. Let Ω be a uniformly n-dimensional region which is a bounded subset of $\mathbb{R}^3$. Technically speaking, Ω is a compact 3-dimensional manifold with boundary (see Section MA-H for the meaning of this term). Consider the following questions:

(1) Given a vector field $\boldsymbol{D}$ such that $\operatorname{div} \boldsymbol{D} = 0$ on Ω, is it possible to find a continuous vector field $\boldsymbol{C}$ such that $\boldsymbol{D} = \operatorname{curl} \boldsymbol{C}$?
(2) Given a vector field $\boldsymbol{H}$ such that $\operatorname{curl} \boldsymbol{H} = 0$ in Ω, is it possible to find a continuous single-valued function ψ such that $\boldsymbol{H} = \operatorname{grad} \psi$?
(3) Given a scalar function ϕ such that $\operatorname{grad} \phi = 0$ in Ω, is $\phi = 0$ in Ω?

These questions have a common form: "Given $\omega \in Z^p(\Omega)$, is $\omega \in B^p(\Omega)$?" where p takes the values $2, 1, 0$, respectively. Equivalently, we can ask: Given $\omega \in Z^p(\Omega)$, is ω cohomologous to zero?

Given an n-dimensional Ω, suppose that, for all p, $C^p(\Omega)$ and $C_p(\Omega)$ are both finite-dimensional. In this case, the fact that ∂_p and d^{p-1} are adjoint operators gives an instant solution to the above questions since, the identity

$$\text{Nullifier}(\operatorname{im} d^{p-1}) = \ker \partial_p,$$

that is,

$$\text{Nullifier}\,(B^p(\Omega)) = Z_p(\Omega)$$

can be rewritten as the following compatibility condition:

$$\omega \in B^p(\Omega) \text{ if and only if } \int_z \omega = 0 \text{ for all } z \in Z_p(\Omega). \tag{1–2}$$

Next, suppose $\omega \in Z^p(\Omega)$ and consider the integral of ω over the coset

$$z + B_p(\Omega) \in H_p(\Omega).$$

Let $c' \in C_{p+1}(\Omega)$ and $b = \partial_{p+1}c'$ an arbitrary element of $B_p(\Omega)$. This gives

$$\begin{aligned}\int_{z+b} \omega &= \int_z \omega + \int_{\partial_{p+1}c'} \omega \quad \text{by linearity}\\ &= \int_z \omega + \int_{c'} d^p\omega \quad \text{by Stokes' theorem}\\ &= \int_z \omega \quad \text{since } \omega \in Z^p(\Omega).\end{aligned}$$

Hence, when $\omega \in Z^p(\Omega)$, the compatibility condition (1–2) depends only on the coset of z in $H_p(\Omega)$. Thus condition (1–2) can be rewritten as

$$\omega \in B^p(\Omega) \iff \omega \in Z^p(\Omega) \text{ and } \int_{z_i} \omega = 0 \text{ for } 1 \le i \le \beta_p(\Omega),$$

where $H_p(\Omega)$ is generated by cosets of the form

$$\sum_{i=1}^{\beta_p(\Omega)} a_i z_i + B_p(\Omega).$$

It turns out that when $C^p(\Omega), C_p(\Omega)$ are finite-dimensional the result of this investigation is true under very general conditions. The result of de Rham which is stated in the next section amounts to saying that

$$H^p(\Omega) \simeq H_p(\Omega),$$

where the isomorphism is obtained through integration. Moreover, $\beta_p(\Omega) = \dim H_p(\Omega)$, and $\beta^p(\Omega) = \dim H^p(\Omega)$ are finite and $\beta^p(\Omega) = \beta_p(\Omega)$. Hence, for an n-dimensional region Ω, given $\omega \in Z^p(\Omega)$, then $z \in B^p(\Omega)$ provided that

$$\int_z \omega = 0$$

over $\beta_p(\Omega)$ independent p-cycles whose cosets in $H_p(\Omega)$ are capable of generating $H_p(\Omega)$. To the uninitiated, this point of view may seem unintuitive and excessively algebraic. For this reason the original statement of de Rham's Theorem and several examples illustrating de Rham's theorem will be considered next.

1D. Nineteenth-Century Problems Illustrating the First and Second Homology Groups

In order to state the theorems of de Rham in their original form the notion of a period is required. Consider a n-dimensional region Ω. The *period* of $\omega \in Z^p(\Omega)$ on $z \in Z_p(\Omega)$ is defined to be the value of the integral

$$\int_z \omega.$$

By Stokes' theorem, the period of ω on z depends only on the coset of z in $H_p(\Omega)$ and the coset of ω in $H^p(\Omega)$. That is,

$$\begin{aligned}\int_{z+\partial_{p+1}c'} \omega + d^{p-1}\omega' &= \int_z \omega + \int_{\partial_{p+1}c'} (\omega + d^{p-1}\omega') + \int_z d^{p-1}\omega' \\ &= \int_z \omega + \int_{c'} d^p\omega + \int_{\partial_p z} \omega' \text{ by Stokes' theorem} \\ &= \int_z \omega \text{ since } \omega \in Z^p(\Omega),\ z \in Z_p(\Omega).\end{aligned}$$

Postponing technicalities pertaining to differentiable manifolds, de Rham's original two theorems can be stated as follows. Let $\tilde{z}_i$, $1 \leq i \leq \beta_p(\Omega)$ be homology classes (cosets in $H_p(\Omega)$) which generate $H_p(\Omega)$. Then:

(1) A closed form whose periods on the $\tilde{z}_i$ vanish is an exact form. That is, $\omega \in B^p(\Omega)$ if $\omega \in Z^p(\Omega)$ and

$$\int_{\tilde{z}_i} \omega = 0, \quad \text{for } 1 \leq i \leq \beta_p(\Omega).$$

(2) Given real a_i, $1 \leq i \leq \beta_p(\Omega)$, there exists a closed form ω such that the period of ω on $\tilde{z}_i$ is a_i, $1 \leq i \leq \beta_p(\Omega)$. That is, given a_i, $1 \leq i \leq \beta_p(\Omega)$, there exists a $\omega \in Z^p(\Omega)$ such that

$$\int_{\tilde{z}_i} \omega = a_i \text{ for } 1 \leq i \leq \beta_p(\Omega).$$

The two theorems above are an explicit way of saying that $H^p(\Omega)$ and $H_p(\Omega)$ are isomorphic. The following examples will illustrate how the de Rham isomorphism between homology and cohomology groups occurs in vector analysis and, whenever possible, the approach will mimic the nineteenth century reasoning.

Example 1.6 Cohomology: $\Omega \subset \mathbb{R}^3$, $H^2(\Omega)$ is of interest. Let Ω be a three-dimensional subset of $\mathbb{R}^3$ and consider a continuous vector field $\boldsymbol{D}$ such that $\operatorname{div} \boldsymbol{D} = 0$ in Ω. When is it possible to find a vector field $\boldsymbol{C}$ such that $\boldsymbol{D} = \operatorname{curl} \boldsymbol{C}$? We consider three cases.

(1) If Ω has no cavities, that is if $\mathbb{R}^3 - \Omega$ is connected, then no such a vector field exists since if $H_2(\Omega) = 0$ then $H^2(\Omega) = 0$.

(2) In order to illustrate that there may be no vector field $\boldsymbol{C}$ if $H_2(\Omega) \neq 0$ consider the following situation. Let S be a unit sphere centered at the origin of $\mathbb{R}^3$, S' a sphere of radius 3 concentric with S, and Ω the spherical shell with S and S' as its boundary, $\partial\Omega = S' - S$. It is clear that a sphere

z of radius 2, centered about the origin and oriented by its unit outward normal is not homologous to zero and that $\beta_2(\Omega) = 1$. Let $\operatorname{div} \boldsymbol{D} = 0$ in Ω, and

$$Q = \int_z \boldsymbol{D} \cdot \boldsymbol{n}\, dS$$

specifies the period of the field $\boldsymbol{D}$ over the nontrivial homology class. In terms of electromagnetism, one can think of S as supporting a nonzero electric charge Q, S' as a perfectly conducting sphere, and regard $\boldsymbol{D}$ as the electric flux density

(3) More generally, the "intuitive" condition for ensuring that such a vector field $\boldsymbol{C}$ exists if $\operatorname{div} \boldsymbol{D} = 0$ can be given as follows [Ste54, Max91]. As mentioned in Example 1.4, Maxwell uses the term periphractic region referring to $H_2(\Omega) \neq 0$ and periphractic number for $\beta_2(\Omega)$. [Max91, Article 22]. Consider again the region Ω of Example 1.4 where the boundary of Ω had $n+1$ connected components S_i, for $0 \leq i \leq n$, S_0 being the boundary of the unbounded component of $\mathbb{R}^3 - \Omega$. In this case $H_2(\Omega)$ is generated by linear combinations of the closed surfaces S_i, and the conditions for ensuring that $\boldsymbol{D} = \operatorname{curl} \boldsymbol{C}$ in Ω if $\operatorname{div} \boldsymbol{D} = 0$ in Ω are

$$\int_{S_i} \boldsymbol{D} \cdot \boldsymbol{n}\, dS = 0 \quad \text{for } 1 \leq i \leq n = \beta_2(\Omega).$$

This is also the answer to be expected by de Rham's Theorem. The integral condition above is satisfied identically on S_0 in this case, since

$$0 = \int_\Omega \operatorname{div} \boldsymbol{D}\, dV = \int_{\partial\Omega} \boldsymbol{D} \cdot \boldsymbol{n}\, dS = \sum_{i=0}^{n} \int_{S_i} \boldsymbol{D} \cdot \boldsymbol{n}\, dS = \int_{S_0} \boldsymbol{D} \cdot \boldsymbol{n}\, dS;$$

this reaffirms that

$$\sum_{i=0}^{n} S_i \sim 0.$$

The case where Ω is not connected is easily handled by applying the above considerations to each connected component of Ω. □

Example 1.7 Cohomology: $\Omega \subset \mathbb{R}^3, H^0(\Omega)$ is of interest. Let Ω be a three-dimensional subset of $\mathbb{R}^3$ and consider a function ϕ such that $\operatorname{grad} \phi = 0$ in Ω. When is it possible to say that $\phi \in B^0(\Omega)$, that is $\phi = 0$? If Ω is connected, then ϕ is determined to within a constant since $\beta_0(\Omega) = 1$ if and only if $\beta^0(\Omega) = 1$.

This problem is the usual one in electrostatics and in this context it is possible to see that ϕ is not necessarily a constant if $\beta_0(\Omega) > 1$. The physical situation is the following. Suppose that there are n connected parts Ω_i' in $\mathbb{R}^3 - \Omega$, each a conducting body supporting an electrical charge Q_i, for $1 \leq i \leq n$. The n charged bodies are inside a conducting shell Ω_0' which supports a charge Q_0. Let

$$\Omega' = \bigcup_{i=0}^{n} \Omega_i'.$$

The electric field vector $\boldsymbol{E} = -\operatorname{grad}\phi$ vanishes inside each conducting body. However, depending on the charges Q_i and hence on the charge

$$-\sum_{i=0}^{n} Q_i$$

somewhere exterior to the problem, it is well-known that constants

$$\phi\big|_{\Omega'_i} = \phi_i, \quad \text{for } 1 \le i \le n,$$

can be assigned arbitrarily. In general, the scalar potential ϕ vanishes only if the constants all vanish; hence

$$\beta_0(\Omega') = n+1.$$

This example can be used to illustrate an additional point. In electrostatics it is customary to let

$$\phi\big|_{\Omega'_0} = 0 \ \text{(datum)}$$

and

$$Q_0 = \sum_{i=1}^{n} Q_i \ \text{(conservation of charge)}.$$

Let $\Omega' = \mathbb{R}^3 - \Omega$ where Ω'_i, for $0 \le i \le n$, are the connected parts of Ω' while Ω'_0 is the unbounded part of Ω'. Using the final equation of Example 1.4, it is clear that

$$n = \beta_2(\Omega) = \beta_0(\mathbb{R}^3 - \Omega) - 1 = \beta_0(\Omega') - 1.$$

Interpreting $\beta_2(\Omega)$ as the number of independent charges in the problem and $\beta_0(\Omega') - 1$ as the number of independent potential differences, this equation says that the number of independent charges equals the number of independent potential differences . □

Example 1.8 Cohomology: Ω a 2-dimensional surface, $H^1(\Omega)$ is of interest. Let Ω be a two-dimensional orientable surface and consider the conjugate versions of the usual integral theorems:

$$\int_c \overline{\operatorname{curl}}\,\chi \cdot \boldsymbol{n}\,dl = \chi(p_2) - \chi(p_1) \quad \text{for } c \in C_1(\Omega) \text{ and } \partial c = p_2 - p_1$$

and

$$\int_c \operatorname{div} \boldsymbol{K}\,dS = \int_{\partial c} \boldsymbol{K} \cdot \boldsymbol{n}\,dl \quad \text{for } c \in C_2(\Omega),$$

where $\boldsymbol{n}$ is the unit vector normal to the curve c. The operation $\overline{\operatorname{curl}}\,\chi$ is defined as $\boldsymbol{n}' \times \operatorname{grad}\chi$, where $\boldsymbol{n}'$ is the unit normal vector to the two-dimensional orientable surface [Ned78, p. 582]. In this case, the operator identity

$$\operatorname{div} \overline{\operatorname{curl}} = 0$$

shows that it is natural to ask the following question. Consider a vector field $\boldsymbol{J}$ on Ω such that $\operatorname{div}\boldsymbol{J} = 0$. When is it possible to write $\boldsymbol{J} = \overline{\operatorname{curl}}\,\chi$ for some single-valued stream function χ? If Ω is simply connected, then it is well-known that $\boldsymbol{K} = \overline{\operatorname{curl}}\,\chi$, i.e. $\beta_1(\Omega) = 0 \Rightarrow \beta^1(\Omega) = 0$. In order to see that it may not be possible to find such a χ if Ω is not simply connected, consider the following situation where Ω is homeomorphic to an annulus. On Ω let $\boldsymbol{K}$ flow outward in

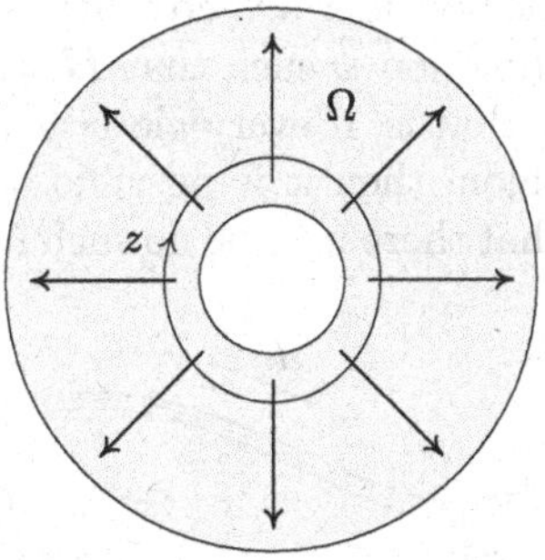

Figure 1.6. Radial surface current on conducting annulus.

the radial direction, and let $z \in Z_1(\Omega)$ be a 1-cycle which encircles the hole (see Figure 1.6). The period of $\boldsymbol{K}$ on the cycle z will be called the current per unit of thickness through z and denoted by I. In this case, relating $\boldsymbol{K}$ to a single-valued stream function χ leads to a contradiction because

$$0 \neq I = \int_z \boldsymbol{K} \cdot \boldsymbol{n}\, dl = \int_z \overline{\operatorname{curl}}\,\chi \cdot \boldsymbol{n}\, dl = \int_{\partial z} \chi = 0 \quad \text{since } \partial z = 0.$$

More generally, consider the surface Ω of Example 1.5 where there are generators z_i of $H_1(\Omega)$, and corresponding cuts c_i where $1 \leq i \leq \beta_1(\Omega) = 2n + k - 1$, so that

$$\Omega^- = \Omega - \left(\bigcup_{i=1}^{\beta_1(\Omega)} c_i \right)$$

is connected and simply connected. Since Ω^- is simply connected it is possible to define a stream function χ^- on Ω^- such that

$$\boldsymbol{K} = \overline{\operatorname{curl}}\,\chi^- \quad \text{on } \Omega^-.$$

Letting the current flowing through z_i be I_i so that

$$\int_{z_i} \boldsymbol{K} \cdot \boldsymbol{n}\, dl = I_i,$$

it is apparent from the integral laws that

$$I_i = \int_{z_i \cap \Omega^-} \overline{\operatorname{curl}}\,\chi^- = (\text{jump in } \chi^- \text{ across } c_i).$$

That is, χ^- is in general multivalued and it is single-valued if and only if all the periods of $\boldsymbol{K}$ on the z_i vanish, that is each I_i must vanish. Hence $\boldsymbol{K} = \overline{\operatorname{curl}}\,\chi$ on Ω for some single-valued χ if and only if $\operatorname{div} \boldsymbol{K} = 0$ and

$$\int_{z_i} \boldsymbol{K} \cdot \boldsymbol{n}\, dl = 0 \ \text{ for } \ 1 \leq i \leq \beta_1(\Omega).$$

See Klein [Kle63] for pictures, interpretations and references to the nineteenth century literature on similar examples. □

Example 1.9 Cohomology: $\Omega \subset \mathbb{R}^3$, $H^1(\Omega)$ is of interest. Let Ω be a three-dimensional subset of $\mathbb{R}^3$ and consider a vector field $\boldsymbol{H}$ such that $\operatorname{curl} \boldsymbol{H} = 0$ in Ω. Is there a single-valued function ψ such that $\boldsymbol{H} = \operatorname{grad} \psi$?

If Ω is simply connected, that is, if every closed curve in Ω can be shrunk to a point in a continuous fashion, then it is possible to find such a single-valued function ψ. In order to see that there may be no such function ψ if Ω is not simply

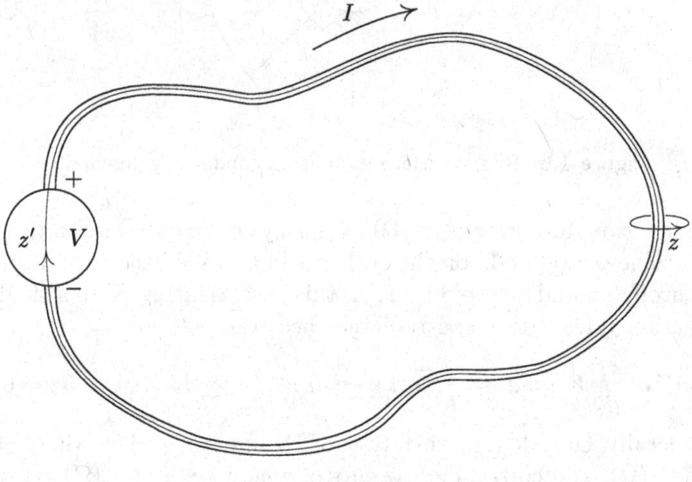

Figure 1.7.

connected, let Ω be the region exterior to a thick resistive wire connected across a battery and $\Omega' = \mathbb{R}^3 - \Omega$ as shown in Figure 1.7. Here, $\beta_1(\Omega) = \beta_1(\Omega') = 1$, $z \in Z_1(\Omega)$, and $z' \in Z_1(\Omega')$ represent nontrivial homology classes of $H_1(\Omega)$ and $H_1(\Omega')$ respectively. Let $S, S' \in C_2(\mathbb{R}^3)$ be a pair of 2-chains which, when considered as sets, are homeomorphic to discs so that

$$\partial S' = z' \text{ and } \partial S = z.$$

Under the assumption of magnetostatics,

$$\operatorname{curl} \boldsymbol{H} = 0 \text{ in } \Omega, \text{ and } \operatorname{curl} \boldsymbol{E} = 0 \text{ in } \Omega'.$$

The periods of $\boldsymbol{H}$ and $\boldsymbol{E}$,

$$\int_z \boldsymbol{H} \cdot \boldsymbol{t}\, dl = I \text{ and } \int_{z'} \boldsymbol{E} \cdot \boldsymbol{t}\, dl = V,$$

are nonzero. However, assuming that $\boldsymbol{E}$ and $\boldsymbol{H}$ can be represented as gradients of single-valued scalar potentials ψ' and ψ, respectively, leads to contradictions since

$$0 \neq I = \int_z \boldsymbol{H} \cdot \boldsymbol{t}\, dl = \int_z \operatorname{grad} \psi \cdot \boldsymbol{t}\, dl = \int_{\partial z} \psi = 0$$

since $\partial z = 0$ and

$$0 \neq V = \int_{z'} \boldsymbol{E} \cdot \boldsymbol{t}\, dl = \int_{z'} \operatorname{grad} \psi' \cdot \boldsymbol{t}\, dl = \int_{\partial z'} \psi' = 0$$

since $\partial z' = 0$. In this case, note that

$$H_1(\Omega - S') = 0 \text{ and } H_1(\Omega' - S) = 0$$

so that the magnetic field can be represented as the gradient of a scalar ψ in $\Omega - S'$ where the scalar has a jump of value I whenever S' is traversed in the direction of its normal. Similarly, the electric field can be represented as the gradient of a scalar ψ' in $\Omega' - S$ where the scalar has a jump of value V whenever S is traversed in the direction of its normal. Note that ψ, ψ' are continuous and single-valued on Ω, and Ω' respectively if and only if $I = 0$ and $V = 0$. Thus it is seen that the irrotational fields $\boldsymbol{H}$ and $\boldsymbol{E}$ in Ω can be expressed in terms of single-valued scalar functions once the cuts S and S' are introduced.

The general intuitive conditions for representing an irrotational vector field $\boldsymbol{H}$ as the gradient of a scalar potential have been studied for a long time. See [Tho69], [Max91, articles 18–20, 421] and [Lam32, articles 47–55, 132–134, and 139–141]. In Maxwell's terminology, acyclic means that Ω is simply connected, cyclosis refers to multiple connectivity, and cyclic constants are periods on generators of $H_1(\Omega)$. Cyclic constants were usually called "Kelvin's constants of circulation" in the nineteenth century literature.

A formal justification for introducing cuts into a space involves duality theorems for homology groups of orientable manifolds which will be considered in Section 3A. For the time being, the general procedure for introducing cuts will be illustrated by trying to generalize the above case involving a battery and a wire. Let Ω be a connected subset of $\mathbb{R}^3$. The first thing to do is to find 2-chains $S'_i \in C_2(\mathbb{R}^3)$, for $1 \leq i \leq n$, which, when considered as surfaces, satisfy the following conditions:

(1) $H_1(\mathbb{R}^3 - \Omega)$ is generated by cosets of the form

$$\sum_{i=1}^{n} a'_i \partial S'_i + B_1(\mathbb{R}^3 - \Omega) \quad \text{for } a_i \in \mathbb{R},$$

and n is chosen so that $n = \beta_1(\mathbb{R}^3 - \Omega)$. Note that $\partial S'_i \notin B_1(\mathbb{R}^3 - \Omega)$, where $1 \leq i \leq \beta_1(\mathbb{R}^3 - \Omega)$.

(2) It turns out that one can also do the reverse, namely find 2-chains $S_i \in C_2(\mathbb{R}^3)$, for $1 \leq i \leq n$, that when considered as surfaces satisfy the following condition: $H_1(\Omega)$ is generated by cosets of the form

$$\sum_{i=1}^{n} a_i \partial S_i + B_1(\Omega) \quad \text{for } a_i \in \mathbb{R}$$

and n is chosen such than $n = \beta_1(\Omega)$. Note that $\partial S_i \neq B_1(\Omega)$ for $1 \leq i \leq \beta_1(\Omega)$.

With luck, the ∂S_i intersect S'_j very few times and likewise for $\partial S'_i$ and S_j. The result

$$\beta_1(\Omega) = \beta_1(\mathbb{R}^3 - \Omega)$$

is apparent at this stage and was known to Maxwell [Max91, Article 18].

If curl $\boldsymbol{H} = 0$ in Ω then by the above construction, there exists a

$$\psi \in C^0\left(\Omega - \left(\bigcup_{i=1}^{\beta^1(\Omega)} S_i'\right)\right)$$

such that

$$\boldsymbol{H} = \operatorname{grad}\psi \quad \text{on } \Omega - \left(\bigcup_{i=1}^{\beta_1(\Omega)} S_i'\right).$$

Furthermore the jump in ψ over the surface S_i' can be deduced from the periods

$$\int_{\partial S_l} \boldsymbol{H}\cdot\boldsymbol{t}\,dl = I_l, \quad \text{for } 1 \le l \le \beta_1(\Omega)$$

by solving a set of linear equations which have trivial solutions if and only if all of the periods vanish.

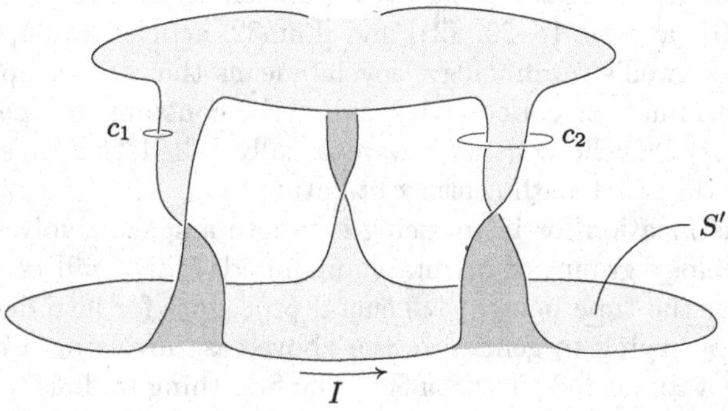

Figure 1.8. A 2-cycle relative to the boundary in the trefoil knot complement.

As a simple example of this procedure, consider a current-carrying trefoil knot $\mathbb{R}^3 - \Omega$ and its complement Ω as shown in Figure 1.8. Here

$$I_i = \int_{\partial S_i} \boldsymbol{H}\cdot\boldsymbol{t}\,dl = (\text{current flowing through } S_i),$$

$$\beta_1(\Omega) = \beta_1(\mathbb{R}^3 - \Omega) = 1,$$

and

$$\boldsymbol{H} = \operatorname{grad}\psi \quad \text{on} \quad \Omega - S'.$$

The jumps in ψ across S_i' are given by I_i. It is clear that the scalar potential will be continuous and single-valued in Ω if and only if $I = 0$.

Although this illustration makes the general procedure look like a silly interpretation of the method of mesh analysis in electrical circuit theory, problems where the $\partial S_i'$ are not necessarily in the same plane may be harder to tackle as are problems where $\beta_0(\Omega) > 1$. The case where Ω may be disconnected is handled by separately treating each connected part of Ω. As a nontrivial mental exercise the reader may convince himself that

$$\beta_1(\Omega) = \beta_1(\mathbb{R}^3 - \Omega) = 2n + k - 1$$

when Ω is the two-dimensional region of example 1.5. This is straightforward when one realizes that generators of $H_1(\Omega)$ can be taken to be the boundaries of cuts in $\mathbb{R}^3 - \Omega$ and generators of $H_1(\mathbb{R}^3 - \Omega)$ can be considered to be boundaries of surfaces which intersect Ω along the cuts c_i, where $1 \leq i \leq \beta_1(\Omega)$. □

1E. Homotopy Versus Homology and Linking Numbers

By some divine justice the homotopy groups of a finite polyhedron or a manifold seem as difficult to compute as they are easy to define.

Raoul Bott [BT82]

While there can be no single-valued scalar potential if the region Ω is not simply connected, it is not true that the cuts must render the region simply connected. One such example has already surfaced in Figure 1.8 where the relative 2-cycle S is a cut for a scalar potential in the scalar potential. Authors in various fields, including Maxwell [Max91] and others [Str41, Lam32, VB89, AL65, HS85], have assumed that the cut must make the (nonconducting) region simply connected. Seeking simple connectivity has led to formulations based on the notion of homotopy. For two-dimensional problems, the assumption leads to the right cuts. The relation between homotopy and homology reveals that in three-dimensional problems such an assumption is not equivalent to the physical conclusion drawn from Ampère's law that the problem of making cuts is one of linking zero current.

While the present objective is to avoid homotopy notions, we briefly consider a geometric and algebraic summary of the first homotopy group. The first homotopy group π_1 of a region V embedded in $\mathbb{R}^3$ is an algebraic classification of all closed loops in V which are topologically different in the sense of continuous deformation described below. In order to illustrate homotopy and make the above description more precise, consider a closed, oriented curve such as the current-carrying trefoil knot shown in Figure 1.9, where V denotes the region complementary to the knot and $\mathbb{R}^3 - V$ is the knot itself. An arbitrary point p in V is chosen as a base for drawing oriented, closed paths $a, b, c, \ldots$ in V. The set of all closed curves based at p can be partitioned into equivalence classes called homotopy classes where two closed paths are homotopic if either curve can be continuously deformed into the other. The path a represents or generates a class $[a]$ which contains all paths homotopic to a. It can be shown that composition of paths induces a product law for homotopy classes, $[a][b] = [ab]$ where we think of ab as traversing a followed by b as shown in Figure 1.9. The loop a^{-1} is a with opposite orientation and generates its own homotopy class $[a^{-1}]$ while the constant, or identity, loop e is a path which can be contracted to the basepoint without encountering the knot. These notions are discussed and extensively illustrated in [Neu79].

It can also be shown that homotopy classes satisfy the following properties:

$$\begin{aligned} &([a][b])[c] = [a]([b][c]) && \text{(associativity)},\\ &[a][e] = [a] = [e][a] && \text{(identity)},\\ &[a][a^{-1}] = [e] = [a^{-1}][a] && \text{(inverse)}. \end{aligned}$$

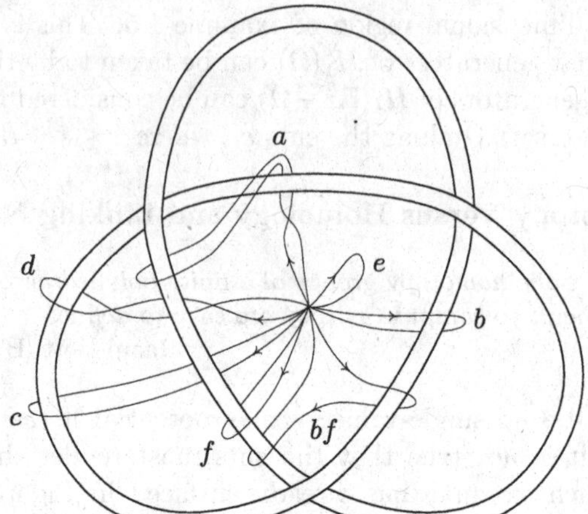

Figure 1.9. Closed loops based at p in the region complementary to the current-carrying knot. Loops a and b are homotopic. So are c and d. Loop e is trivial since it can be contracted to p. Product bf is shown.

Hence, the set of all homotopy classes in V forms a group, written π_1, where the group law is written as a multiplication. Note that a region is said to be simply connected if all loops in the region are homotopic to the identity, in which case π_1 is trivial. Formally, a multiply connected region is one which is not simply connected in this sense.

Particularly significant to π_1 are the homotopy classes of the form $[xyx^{-1}y^{-1}]$, called commutators. If π_1 is commutative, commutators are equal to the identity, that is xy is deformable to yx so that both represent the same homotopy class, $[x][y] = [y][x]$. In general π_1 is not commutative so that commutators are not equal to the identity and π_1 has a commutator subgroup $[\pi_1, \pi_1]$ generated by all possible commutator products.

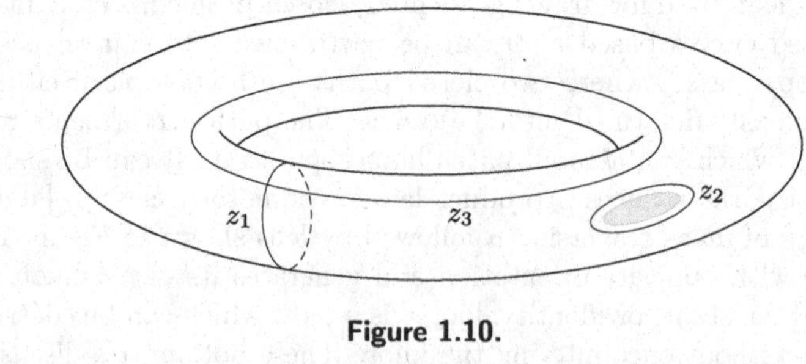

Figure 1.10.

Example 1.10 Fundamental group of the torus. In reference to Figure 1.10 we note that on the torus π_1 is generated by z_1 and z_3, before adding the puncture shown enclosed by z_2. In this case the commutator subgroup is trivial so that π_1 is commutative. On the other hand, π_1 of the punctured torus has a nontrivial commutator subgroup since the commutator is homotopic to the boundary of

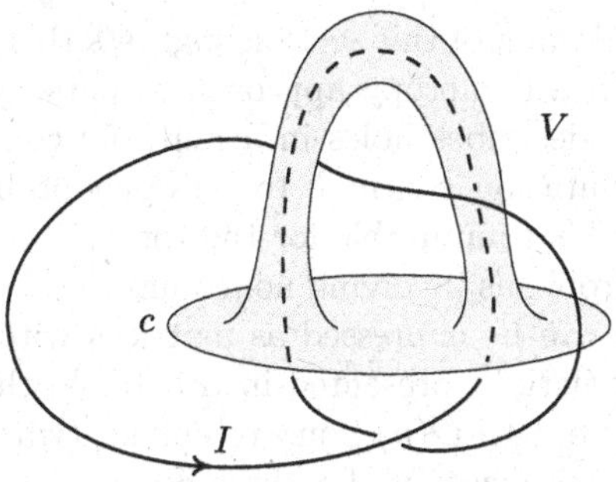

Figure 1.11. c is in the commutator subgroup of π_1 for the trefoil knot complement V. c links zero net current and is the boundary of a surface (a disc with a handle) in V.

the hole. Note that in homology, the effect of puncturing the torus becomes apparent only in the second homology group. □

The formal relationship between the first homotopy and homology groups is a homomorphism, $\pi_1(V) \to H_1(V)$. The Poincaré isomorphism theorem states that the kernel of the homomorphism is $[\pi_1, \pi_1]$ so that there is an isomorphism

$$\pi_1/[\pi_1, \pi_1] \simeq H_1$$

for any region V. When $[\pi_1, \pi_1]$ is nontrivial, commutators are nontrivial closed loops which, by virtue of the Poincaré isomorphism, are zero-homologous. It can be shown that this is the case [Sti93, GH81], thus commutators are boundaries of surfaces which lie entirely in V. The homology classes of 2-chains bounded by zero-homologous paths can be represented by orientable manifolds in V [Sti93]. Hence no current can be linked by a commutator and surfaces bounded by commutators are unrelated to surfaces used in Ampère's law. A proof of the π_1-H_1 relation can be found in [GH81] and is discussed in the context of Riemann surfaces and complex analysis in [Spr57]. A discussion of the consequences of the π_1-H_1 relation for computational methods can be found in [Cro78].

Figure 1.11 illustrates a commutator element and the surface bounded by the commutator for the current-carrying trefoil knot where c denotes a class in $[\pi_1, \pi_1]$. Since $c \in [\pi_1, \pi_1]$ and $[\pi_1, \pi_1]$ is the kernel of the Poincaré map, c is zero-homologous meaning that a surface S such that $c = \partial S$ exists. By Ampère's Law we then have

$$\int_c \boldsymbol{H} \cdot d\boldsymbol{r} = \int_S \boldsymbol{J} \cdot \boldsymbol{n}\, ds = 0,$$

because S lies in V and $\boldsymbol{J} = 0$ in V. It follows that $\text{Link}(c, c') = 0$ for $c \in [\pi_1, \pi_1]$ and $c' \in H_1(\mathbb{R}^3 - V)$.

To make the nonconducting region simply connected requires elimination of all nontrivial closed paths in the region. In problems where $[\pi_1, \pi_1]$ is nontrivial, there exist commutator loops which are zero-homologous, but since commutators link zero current, they are unimportant in light of Ampère's law; and they are irrelevant to finding cuts which make the scalar potential single-valued.

The quote at the beginning of this section suggests that there exist other fundamental problems with a homotopy approach to cuts. While π_1 is appealing because it "accurately" describes holes in a multiply connected region, practical aspects of its computation continue to be open problems in mathematics [DR91, Sti93]. While π_1 is computable for the torus, it is difficult in general to resolve basic decision problems involving noncommutative groups. In homology, groups are abelian and can be expressed as matrices with integer entries. Furthermore, for the constructions presented in this book, the matrices associated with homology are sparse with $\mathcal{O}(n)$ nonzero entries (where n is the number of tetrahedra in the tetrahedralization of a three-dimensional domain) and can be computed with graph-theoretical techniques. This is made clear in [BS90] and [Rot71] for electrical circuits and is discussed in Chapter 6.

1F. Chain and Cochain Complexes

Chain and cochain complexes are the setting for homology theory. Algebraically speaking, a chain complex $C_* = \{C_p, \partial_p\}$ is a sequence of modules C_i over a ring R and a sequence of homomorphisms

$$\partial_p : C_p \to C_{p-1}$$

such that

$$\partial_{p-1}\partial_p = 0.$$

For our purpose, the ring R will be $\mathbb{R}$ or $\mathbb{Z}$ in which case the modules C_p are vector spaces or abelian groups respectively. A familiar example is the chain complex

$$C_*(\Omega; \mathbb{R}) = \{C_p(\Omega; \mathbb{R}), \partial_p\}$$

considered up to now. Similarly, one has the chain complex

$$C_*(\Omega; \mathbb{Z}) = \{C_p(\Omega; \mathbb{Z}), \partial_p\}$$

when the coefficient group is $\mathbb{Z}$.

Cochain complexes are defined in a way similar to chain complexes except that the arrows are reversed. That is, a cochain complex $C^* = \{C^p, d^p\}$ is a sequence of modules C^p and homomorphisms

$$d^p : C^p \to C^{p+1}$$

such that

$$d^{p+1}d^p = 0.$$

An example of a cochain complex is

$$C^*(\Omega; \mathbb{R}) = \{C^p(\Omega; \mathbb{R}), d^p\}$$

which has been considered in the context of integration. When the coefficient group is not mentioned, it is understood to be $\mathbb{R}$.

From the definition of chain and cochain complexes, homology and cohomology are defined as follows. For homology,

$$B_p = \operatorname{im} \partial_{p+1} \text{ and } Z_p = \ker \partial_p,$$

so that

$$B_p \subset Z_p \text{ and } H_p = Z_p/B_p,$$

where

$$\beta_p = \operatorname{Rank} H_p.$$

For cohomology,

$$B^p = \operatorname{im} d^{p-1} \text{ and } Z^p = \ker d^p,$$

so that

$$B^p \subset Z^p \text{ and } H^p = Z^p/B^p,$$

where

$$\beta^p = \operatorname{Rank} H^p.$$

When dealing with chain and cochain complexes, it is often convenient to suppress the subscript on ∂_p and the superscript on d^p and let ∂ and d be the boundary and coboundary operators in the complex where their interpretation is clear form context.

The reader should realize that the introduction has thus far aimed to motivate the idea of chain and cochain complexes and the resulting homology and cohomology. Explicit methods for setting up complexes and computing homology from triangulations or cell decompositions can be found in [Mas80, Gib81, Wal57, GH81], while computer programs to compute Betti numbers and other topological invariants have existed for over four decades [Pin66]. In contrast to the vast amount of literature on homology theory, there seems to be no systematic exposition on its role in boundary value problems of electromagnetics, though early papers by Bossavit [Bos81, Bos82], Bossavit and Verité [BV82, BV83], Milani and Negro [MN82], Brown [Bro84], Nedelec [Ned78], and Post [Pos78, Pos84] were valuable first steps.

Just as groups, fields and vector spaces are examples of algebraic structures, complexes are a type of algebraic structure and as such it is useful to consider mappings between complexes. In the case of a chain complex the useful mappings to consider are the ones which have nice properties when it comes to homology. Such mappings, called chain homomorphisms, are defined as follows. Given two complexes

$$C_* = \{C_p, \partial_p\} \text{ and } C'_* = \{C'_p, \partial'_p\}$$

a chain homomorphism

$$f_* : C_* \to C'_*$$

is a sequence of homomorphisms $\{f_p\}$ such that

$$f_p : C_p \to C'_p$$

and

$$\partial'_p f_p = f_{p-1}\partial_p.$$

That is, for each p, the diagram

$$\begin{array}{ccc} \downarrow & & \downarrow \\ C_p & \xrightarrow{f_p} & C'_p \\ \partial_p \downarrow & & \downarrow \partial'_p \\ C_{p-1} & \xrightarrow{f_{p-1}} & C'_{p-1} \\ \downarrow & & \downarrow \end{array}$$

is commutative. In the case of cochain complexes, cochain homomorphisms are defined analogously.

In order to illustrate chain and cochain homomorphisms, consider a region Ω and a closed and bounded subset S. Since $C_p(S) \subset C_p(\Omega)$ for all p and the boundary operator ∂' in the complex $C_*(S)$ is the restriction of the boundary operator ∂ in $C_*(\Omega)$ to $C_*(S)$, where $C_*(S)$ is a subcomplex of $C_*(\Omega)$. There is a chain homomorphism

$$i_* : C_*(S) \to C_*(\Omega),$$

where

$$i_p : C_p(S) \to C_p(\Omega)$$

is an inclusion. Obviously $i_{p-1}\partial'_p c = \partial_p i_p c$ for all $c \in C_p(S)$ as required. Similarly, considering the restriction of a p-form on Ω to one on S for all values of p, there is a cochain homomorphism

$$r^* : C^*(\Omega) \to C^*(S)$$

where

$$r^p : C^p(\Omega) \to C^p(S).$$

If the coboundary operator (exterior derivative) in $C^*(\Omega)$ is d and the corresponding coboundary operator in $C^*(S)$ is d' then

$$d'^p r^p \omega = r^{p-1} d^p \omega \text{ for all } \omega \in C^p(\Omega)$$

as required.

Chain Complexes in Electrical Circuit Theory. The notion of a complex is a fundamental idea in electrical circuit theory. Let A be the incidence matrix of branches with vertices, and B the loop matrix of a network. Since $AB^T = 0$ there is a chain complex

$$0 \to \{\text{meshes}\} \xrightarrow{B^T} \{\text{branches}\} \xrightarrow{A} \{\text{nodes}\} \to 0$$

in which 2-chains are linear combinations of mesh currents and 1-chains are linear combinations of branch currents. In addition, the transpose of this complex, results in a cochain complex. Since $BA^T = 0$, there is a cochain complex

$$0 \leftarrow \{\text{meshes}\} \xleftarrow{B} \{\text{branches}\} \xleftarrow{A^T} \{\text{nodes}\} \leftarrow 0$$

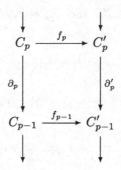

where 0-cochains are linear combinations of potentials at nodes of the network, and 1-cochains are linear combinations of branch voltages [BB69, Sect. 2.2]. Furthermore, if the network is planar and A is the reduced incidence matrix obtained by ignoring one node on each connected component of the network graph, then the homology of the complex is trivial. Kirchhoff's laws can be expressed as

$$\text{Kirchhoff Voltage Law: } v \in \operatorname{im} A^T \quad \text{(or } v \text{ is a 1-coboundary)}$$

$$\text{Kirchhoff Current Law: } i \in \ker A \quad \text{(or } i \text{ is a 1-cycle)}$$

so that if $v = A^T e$ for some set of nodal potentials e, then Tellegen's Theorem is easily deduced:

$$0 = (e, Ai) = (A^T e, i) = (v, i).$$

Thus Tellegen's Theorem is an example of orthogonality between cycles and coboundaries. This view of electrical network theory is usually attributed to Weyl [Wey23] (see also [Sle68, Fla89, Sma72]). Systematic use of homology theory in electrical network theory can be found in [Chi68] and in the work of Roth [Rot55b, Rot55a, Rot59, Rot71]. Kron [Kro59] generalized electrical network theory by introducing branch relations associated with k-dimensional ones. An explanation of Kron's method as well as references to additional papers by Kron can be found in [BLG70].

The interplay between continuum and network models through the use of complexes is developed in [Bra66, Ton77, KI59, MIK59]. Examples of cochain complexes for differential operators encountered in the work of Tonti and Branin are, for vector analysis in three dimensions where informally, the complex is

$$0 \to \left\{\begin{matrix}\text{scalar}\\ \text{functions}\end{matrix}\right\} \xrightarrow{\text{grad}} \left\{\begin{matrix}\text{field}\\ \text{vectors}\end{matrix}\right\} \xrightarrow{\text{curl}} \left\{\begin{matrix}\text{flux}\\ \text{vectors}\end{matrix}\right\} \xrightarrow{\text{div}} \left\{\begin{matrix}\text{volume}\\ \text{densities}\end{matrix}\right\} \to 0$$

with

$$\operatorname{curl}(\operatorname{grad}) = 0$$
$$\operatorname{div}(\operatorname{curl}) = 0.$$

Similarly, in two dimensions,

$$0 \to \left\{\begin{matrix}\text{scalar}\\ \text{functions}\end{matrix}\right\} \xrightarrow{\text{grad}} \left\{\begin{matrix}\text{field}\\ \text{vectors}\end{matrix}\right\} \xrightarrow{\text{curl}} \{\text{densities}\} \to 0.$$

The two complexes above are special cases of the complex $C^*(\Omega)$ considered thus far; however, when there is no mention of the domain Ω over which functions are to be defined, it is impossible to say anything about the homology of the complex. Hence, unless an explicit dependence on the domain Ω is recognized in the definition of the complex, it is virtually impossible to say anything concrete about global aspects of solvability conditions, gauge transformations or complementary variational principles, since these aspects depend on the cohomology groups of the complex which in turn depend on the topology of the domain Ω. Furthermore, imposing boundary conditions on some subset $S \subset \partial\Omega$ necessitates the consideration of relative cohomology groups to resolve questions of solvability and gauge ambiguity. Again the situation becomes hopelessly complicated unless a complex which depends explicitly on Ω and S is defined. The cohomology groups of this complex, which are called the relative cohomology groups of Ω modulo S are the ones required to describe the global aspects of the given problem. Relative homology and cohomology groups are considered in the next section.

1G. Relative Homology Groups

Relative chain, cycle, boundary and homology groups of a region Ω modulo a subset S will now be considered. Relative homology groups are the generalization of ordinary homology groups necessary in order to describe the topological aspects of cochains (forms) subject to boundary conditions.

Consider a region Ω and the chain complex $C_*(\Omega) = \{C_p(\Omega), \partial_p\}$ associated with it. Let S be a compact subset of Ω and $C_*(S) = \{C_p(S), \partial'_p\}$ be the chain complex associated with S. The boundary operator of $C_*(S)$ is the boundary operator on $C_*(\Omega)$ with a restricted domain. Furthermore,

$$C_p(S) \subset C_p(\Omega) \quad \text{for all } p.$$

It is useful to define the quotient group of p-chains on Ω modulo S

$$C_p(\Omega, S) = C_p(\Omega)/C_p(S)$$

when one wants to consider p-chains on Ω while disregarding what happens on some subset S. In this way, the elements of $C_p(\Omega, S)$ are cosets of the form

$$c + C_p(S), \quad \text{where } c \in C_p(\Omega).$$

Although this definition makes sense with any coefficient ring R, in the case of coefficients in $\mathbb{R}$ the definition of $C_p(\Omega, S)$ is made intuitive if one defines $C^p(\Omega, S)$ to be the subset of $C^p(\Omega)$ where the support of $\omega \in C^p(\Omega, S)$ lies in $\Omega - S$. In this case it is possible to salvage the idea that integration should be a bilinear pairing between $C_p(\Omega, S)$ and $C^p(\Omega, S)$. That is,

$$\int : C^p(\Omega, S) \times C_p(\Omega, S) \to \mathbb{R}$$

should be such that

$$\int_c \omega = 0 \text{ for all } \omega \in C^p(\Omega, S) \quad \text{implies} \quad c \in C_p(S)$$

and

$$\int_c \omega = 0 \text{ for all } c \in C_p(\Omega, S) \quad \text{implies} \quad \omega = 0.$$

Note that when S is the empty set, the definitions of relative chain and cochain groups reduce to those of their absolute counterparts.

Returning to the general case where the chains could be considered with coefficients in rings such as $\mathbb{R}$ or $\mathbb{Z}$, the induced boundary operator

$$\partial''_p : C_p(\Omega)/C_p(S) \to C_{p-1}(\Omega)/C_{p-1}(S)$$

makes the following definitions appropriate:

$$Z_p(\Omega, S) = \ker\big(C_p(\Omega, S) \xrightarrow{\partial''_p} C_{p-1}(\Omega, S)\big)$$

is the group of *relative p-cycles* of Ω modulo S, and

$$B_p(\Omega, S) = \mathrm{im}\big(C_{p+1}(\Omega, S) \xrightarrow{\partial''_{p+1}} C_p(\Omega, S)\big)$$

is the group of *relative p-boundaries* of Ω modulo S. Intuitively, relative cycles and boundaries can be interpreted as follows. Given $z, b \in C_p(\Omega)$,

$$\begin{aligned} z + C_p(S) &\in Z_p(\Omega, S) \quad \text{if } \partial_p z \in i_{p-1}\left(C_{p-1}(S)\right), \\ b + C_p(S) &\in B_p(\Omega, S) \quad \text{if } \partial_{p+1} c - b \in i_p\left(C_p(S)\right), \end{aligned}$$

for some $c \in C_{p+1}(\Omega)$. Hence, z is a relative p-cycle if its boundary lies in the subset S while b is a relative p-boundary if it is homologous to some p-chain on S.

From the definition of ∂''_p, it is apparent that

$$\partial''_p \partial''_{p+1} = 0.$$

Hence

$$B_p(\Omega, S) \subset Z_p(\Omega, S);$$

the *pth relative homology group of* Ω *modulo* S is defined as

$$H_p(\Omega, S) = Z_p(\Omega, S)/B_p(\Omega, S)$$

and the *relative pth Betti number* of Ω modulo S as

$$\beta_p(\Omega, S) = \mathrm{Rank} H_p(\Omega, S).$$

By defining $C_p(\Omega, S) = 0$ for $p < 0$ and $p > n$, the above definitions make it apparent that

$$C_*(\Omega, S) = \left\{C_p(\Omega, S), \partial''_p\right\}$$

is a complex. Furthermore, if

$$j_p : C_p(\Omega) \to C_p(\Omega, S)$$

is the homomorphism which takes a $c \in C_p(\Omega)$ into a coset of $C_p(\Omega, S)$ according to the rule

$$j_p(c) = c + C_p(S),$$

then the collection of homomorphisms $j_* = \{j_p\}$ is a chain homomorphism

$$j_* : C_*(\Omega) \to C_*(\Omega, S),$$

since

$$\partial''_p j_p(c) = j_{p-1} \partial_p(c) \text{ for all } c \in C_p(\Omega).$$

Though the definitions leading to relative homology groups seem formidable at first sight, they are actually quite a bit of fun as the following example shows.

Example 1.11 Two-dimensional example of relative homology. In this example the relative homology groups associated with the cross-section of a coaxial cable are considered. The usefulness of relative homology groups will become apparent in later sections once relative cohomology groups have been introduced. Consider a piece of coaxial cable of elliptic cross-section, let Ω be the "insulator" as shown in Figure 1.12, and consider 1-chains z, z', z'' and 2-chains c, c' as shown in Figure 1.12. From the picture we see that z, z', z'' represent nontrivial cosets in $Z_1(\Omega, \partial\Omega)$ but

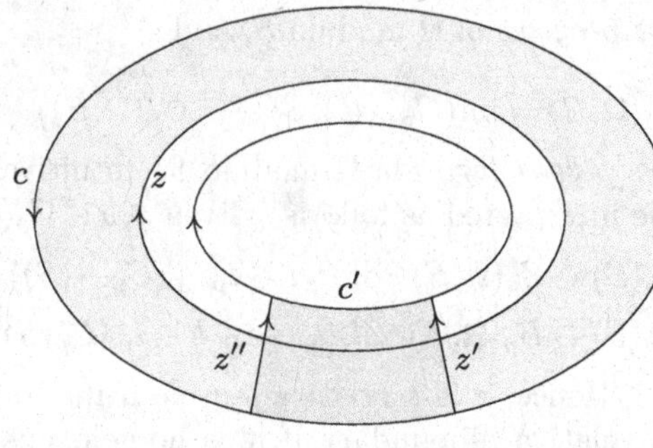

Figure 1.12. Cycles and relative cycles on elliptical annulus.

$$j_1(z) \sim 0 \text{ in } H_1(\Omega, \partial\Omega) \quad \text{since } \partial c - z \in i_1\left(C_1(\partial\Omega)\right),$$

and

$$j_1(z') \sim j_1(z'') \text{ in } H_1(\Omega, \partial\Omega) \quad \text{since } \partial c' - z' + z'' \in i_1\left(C_1(\partial\Omega)\right).$$

However, it is apparent that $j_1(z')$ is not homologous to zero in $H_1(\Omega, \partial\Omega)$ and that $\beta_1(\Omega, \partial\Omega) = 1$ so that the cosets of $H_1(\Omega, \partial\Omega)$ are $az' + B_1(\Omega, \partial\Omega)$ for $a \in \mathbb{R}$.

Next consider the other relative homology groups. Obviously $\beta_0(\Omega, \partial\Omega) = 0$, since any point in Ω can be joined to the boundary by a curve which lies in Ω. Furthermore, considering Ω as a 2-chain in $C_2(\Omega, \partial\Omega)$ it is apparent that

$$\Omega \in Z_2(\Omega, \partial\Omega) \text{ since } \partial_2(\Omega) \in i_1\left(C_1(\partial\Omega)\right)$$

hence, since $B_2(\Omega, \partial\Omega) = 0$, Ω is a nontrivial generator of $H_2(\Omega, \partial\Omega)$ and since the region is planar, it is plausible that there are no other independent generators of $H_2(\Omega, \partial\Omega)$. Thus $\beta_2(\Omega, \partial\Omega)$ and the cosets of $H_2(\Omega, \partial\Omega)$ are $a\Omega$ for $a \in \mathbb{R}$.

In the light of the previous examples the absolute homology groups of the region Ω are obvious once one notices that $\beta_0(\Omega) = 1$, the 1-cycle z is the only independent generator of $H_1(\Omega)$ hence $\beta_1(\Omega) = 1$, and $\beta_2(\Omega) = 0$ since $Z_2(\Omega) = 0$. Hence, in summary,

$$\begin{aligned}
\beta_0(\Omega) &= \beta_2(\Omega, \partial\Omega) = 1, \\
\beta_1(\Omega) &= \beta_1(\Omega, \partial\Omega) = 1, \\
\beta_2(\Omega) &= \beta_0(\Omega, \partial\Omega) = 0.
\end{aligned}$$

In order to exercise the newly acquired concepts, suppose that the capacitance of the cable was to be determined by a direct variational method. In this case it is convenient to exploit the inherent symmetry to reduce the problem to one-quarter of the annulus. Consider the diagram shown in Figure 1.13. It is apparent that for $a_1, a_2 \in \mathbb{R}$, the cosets of $H_1(\Omega', S_1)$ and $H_1(\Omega', S_2)$ look like

$$a_1 z_1 + B_1(\Omega', S_1) \text{ and } a_2 z_2 + B_1(\Omega', S_2)$$

respectively, and that

$$Z_2(\Omega', S_1) = 0 = Z_2(\Omega', S_2).$$

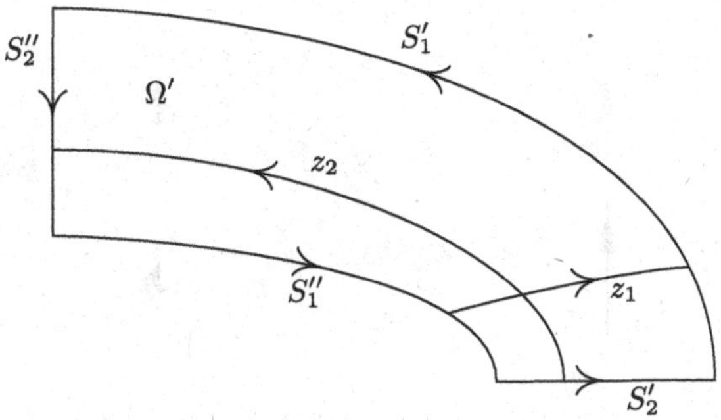

Figure 1.13. Cycles and relative cycles of elliptical annulus with respect to symmetry of the full annulus. The boundary $\partial\Omega'$ equals $S_1 + S_2$, where $S_1 = S_1' + S_1''$ and $S_2 = S_2' + S_2''$.

Hence

$$\beta_0(\Omega', S_1) = 0 = \beta_2(\Omega', S_2),$$
$$\beta_1(\Omega', S_1) = 1 = \beta_1(\Omega', S_2),$$
$$\beta_2(\Omega', S_1) = 0 = \beta_0(\Omega', S_2). \qquad \square$$

Let the pth relative homology group of Ω modulo a subset S with coefficients in $\mathbb{Z}$, be denoted by

$$H_p(\Omega, S; \mathbb{Z}).$$

This is an abelian group by construction. By the structure theorem for finitely generated abelian groups (see [Jac74, Theorem 3.13] or [Gib81, Theorem A.26 and Corollary A.27]), this relative homology group is isomorphic to the direct sum of a free abelian group F on $\beta_p(\Omega, S)$ generators and a torsion group T on $\tau_p(\Omega, S)$ generators, where $\tau_p(\Omega, S)$ is called the pth torsion number of Ω modulo S. When homology is computed with coefficients in $\mathbb{R}$ one obtains all the information associated with the free subgroup F and no information about the torsion subgroup T. It turns out that $\tau_p(\Omega, S) = 0$ if Ω is a subset of $\mathbb{R}^3$ and S is the empty set. In other words, for subsets of $\mathbb{R}^3$ the torsion subgroups of the homology groups

$$H_p(\Omega; \mathbb{Z}), \quad \text{for } 0 \le p \le 3,$$

are trivial (see, for example, [Mas80, Ch. 9, Ex. 6.6] for details). The relationship between $H_p(\Omega, S; \mathbb{R})$ and $H_p(\Omega, S; \mathbb{Z})$ is important since problems in vector analysis are resolved by knowing the structure of $H_p(\Omega, S; \mathbb{R})$ while it is convenient to use integer coefficients in numerical computations and determine $H_p(\Omega, S; \mathbb{Z})$. When $H_p(\Omega, S; \mathbb{Z})$ is found, the absolute homology groups with coefficients in $\mathbb{R}$ are easily deduced and relative homology groups with coefficients in $\mathbb{R}$ are deduced by throwing away torsion information. The following example illustrates a relative homology group with nontrivial torsion subgroup.

Example 1.12 Torsion phenomena in relative homology. Consider a Möbius band which is obtained by identifying the sides of a square I^2 as shown in Figure 1.14. Let Ω be the Möbius band and $S = z_a + z_b$ be the 1-chain along the edge of the band. Regarding S as a set, the homology groups $H_1(\Omega, \mathbb{Z})$ and

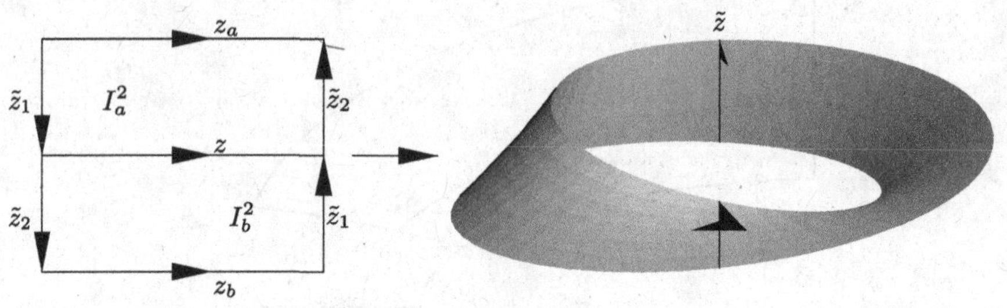

Figure 1.14. Möbius band. Note that $I^2 = I_a^2 + I_b^2$ and $\tilde{z} = \tilde{z}_1 + \tilde{z}_2$.

$H_1(\Omega, S; \mathbb{Z})$ will be deduced. The cosets of $H_1(\Omega, \mathbb{Z})$ are $az + B_1(\Omega, \mathbb{Z})$ with $a \in \mathbb{Z}$ so that $\beta_1(\Omega) = 1$.

In contrast, the situation is different when the relative homology group is considered. Observe that $j_1(\tilde{z})$ is not homologous to zero in $H_1(\Omega, S; \mathbb{Z})$, that is $\partial\tilde{z} \in C_0(S, \mathbb{Z})$ but there is no $c \in C_2(\Omega, \mathbb{Z})$ such that $\partial c - \tilde{z} \in C_1(S, \mathbb{Z})$. However, the square I^2 from which Ω was obtained has boundary

$$\partial_2(I^2) = 2\tilde{z} + z_b - z_a;$$

hence

$$\partial_2(I^2) - 2\tilde{z} \in i_1\left(C_1(S, Z)\right),$$

or

$$j_1(2\tilde{z}) \sim 0 \text{ in } H_1(\Omega, S; \mathbb{Z}).$$

Thus $\tilde{z}$ is an element of the torsion subgroup of the relative homology group since it is not homologous to zero, although a multiple of $\tilde{z}$ is zero-homologous. Similarly $j_1(z) \not\sim 0$ in $H_1(\Omega, S, \mathbb{Z})$ and $j_1(2z) \sim 0$ in $H_1(\Omega, S, \mathbb{Z})$. In order to see this one may imagine the Möbius band to be made out of paper and is cut along the 1-cycle z to yield a surface Ω'. The surface Ω' is orientable and

$$\partial_2(\Omega') = 2z - z_a - z_b$$

or

$$2z - \partial_2(\Omega') \in C_1(S, \mathbb{Z}).$$

Hence z and $\tilde{z}$ are nontrivial generators of $H_1(\Omega, S, \mathbb{Z})$. However $z \sim \tilde{z}$ since, referring back to the picture, it is apparent that

$$\partial_2(I_a^2) = \tilde{z}_1 + z + \tilde{z}_2 - z_a = \tilde{z} + z - z_a,$$
$$\partial_2(I_b^2) = \tilde{z}_1 - z + \tilde{z}_2 + z_b = \tilde{z} - z + z_b,$$

so that

$$\tilde{z} - (-z) - \partial_2(I_a^2) \in C_1(S; \mathbb{Z}) \quad \text{and} \quad \tilde{z} - (z) - \partial_2(I_b^2) \in C_1(S; \mathbb{Z}).$$

That is,

$$j_1(\tilde{z}) \sim j_1(z) \quad \text{and} \quad j_1(\tilde{z}) \sim -j_1(z) \text{ in } H_1(\Omega, S, \mathbb{Z}).$$

Thus, it is plausible that

$$H_1(\Omega, S; \mathbb{Z}) \simeq \mathbb{Z}/2$$

where $\mathbb{Z}$ denotes the integers modulo 2. In summary,

$$\beta_1(\Omega, S) = 0, \quad \beta_1(\Omega) = 1, \quad \tau_1(\Omega, S) = 1, \quad \tau_1(\Omega) = 0. \qquad \square$$

Before considering the role of relative homology groups in resolving topological problems of vector analysis it is useful to consider the long exact homology sequence since it is the key to understanding relative homology.

1H. The Long Exact Homology Sequence

In the context of this book, the long exact homology sequence is a result which enables one to visualize a full set of generators for the homology of a region Ω modulo a closed subset S in situations where intuition can be trusted only with the absolute homology groups of Ω and S. To see how the long exact homology sequence comes about, consider the three complexes

$$C_*(\Omega) = \{C_p(\Omega), \partial_p\},$$
$$C_*(S) = \{C_p(S), \partial'_p\},$$
$$C_*(\Omega, S) = \{C_p(\Omega, S), \partial''_p\},$$

and the two chain homomorphisms

$$i* = \{i_p\}, \qquad j_* = \{j_p\},$$

$$0 \to C_*(S) \xrightarrow{i_*} C_*(\Omega) \xrightarrow{j_*} C_*(\Omega, S) \to 0$$

where the inclusion map i_p takes a p-chain on S and sends it to a p-chain which coincides with it on S and vanishes elsewhere on $\Omega - S$ while j_p takes a $c \in C_p(\Omega)$ and sends it into the coset $c + C_p(S)$ in $C_p(\Omega, S)$. It is clear that i_* is injective, j_* is surjective and that $\operatorname{im} i_* = \ker j_*$. Such a sequence of three complexes and chain homomorphisms i_*, j_* is an example of a "short exact sequence of complexes". When written out in full, the sequence is:

$$\begin{array}{ccccccccc}
 & & \downarrow & & \downarrow & & \downarrow & & \\
0 & \longrightarrow & C_{p+1}(S) & \xrightarrow{i_{p+1}} & C_{p+1}(\Omega) & \xrightarrow{j_{p+1}} & C_{p+1}(\Omega, S) & \longrightarrow & 0 \\
 & & \downarrow \partial'_{p+1} & & \downarrow \partial_{p+1} & & \downarrow \partial''_{p+1} & & \\
0 & \longrightarrow & C_p(S) & \xrightarrow{i_p} & C_p(\Omega) & \xrightarrow{j_p} & C_p(\Omega, S) & \longrightarrow & 0 \\
 & & \downarrow \partial'_p & & \downarrow \partial_p & & \downarrow \partial''_p & & \\
0 & \longrightarrow & C_{p-1}(S) & \xrightarrow{i_{p-1}} & C_{p-1}(\Omega) & \xrightarrow{j_{p-1}} & C_{p-1}(\Omega, S) & \longrightarrow & 0 \\
 & & \downarrow \partial'_{p-1} & & \downarrow \partial_{p-1} & & \downarrow \partial''_{p-1} & &
\end{array}$$

It is a fundamental and purely algebraic result (see [Jac80, Vol. II, Sect. 6.3, Theorem 6.1]) that given such a short exact sequence of complexes, there is a long exact sequence in homology. This means that if $\tilde{\imath}_p$ and $\tilde{\jmath}_p$ are the homomorphisms which i_p and j_p induce on homology, and δ_p is a map on homology classes which takes $z \in Z_p(\Omega, S)$ into $z' \in Z_{p-1}(S)$ according to the rule

$$(z' + B_{p-1}(S)) = (i_{p-1})^{-1}\partial_p(j_p)^{-1}(z + B_p(\Omega, S)),$$

then the diagram

$$\begin{array}{l} 0 \longrightarrow H_n(S) \xrightarrow{\tilde{\imath}_n} \cdots \longrightarrow H_{p+1}(\Omega, S) \longrightarrow \\ \xrightarrow{\delta_{p+1}} H_p(S) \xrightarrow{\tilde{\imath}_p} H_p(\Omega) \xrightarrow{\tilde{\jmath}_p} H_p(\Omega, S) \longrightarrow \\ \xrightarrow{\delta_p} H_{p-1}(S) \xrightarrow{\tilde{\imath}_{p-1}} H_{p-1}(\Omega) \xrightarrow{\tilde{\jmath}_{p-1}} H_{p-1}(\Omega, S) \xrightarrow{\delta_{p-1}} \\ \cdots \xrightarrow{\tilde{\jmath}_0} H_0(\Omega, S) \longrightarrow 0 \end{array}$$

satisfies three conditions:

$$\ker \tilde{\imath}_p = \operatorname{im} \delta_{p+1}, \quad \ker \tilde{\jmath}_p = \operatorname{im} \tilde{\imath}_p, \quad \ker \delta_p = \operatorname{im} \tilde{\jmath}_p.$$

This result is valid for the coefficient groups $\mathbb{Z}$ and $\mathbb{R}$. Here we consider the case where homology is computed with coefficients in $\mathbb{R}$ so that all of the homology groups are vector spaces. Let

$$H_p(\Omega, S) \simeq \left(\frac{H_p(\Omega, S)}{\ker \delta_p}\right) \oplus \ker \delta_p.$$

The two summands can be interpreted using the conditions above. The first satisfies

$$\frac{H_p(\Omega, S)}{\ker \delta_p} \simeq \delta_p^{-1}(\operatorname{im} \delta_p) \simeq \delta_p^{-1}(\ker \tilde{\imath}_{p-1})$$

by the first condition, and the second satisfies

$$\ker \delta_p = \operatorname{im} \tilde{\jmath}_p \simeq \tilde{\jmath}_p\left(\frac{H_p(\Omega)}{\ker \tilde{\jmath}_p}\right) \simeq \tilde{\jmath}_p\left(\frac{H_p(\Omega)}{\tilde{\imath}_p(H_p(S))}\right)$$

by the second condition. Thus, $H_p(\Omega, S)$ can be rewritten as

$$H_p(\Omega, S) \simeq \delta_p^{-1}(\ker \tilde{\imath}_{p-1}) \oplus \tilde{\jmath}_p\left(\frac{H_p(\Omega)}{\tilde{\imath}_p(H_p(S))}\right). \tag{1–3}$$

By this identity it is usually easy to deduce a set of generators of $H_p(\Omega, S)$ if $H_p(S)$, $H_p(\Omega)$, $H_{p-1}(S)$, and $H_{p-1}(\Omega)$ are known. The generators for these groups can be found by the following three-step procedure:

(1) Find a basis for the vector space V_p defined by

$$H_p(\Omega) = (\operatorname{im} \tilde{\imath}_p) \oplus V_p.$$

Hence, $\tilde{\jmath}_p(V_p)$ gives $\beta_p(\Omega) - \dim \operatorname{im} \tilde{\imath}_p$ generators of $H_p(\Omega, S)$.

(2) Find a basis for $\ker \tilde{\imath}_{p-1}$ from the basis for $H_{p-1}(S)$ so that the

$$\dim \ker \tilde{\imath}_{p-1}$$

remaining generators of $H_p(\Omega, S)$ can be deduced from

$$\delta_p^{-1}\left(\ker \tilde{\imath}_{p-1}\right).$$

This is done as follows: Let $\tilde{z}_i$ be a basis for $\ker \tilde{\imath}_{p-1}$ and find a set of z_i, for $1 \leq i \leq \dim \ker \tilde{\imath}_{p-1}$, such that

$$z_i = j_p(\partial_p)^{-1} i_{p-1} \tilde{z}_i.$$

(3) $H_p(\Omega, S) = (j_p^* V_p) \oplus \delta_p^{-1}\left(\ker \tilde{\imath}_{p-1}\right)$, where a basis results from steps 1 and 2. Furthermore

$$\beta_p(\Omega, S) = \beta_p(\Omega) - \dim \operatorname{im} \tilde{\imath}_p + \dim \ker \tilde{\imath}_{p-1}.$$

Although this procedure is algebraic, it enables one to proceed in a systematic way in complicated problems, as illustrated in the following example.

Example 1.13 Embedded surfaces and relative homology. Recalling the 2-dimensional surface with n "handles" and k "holes" which was considered in Example 1.5, we will deduce the relative homology groups $H_1(\Omega, \partial\Omega)$ and $H_2(\mathbb{R}^3, \Omega)$.

For $H_1(\Omega, \partial\Omega)$, consider the long exact homology sequence for the pair $(\Omega, \partial\Omega)$:

$$0 \xrightarrow{\tilde{\imath}_2} H_2(\Omega) \xrightarrow{j_2} H_2(\Omega, \partial\Omega) \rightarrow$$

$$\xrightarrow{\delta_2} H_1(\partial\Omega) \xrightarrow{\tilde{\imath}_1} H_1(\Omega) \xrightarrow{j_1} H_1(\Omega, \partial\Omega) \rightarrow$$

$$\xrightarrow{\delta_1} H_0(\partial\Omega) \xrightarrow{\tilde{\imath}_0} H_0(\Omega) \xrightarrow{j_0} H_0(\Omega, \partial\Omega) \longrightarrow 0.$$

Following the three-step procedure for homology generators $H_1(\Omega, \partial\Omega)$ is obtained as follows.

(1) $\operatorname{im} \tilde{\imath}_1$ and V_1 are readily seen to be of the form $\sum_{j=1}^{k-1} a_{2n+j} z_{2n+j} + B_1(\Omega)$ and $\sum_{j=1}^{2n} a_j z_j + B_1(\Omega)$, respectively. That is, $j_1(z_{2n+j})$, with $1 \leq j \leq k-1$, are homologous to zero in $H_1(\Omega, \partial\Omega)$, while $j_1(z_j)$, for $1 \leq j \leq 2n$, are not homologous to zero in $H_1(\Omega, \partial\Omega)$.

(2) The kernel $\ker \tilde{\imath}_0$ is of the form

$$\sum_{j=1}^{k-1} a_j(p_j - p_k) + B_0(\partial\Omega),$$

while the point p_k can be used to generate $H_0(\Omega)$. Thus the curves $j_1(C_{2n+j})$, for $1 \leq j \leq k-1$, from Example 1.5 can be used as $k-1$ additional generators in $H_1(\Omega, \partial\Omega)$ since

$$\partial c_{2n+j} = p_j - p_k \quad \text{for } 1 \leq j \leq k-1.$$

(3) Looking at the definitions of the c_j, for $1 \le j \le 2n+k-1$, it is clear that

$$\sum_{j=1}^{2n} a_j c_j + B_1(\Omega, \partial\Omega) = \operatorname{im} \tilde{\jmath}_1$$

and

$$\sum_{j=2n+1}^{2n+k-1} a_j c_j + B_1(\Omega, \partial\Omega) = \delta_1^{-1}(\ker \tilde{\imath}_0).$$

Thus, the cosets of $H_1(\Omega, \partial\Omega)$ look like

$$\sum_{j=1}^{2n+k+1} a_j c_j + B_1(\Omega, \partial\Omega);$$

that is to say, the chains along which cuts were made to obtain a simply connected surface yield a set of generators for $H_1(\Omega, \partial\Omega)$, and

$$\beta_1(\Omega) = 2n+k-1 = \beta_1(\Omega, \partial\Omega).$$

For $H_2(\mathbb{R}^3, \Omega)$, consider part of the long exact homology sequence for the pair $(\mathbb{R}^3, \Omega)$:

$$\cdots \xrightarrow{\tilde{\imath}_2} H_2(\mathbb{R}^3) \xrightarrow{\tilde{\jmath}_2} H_2(\mathbb{R}^3, \Omega) \xrightarrow{\delta_2} H_1(\Omega) \xrightarrow{\tilde{\imath}_1} H_1(\mathbb{R}^3) \xrightarrow{\tilde{\jmath}_1} \cdots$$

Given that

$$H_p(\mathbb{R}^3) \simeq \begin{cases} \mathbb{R} & \text{if } p = 0, \\ 0 & \text{if } p \neq 0, \end{cases}$$

the part of the long exact sequence displayed above reduces to:

$$0 \xrightarrow{\tilde{\jmath}_2} H_2(\mathbb{R}^3, \Omega) \xrightarrow{\delta_2} H_1(\Omega) \xrightarrow{\tilde{\imath}_1} 0.$$

Hence δ_2 is an isomorphism since the sequence is exact. It is instructive to deduce $H_2(\mathbb{R}^3, \Omega)$ by the three-step procedure for homology generators:

(1) The first step can be ignored, since $H_2(\mathbb{R}^3) \simeq 0$ implies that $\operatorname{im} \tilde{\jmath}_2 = 0$.

(2) Since $H_2(\mathbb{R}^3, \Omega) \simeq H_1(\Omega)$ take generators z_i of $H_1(\Omega)$, for $1 \le i \le \beta_1(\Omega)$, and consider $\delta_2^{-1} z_i$. In other words, relative 2-cycles $j_2(S_i) \in Z_2(\mathbb{R}^3, \Omega)$ must be found such that

$$j_2(S_i) = j_2(\partial_2^{-1}) i_1 z_i \quad \text{for } 1 \le i \le \beta_1(\Omega)$$

can be used to generate a basis vector of $H_1(\mathbb{R}^3, \Omega)$. By individually considering the "handles" and "holes" of Ω, we easily see that such a set can be found for the jth handle and the jth hole, as illustrated in Figure 1.15.

There is nothing to do at this stage. The cosets of $H_2(\mathbb{R}^3, \Omega)$ look like

$$\sum_{i=0}^{2n+k-1} a_i S_i + B_2(\mathbb{R}^3, \Omega) \quad \text{for } a_i \in \mathbb{R}$$

and

$$\beta_2(\mathbb{R}^3, \Omega) = 2n+k-1 = \beta_1(\Omega).$$

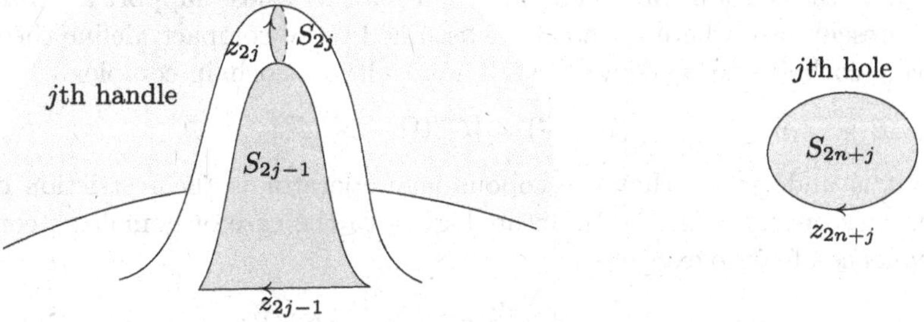

Figure 1.15. Regions bounded by homology generators of handles on the body in Figure 1.4. For the handle $1 \le j \le n$ and for hole $1 \le j \le k-1$.

It is useful to realize that the same arguments hold if Ω is knotted or has several connected components with the exception that representatives of generators of $H_2(\mathbb{R}^3, \Omega)$ may not be homeomorphic to discs. □

Instead of considering more examples of relative homology groups, a heuristic argument will now be considered in order to illustrate the use of relative homology groups in vector analysis.

1I. Relative Cohomology and Vector Analysis

Given a region Ω, one can form a vector space $C_c^p(\Omega)$ by considering linear combinations of p-cochains (p-forms) which have compact support in Ω. Since the coboundary operator (exterior derivative) applied to a p-cochain of compact support yields a $p+1$-cochain of compact support, one can define a complex

$$C_c^*(\Omega) = \{C_c^p(\Omega), d^p\}$$

and, by virtue of the fact that one has a complex, the cocycle, coboundary, cohomology groups, and Betti numbers can be defined as usual:

$$\begin{aligned} Z_c^p(\Omega) &= \ker\big(C_c^p(\Omega) \xrightarrow{d^p} C_c^{p+1}(\Omega)\big), \\ B_c^p(\Omega) &= \operatorname{im}\big(C_c^{p-1}(\Omega) \xrightarrow{d^{p-1}} C_c^p(\Omega)\big), \\ H_c^p(\Omega) &= Z_c^p(\Omega)/B_c^p(\Omega), \\ \beta_c^p(\Omega) &= \operatorname{Rank}\left(H_c^p(\Omega)\right). \end{aligned}$$

In general, if Ω is a compact region then the cohomology of the complexes $C_c^*(\Omega)$ and $C^*(\Omega)$ are identical. However, if Ω is an open set then the cohomology of the set of complexes will in general be different from $C_c^*(\Omega)$ since the cochains with compact support have restrictions on the boundary of the set.

In order to formulate the idea of relative cohomology, let Ω be a compact region, S a compact subset, and consider the complexes $C_c^*(\Omega) = \{C_c^p(\Omega), d^p\}$ and $C_c^*(S) = \{C_c^p(S), d^p\}$. It is understood that the coboundary operator in the latter complex is the coboundary operator of the first complex except that the domain is restricted. In the heuristic motivation for relative homology groups it was mentioned that in order to regard integration as a bilinear pairing between

$C^p(\Omega, S)$ and $C_p(\Omega, S)$, the definition of $C_p(\Omega, S)$ makes sense if $C^p(\Omega, S)$ is taken to be the set of linear combinations of p-forms whose support lies in $\Omega - S$. In the present case where Ω and S are assumed to be compact, define the set of relative p-cochains to be $C_c^p(\Omega - S)$. The result is a cochain complex

$$C_c^*(\Omega - S) = \{C_c^p(\Omega - S), d^p\},$$

where it is understood that the coboundary operator is the restriction of the coboundary operator in $C_c^*(\Omega)$. In analogy with the case of homology, consider the sequence of complexes

$$0 \longrightarrow C_c^*(\Omega - S) \xrightarrow{e^*} C_c^*(\Omega) \xrightarrow{r^*} C_c^*(S) \longrightarrow 0$$

and the cochain homomorphisms

$$e^* = \{e^p\}, \quad r^* = \{r^p\}.$$

Here e^p takes a p-cochain on $\Omega - S$ and extends it by 0 to the rest of Ω, while r^p takes an p-cochain on Ω and gives its restriction to S. Although this sequence of complexes fails to be exact at $C_c^*(\Omega)$, (that is $\operatorname{im} e^* \neq \ker r^*$), a limiting argument shows that there is still a long exact sequence in cohomology (see [Spi79, p. 589, Theorems 12 and 13]). Furthermore, there exist relative de Rham isomorphisms (see Duff [Duf52] for the basic constructions).

Instead of trying to develop the idea that the coboundary operator in the complex $C_c^*(\Omega - S)$ is adjoint to the boundary operator in the complex $C_*(\Omega, S)$, and trying to justify a relative de Rham isomorphism, familiar examples of the relative isomorphism will be considered. These examples will serve to solidify the notion of relative homology and cohomology groups and relative de Rham isomorphism so that an intuitive feel can be developed before a more concise formalism is given in the next chapter. When considering relative chains on $C_c^*(\Omega - S)$ there are certain boundary conditions which cochains must satisfy when S is approached from within $\Omega - S$. Although these conditions are transparent in the formalism of differential forms, they will be stated often in the following examples, without proof, since in specific instances they are easily deduced by using the integral form of Maxwell's Equations.

In these examples, the relative de Rham isomorphism is understood to mean

$$H_p(\Omega, S) \simeq H_c^p(\Omega - S) \text{ for all } p.$$

Also, two forms $\omega_1, \omega_2 \in Z_c^p(\Omega - S)$ are said to be cohomologous in the relative sense if

$$\omega_1 - \omega_2 \in B_c^p(\Omega - S).$$

As usual, this forms an equivalence relation where the condition above is written as

$$\omega_1 \sim \omega_2.$$

The notion of a relative period is defined as follows. If $\omega \in Z_c^p(\Omega - S)$ and $z \in Z_p(\Omega, S)$, the integral $\int_z \omega$ is called the *relative period* of ω on z; here, by Stokes' theorem, it is easily verified that the period depends only on the cohomology and homology classes of ω and z respectively. Thus the relative de Rham theorem should be interpreted as asserting that integration induces a nondegenerate bilinear pairing

$$\int : H_p(\Omega, S) \times H_c^p(\Omega - S) \to \mathbb{R}$$

where the values of this bilinear pairing can be deduced from evaluating the periods of basis vectors of $H_c^p(\Omega - S)$ on basis vectors of $H_p(\Omega, S)$. In most cases these periods have the interpretation of voltages, currents, charges, or fluxes.

One of the virtues of axiomatic homology theory is that one can show that once a method of computing homology for a certain category of spaces, such as manifolds, has been devised, the resulting homology groups are unique up to an isomorphism. Thus, for example, the de Rham isomorphism can be regarded as a consequence of devising a method of computing cohomology with differential forms and simplicial complexes, and showing that both methods satisfy the requirements of the axiomatic theory in the case of differentiable manifolds.

Example 1.14 Electrostatics: Visualizing $H_1(\Omega, S)$ in 3 dimensions. Consider a compact subset Ω of $\mathbb{R}^3$. On the boundary $\partial\Omega$ let $\partial\Omega$ be a union $S_1 \cup S_2$, where $S_1 \cap S_2$ has no area. There is a vector field $\boldsymbol{E}$ in Ω with a constraint on S_1 such that:

$$\operatorname{curl} \boldsymbol{E} = 0 \quad \text{in } \Omega,$$
$$\boldsymbol{n} \times \boldsymbol{E} = 0 \quad \text{on } S_1.$$

No constraint is specified on S_2. Elements of $C_c^1(\Omega - S_1)$ are associated with vector fields whose components tangential to S_1 vanish, hence $\boldsymbol{E}$ can be associated with an element of $Z_c^1(\Omega - S_1)$.

Considering the long exact homology sequence for the pair (Ω, S_1), one has

$$H_1(\Omega, S_1) = \delta_1^{-1}(\ker i_0) \oplus \tilde{\jmath}_1 \left(\frac{H_1(\Omega)}{\tilde{\imath}_1 (H_1(S_1))} \right),$$

where the relevant portion of the long exact homology sequence is

$$\cdots \xrightarrow{\delta_2} H_1(S_1) \xrightarrow{\tilde{\imath}_1} H_1(\Omega) \xrightarrow{\tilde{\jmath}_1} H_1(\Omega, S_1) \xrightarrow{\delta_1} H_0(S_1) \xrightarrow{\tilde{\imath}_0} H_0(\Omega) \longrightarrow \cdots$$

Let c_i, for $1 \le i \le \beta_1(\Omega, S)$, be a set of curves which are associated with the generators of $H_1(\Omega, S_1)$. These curves can be arranged into two groups according to the three-step procedure on page 38:

(1) There are dim im $\tilde{\jmath}_1$ generators of $H_1(\Omega, S_1)$ which are homologous in the absolute sense to generators of $H_1(\Omega)$. These generators can be associated with closed curves c_i, for $1 \le i \le \dim \operatorname{im} \tilde{\imath}_1$.

Thinking of electromagnetism, the period

$$\int_{c_i} \boldsymbol{E} \cdot \boldsymbol{t} \, dl = V_i$$

is equal to the rate of change of magnetic flux which links c_i. Although there is a static problem in Ω, the periods are associated with magnetic circuits in $\mathbb{R}^3 - \Omega$. It is usually wise to set these periods equal to zero if possible.

(2) There are $\dim\ker\tilde{\imath}_0$ remaining generators of $H_1(\Omega, S_1)$ which can be associated with simple open curves whose end points lie in distinct points of S_1. In other words, if

$$c_{(i+\dim\operatorname{im}\tilde{\jmath}_1)} \quad \text{for } 1 \le i \le \dim\ker\tilde{\imath}_0,$$

is such a set of curves, they can be defined (assuming Ω is connected) so that

$$\partial c_{(i+\dim\operatorname{im}\tilde{\jmath})} = p_i - p_0,$$

where p_0 is a datum node lying in some connected component of S_1 and each p_i lies in some distinct connected component of S_1. That is, there is one p_i in each connected component of S_1.

In this case, the electromagnetic interpretation is that the period

$$\int_{c_i} \boldsymbol{E} \cdot \boldsymbol{t}\, dl = V_i$$

is associated with potential differences between connected components of S_1.

Now suppose that the periods of $\boldsymbol{E}$ on generators of the first group vanish. In this case, it is seen from the long exact sequence that the period of $\boldsymbol{E}$ vanishes on all generators of $H_1(\Omega)$ since the tangential components of $\boldsymbol{E}$ vanish on S_1. Hence, $\boldsymbol{E}$ may be expressed as the gradient of a single-valued scalar ϕ. Furthermore, the scalar is a constant on each connected component of S_1. That is, $\boldsymbol{E} = \operatorname{grad}\phi$ in Ω, where $\phi = \phi(p_i)$ on the ith component of S_1.

When $\boldsymbol{E}$ is expressed in this form, the periods of $\boldsymbol{E}$ on the generators of $H_1(\Omega, S_1)$ which lie in the second group are easy to calculate:

$$V_{(i+\dim\operatorname{im}\tilde{\jmath}_1)} = \int_{c_{(i+\dim\operatorname{im}\tilde{\jmath}_1)}} \operatorname{grad}\phi \cdot dl = \phi(p_i) - \phi(p_0)$$

since

$$\partial c_{(\dim\operatorname{im}\tilde{\jmath}_1+1)} = p_i - p_0. \qquad \square$$

Example 1.15 Magnetostatics: Visualizing $H_2(\Omega, S)$ in 3 dimensions. As in the previous example, let Ω be a compact subset of $\mathbb{R}^3$ such that $\partial\Omega = S_1 \cup S_2$ where $S_1 \cap S_2$ has no area. Let $\boldsymbol{B}$ be a vector field in Ω so that $\operatorname{div}\boldsymbol{B} = 0$ in Ω and $\boldsymbol{B}\cdot\boldsymbol{n} = 0$ on S_1. There is no constraint on $\boldsymbol{B}$ with respect to S_2. Elements of $C_c^2(\Omega - S_1)$ can be identified with vector fields whose component normal to S_1 vanishes, so $\boldsymbol{B}$ can be associated with an element of $Z_c^2(\Omega - S_1)$. We also let Σ_i, where $1 \le i \le \beta_2(\Omega, S_1)$, be a set of surfaces associated with a basis of $H_2(\Omega, S_1)$.

The long exact homology sequence for the pair (Ω, S_1) is

$$H_2(\Omega, S_1) = \delta_2^{-1}(\ker\tilde{\imath}_1) \oplus \tilde{\jmath}_2\left(\frac{H_2(\Omega)}{\tilde{\imath}_2(H_2(S_1))}\right)$$

where the relevant portion of the long exact homology sequence is

$$\cdots \xrightarrow{\delta_3} H_2(S_1) \xrightarrow{\tilde{\imath}_2} H_2(\Omega) \xrightarrow{\tilde{\jmath}_2} H_2(\Omega, S_1) \xrightarrow{\delta_2} H_1(S_1) \xrightarrow{\tilde{\imath}_1} H_1(\Omega) \xrightarrow{\tilde{\jmath}_1} \cdots$$

Using the three-step procedure from page 38, surfaces Σ_i can be arranged into two groups:

(1) There are dim im $\tilde{\jmath}_2$ generators of $H_2(\Omega, S_1)$ which are homologous in the absolute sense to generators of $H_2(\Omega)$. Thus let $\partial\Sigma_i = 0$, for $1 \leq i \leq$ dim im $\tilde{\jmath}_2$ and associated to these Σ_i is a basis of im $\tilde{\jmath}_2$ in $H_2(\Omega, S_1)$.
(2) There are dim ker $\tilde{\imath}_1$ remaining generators of $H_2(\Omega, S_1)$ whose image under δ_2 form in $H_1(S_1)$ a basis for ker $\tilde{\imath}_1$. Hence let

$$\partial\Sigma_{\dim\,\mathrm{im}\,\tilde{\jmath}_2+i} = z_1 \quad \text{for} \quad 1 \leq i \leq \dim\,\ker\tilde{\imath}_1$$

where the z_i are associated with ker $\tilde{\imath}_1$.

Considering the periods of $\boldsymbol{B}$ on the generators of $H_2(\Omega, S_1)$ in the first group, it is clear that if

$$\boldsymbol{B} = \mathrm{curl}\,\boldsymbol{A} \quad \text{in } \Omega$$

then the periods must vanish because

$$\Phi_i = \int_{\Sigma_i} \boldsymbol{B}\cdot\boldsymbol{n}\,dS = \int_{\partial\Sigma_i} \boldsymbol{A}\cdot\boldsymbol{t}\,dl = 0,$$

since

$$\partial\Sigma_i = 0 \quad \text{for } 1 \leq i \leq \dim\,\mathrm{im}\,\tilde{\jmath}_2.$$

While this is a restriction, it is still natural to formulate the problem in terms of a vector potential since the nonzero periods of $\boldsymbol{B}$ on the generators of the first group can only be associated with distributions of magnetic monopoles in $\mathbb{R}^3 - \Omega$. Assuming that the periods of $\boldsymbol{B}$ on the generators of the first group vanish, and $\boldsymbol{B}$ is related to a vector potential $\boldsymbol{A}$, the periods of $\boldsymbol{B}$ on the generators of $H_2(\Omega, S_1)$ which lie in the second group can easily be expressed in terms of the vector potential:

$$\Phi_{\dim\,\mathrm{im}\,\tilde{\jmath}_2+i} = \int_{\Sigma_{1+\dim\,\mathrm{im}\,\tilde{\jmath}_2}} \boldsymbol{B}\cdot\boldsymbol{n}\,dS = \int_{\partial\Sigma_{\dim\,\mathrm{im}\,\tilde{\jmath}_2+i}} \boldsymbol{A}\cdot\boldsymbol{t}\,dl = \int_{z_i} \boldsymbol{A}\cdot\boldsymbol{t}\,dl,$$

since

$$\partial\Sigma_{\dim\,\mathrm{im}\,\tilde{\jmath}_2+i} = z_i \quad \text{for } 1 \leq i \leq \dim\,\ker\tilde{\imath}_1.$$

It is worthwhile to consider how the tangential components of $\boldsymbol{A}$ are to be prescribed on S_1 so that $\boldsymbol{B}\cdot\boldsymbol{n} = 0$ and the periods of $\boldsymbol{A}$ can be prescribed. One cannot impose curl $\boldsymbol{A}\cdot\boldsymbol{n} = 0$ on S_1 by forcing $\boldsymbol{n}\times\boldsymbol{A} = 0$ on S_1 because this would imply

$$\Phi_{\dim\,\mathrm{im}\,\tilde{\jmath}_2+i} = \int_{z_i} \boldsymbol{A}\cdot\boldsymbol{t}\,dl = 0,$$

since

$$z_i \in C_c^1(S_1).$$

Instead, as in the analogous case of electrostatics, one has to let

$$\boldsymbol{n}\times\boldsymbol{A} = \boldsymbol{n}\times\mathrm{grad}\,\psi = \overline{\mathrm{curl}}\,\psi \quad \text{on } S_1,$$

where ψ is a multivalued function of the coordinates on S_1. Following the reasoning in Example 1.5, this function can be made single-valued on

$$S_1 - \bigcup_{j=i}^{\beta_1(S_1,\partial S_1)} d_i,$$

where the d_i are a set of curves associated with the generators of $H_1, (S_1, \partial S_1)$ and the periods of the multivalued function ψ are given by specifying the jumps of ψ on d_j, denoted by $[\psi]_{d_j}$, where $1 \le j \le \beta_1(S_1, \partial S_1)$. To see how this is done, consider the following portion of the long exact homology sequence for the pair (Ω, S_1):

$$\cdots \longrightarrow H_2(\Omega) \xrightarrow{\tilde{\jmath}_2} H_2(\Omega, S_1) \xrightarrow{\delta_2} H_1(S_1) \xrightarrow{\tilde{\imath}_1} H_1(\Omega) \xrightarrow{\tilde{\jmath}_1} H_1(\Omega, S_1) \xrightarrow{\delta_1} \cdots$$

Using the reasoning of the three-step procedure one has

$$H_1(S_1) \simeq (\tilde{\imath}_1)^{-1} (\ker \tilde{\jmath}_1) \oplus \delta_2 \left(\frac{H_2(\Omega, S_1)}{\tilde{\jmath}_2 (H_2(\Omega))} \right).$$

Thus the generators of $H_2(S_1)$ can be arranged into two groups where $\beta_1(S_2)$ curves z_i, for $1 \le i \le \dim \operatorname{im} \delta_2$, can be associated with boundaries of generators of $H_2(\Omega, S_1)$. In addition, $z_{i+\dim \operatorname{im} \delta_2}$ curves, for $1 \le i \le \dim \ker \tilde{\jmath}_1$, which are homologous in $H_1(\Omega)$, are associated to a set of generators of $\operatorname{im} \tilde{\imath}_1$.

One can define an intersection number matrix $\mathcal{I}$ whose ijth entry is the number of oriented intersections of curve z_i with curve d_j. Then, if p_j is the period of $\operatorname{grad} \psi$ on z_j,

$$\begin{aligned} p_i = \int_{z_i} \boldsymbol{A} \cdot \boldsymbol{t}\, dl &= \int_{z_i} \operatorname{grad} \psi \cdot \boldsymbol{t}\, dl \\ &= \sum_{j=1}^{\beta_1(S_1, \partial S_1)} \mathcal{I}_{ij} [\psi]_{d_j} \quad \text{for } 1 \le i \le \beta_1(S_1), \end{aligned}$$

where $p_i = \Phi_{i+\dim \operatorname{im} \tilde{\jmath}_2}$, for $1 \le i \le \dim \operatorname{im} \delta_2$, and the remaining p_i are prescribed arbitrarily. Assuming, as before, that the matrix with entries m_{ij} is nonsingular, the above system of linear equations can be inverted to give the jumps in the scalar ψ in terms of the dim im δ_2 periods of the vector potential and dim ker $\tilde{\jmath}_1$ other arbitrary constants. This technique generalizes and simplifies that of [MN82]. The assumption that the matrix with entries m_{ij} is square and nonsingular is a consequence of the Lefschetz duality theorem which will be considered in the next chapter. □

1J. A Remark on the Association of Relative Cohomology Groups with Perfect Conductors

The introduction of relative cohomology groups in order to describe the electromagnetic field outside of a good conductor follows logically from two assumptions:

(1) The normal component of the magnetic flux density vector $\boldsymbol{B}$ and the tangential components of the electric field intensity vector $\boldsymbol{E}$, are continuous across the interface between two media. One of the two media can be a perfect conductor.
(2) The fields $\boldsymbol{E}$ and $\boldsymbol{B}$ vanish inside a perfect conductor.

In practice, we do not derive the exact conditions under which these assumptions hold on a case-by-case basis, but we do have general criteria for deciding when these assumptions are valid. The first assumption follows from the assumption that the electromagnetic field is finite at every point in space and that the length scales of interest are large compared to the size of an atom. Otherwise, the notion of an interface is open to question. The second assumption pertaining to the vanishing of fields in a perfect conductor is a little more tricky since a precise justification depends on the notion of a skin depth, denoted by δ. The general rule of thumb is that the fields inside a conductor are considered to be negligible if the local radius of curvature is much larger than δ. This criterion is simple enough to understand if we are given δ, but there is a hitch. The skin depth depends on the frequency of the excitation and the material properties of the medium. Hence, the criterion applies to the Fourier transform of the electromagnetic field with respect to time. Specifically, for a nonmagnetic medium of conductivity σ and a Fourier component of frequency f, we have in MKS units,

$$\delta = \frac{1}{2\pi\sqrt{\frac{\sigma f}{10^7}}}.$$

Although this completes the criteria for deciding whether or not a given conductor behaves like a perfect conductor for some excitation, it is useful to give a few numbers by way of example. For copper, $\sigma = 5.6 \times 10^7$ ohms/meter, so

$$\delta = \frac{1}{2\pi\sqrt{5.6f}}.$$

To relate this to engineering applications involving copper wires, consider the following table, which lists operating frequency and skin depth of several familiar electromagnetic systems:

Application	f	δ
Power system	60 Hz	9 mm
AM radio	1 MHz	67 μm
FM radio / PC motherboard	100 MHz	6.7 μm
Cellular phone / VLSI circuit	2 GHz	1.5 μm

As a rule, if the conductors are much thicker than the frequency-dependent skin depth, we can assume that the tangential electric fields and the normal magnetic fields are negligible.

The axiomatic method has many advantages over honest work.

Bertrand Russell

May one plow with an ox and an ass together? The like of you may write everything and prove everything in quaternions, but in the transition period the bilingual method may help to explain the more perfect.

James Clerk Maxwell, in a letter to P. G. Tait.

2
Quasistatic Electromagnetic Fields

The purpose of this chapter is to articulate the notion of a quasistatic electromagnetic system, and develop the topological aspects of the boundary value problems encountered in the analysis of quasistatic systems. The topological approach gives a new perspective on variational formulations which form the basis of finite element analysis.

2A. The Quasistatic Limit of Maxwell's Equations

Maxwell's Equations. Let S be a surface with boundary, V a volume in $\mathbb{R}^3$, and note that ∂ denotes the boundary operator. The integral versions of Maxwell's equations are as follows:

(2–1) $$\int_{\partial S} \boldsymbol{E} \cdot dl = -\frac{d}{dt}\int_S \boldsymbol{B} \cdot dS \qquad \text{(Faraday's Law)}$$

(2–2) $$\int_{\partial V'} \boldsymbol{B} \cdot dS = 0 \qquad \text{(Gauss' Law for magnetic charge)}$$

(2–3) $$\int_{\partial S'} \boldsymbol{H} \cdot dl = \frac{d}{dt}\int_{S'} \boldsymbol{D} \cdot dS + \int_{S'} \boldsymbol{J} \cdot dS \qquad \text{(Ampère's Law)}$$

(2–4) $$\int_{\partial V} \boldsymbol{D} \cdot dS = \int_V \rho\, dV \qquad \text{(Gauss' Law)}$$

where

$\boldsymbol{E}$ = Electric field intensity vector,
$\boldsymbol{B}$ = Magnetic field flux density vector,
$\boldsymbol{H}$ = Magnetic field intensity vector,
$\boldsymbol{D}$ = Electric field flux density vector,

and the current and charge sources are described by

$$\boldsymbol{J} = \text{Electric current flux density vector,}$$
$$\rho = \text{Electric charge density.}$$

If we let $S' = \partial V$ in Ampère's Law (2–3) and remember that $\partial\partial = 0$, then the field vector can be eliminated between Ampère's and Gauss' law, (2–3) and (2–4) to reveal a statement of charge conservation:

$$0 = \frac{d}{dt}\int_V \rho \, dV + \int_{\partial V} \boldsymbol{J} \cdot d\boldsymbol{S}.$$

This shows that conservation of charge is implicit in Maxwell's equations.

When the surfaces and volumes S, S', V, V' are stationary with respect to the inertial reference frame of the field vectors, one can use the standard theorems of vector calculus to rewrite Maxwell's equations as follows:

$$\text{curl}\, \boldsymbol{E} = -\frac{\partial \boldsymbol{B}}{\partial t}, \tag{2–5}$$

$$\text{div}\, \boldsymbol{B} = 0, \tag{2–6}$$

$$\text{curl}\, \boldsymbol{H} = \frac{\partial \boldsymbol{D}}{\partial t} + \boldsymbol{J}, \tag{2–7}$$

$$\text{div}\, \boldsymbol{D} = \rho, \tag{2–8}$$

$$\text{div}\, \boldsymbol{J} + \frac{\partial \rho}{\partial t} = 0. \tag{2–9}$$

Equations (2–5)–(2–8) are the differential versions of Maxwell's equations. Equation (2–9) is the differential version of the conservation of charge and can be obtained independently by taking the divergence of Equation (2–7) and substituting in (2–9), remembering that div curl = 0.

The differential versions of Maxwell's equations are much less general than the integral laws for three main reasons:

(1) They need to be modified for use in problems involving noninertial reference frames. Differential forms are an essential tool for this modification.
(2) The differential laws do not contain "global topological" information which can be deduced from the integral laws. The machinery of cohomology groups is the remedy for this problem.
(3) The differential laws assume differentiability of the field vectors with respect to spatial coordinates and time. However, there can be discontinuities in fields across medium interfaces, so it is not possible to deduce the proper interface conditions from the differential laws.

In order to get around the third obstacle, we must go back to the integral laws and derive so-called interface conditions which give relations for the fields across material or media interfaces (Figure 2.1). Here we simply state the interface

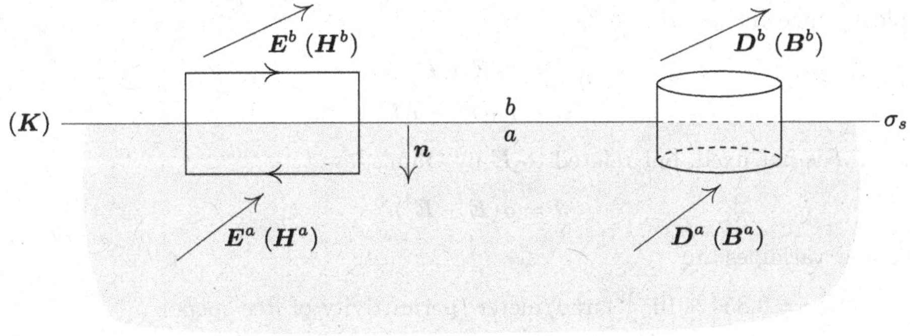

Figure 2.1.

conditions:

(2–10) $$n \times (E^a - E^b) = 0,$$

(2–11) $$n \cdot (B^a - B^b) = 0,$$

(2–12) $$n \times (H^a - H^b) = K,$$

(2–13) $$n \cdot (D^a - D^b) = \sigma_s,$$

where the superscripts refer to limiting values of the field at on the interface when the interface is approached from the side of the indicated medium. The source vectors do not really exist in nature but rather represent a way of modeling current distribution in the limit of zero skin depth (that is, when $\omega = 2\pi f \to \infty$ or $\sigma \to 0$). This approximation will be discussed soon. From a limiting case of the integral laws one has a statement of conservation of surface charge:

$$n \cdot (J^a - J^b) + \operatorname{div}_S K + \frac{\partial \sigma_S}{\partial t} = 0,$$

where div_S is the divergence operator in the interface.

In summary, one can say that the integral version of Maxwell's equations is equivalent to three distinct pieces of information:

(1) the differential version in regions where constitutive laws are continuous, and the field vectors are continuous and have the appropriate degree of differentiability;
(2) interface conditions where constitutive laws are discontinuous;
(3) global topological information which is lost when problems are specified on subsets of Euclidean space—these are the lumped parameters of circuit theory.

Constitutive Laws. In general, Maxwell's equations are insufficient for determining the electromagnetic field since there are six independent equations in twelve unknowns, namely the components of E, B, D, and H. The first step to closing this gap is to introduce constitutive laws. For stationary media, these typically take the form

$$\boldsymbol{D} = \varepsilon_0 \boldsymbol{E} + \boldsymbol{P},$$
$$\boldsymbol{B} = \mu_0(\boldsymbol{H} + \boldsymbol{M})$$

and, if $\boldsymbol{J}$ is not fixed, but related to $\boldsymbol{E}$ by Ohm's law,

$$\boldsymbol{J} = \sigma(\boldsymbol{E} + \boldsymbol{E}^i).$$

The new variables are

$\varepsilon_0 = 8.854 \times 10^{-12}$farad/meter (permittivity of free space),

$\mu_0 = 4\pi \times 10^{-7}$henry/meter (permeability of free space),

$\sigma =$ conductivity,

$\boldsymbol{P} =$ polarization,

$\boldsymbol{M} =$ magnetization,

$\boldsymbol{E}^i =$ external impressed field due to forces of chemical origin.

Generally the functional relationships

$$\boldsymbol{P} = \boldsymbol{P}(\boldsymbol{E}, \boldsymbol{r}), \quad \boldsymbol{M} = \boldsymbol{M}(\boldsymbol{H}, \boldsymbol{r})$$

are assumed, though sometimes it makes sense to have $\boldsymbol{M} = \boldsymbol{M}(\boldsymbol{B}, \boldsymbol{r})$.

There are several ways to characterize media:

(1) Lossless: $\partial P_i/\partial E_j = \partial P_j/\partial E_i$, $\partial M_i/\partial H_j = \partial M_j/\partial H_i$ when referred to Cartesian coordinates.
(2) Homogeneous: $\boldsymbol{P} = \boldsymbol{P}(\boldsymbol{E})$, $\boldsymbol{M} = \boldsymbol{M}(\boldsymbol{H})$.
(3) Isotropic: $\boldsymbol{P} = \varepsilon_0 \chi_e(|\boldsymbol{E}|^2, \boldsymbol{r})\boldsymbol{E}$, $\boldsymbol{M} = \chi_m(|\boldsymbol{H}|^2, \boldsymbol{r})\boldsymbol{H}$ where χ_e and χ_m are called the electric and magnetic susceptibilities, respectively.
(4) Linear: $\boldsymbol{P}$ is a linear function of $\boldsymbol{E}$, and $\boldsymbol{M}$ is a linear function of $\boldsymbol{H}$.

Occasionally the constitutive laws are replaced by linear differential operators:

$$\boldsymbol{P} = \varepsilon_0 \left(\sum_{i=0}^{2} \chi_i \frac{d^i}{dt^i} \right) \boldsymbol{E}.$$

This leads into the topic of dispersion relations in optics and other interesting phenomena which we will sidestep.

Remark on the concept of energy. The condition for lossless media is equivalent to saying that the integrals

$$\int_{\boldsymbol{E}_0}^{\boldsymbol{E}_1} \boldsymbol{D}(\boldsymbol{E}) \cdot d\boldsymbol{E}, \quad \int_{\boldsymbol{H}_0}^{\boldsymbol{H}_1} \boldsymbol{B}(\boldsymbol{H}) \cdot d\boldsymbol{H}$$

are independent of path connecting initial and final states (at each point in physical space). Often integration by parts or some monotonicity assumption then shows that the values of the integrals

$$w_e = \int_{\boldsymbol{D}(\boldsymbol{E}_0)}^{\boldsymbol{D}(\boldsymbol{E}_1)} \boldsymbol{E}(\boldsymbol{D}) \cdot d\boldsymbol{D}, \quad w_m = \int_{\boldsymbol{B}(\boldsymbol{H}_0)}^{\boldsymbol{B}(\boldsymbol{H}_1)} \boldsymbol{H}(\boldsymbol{B}) \cdot d\boldsymbol{B}$$

representing "electric and magnetic energy densities" are defined independent of how final states are achieved. When this is so, we define total energies by:

$$W_e = \int_V w_e \, dV, \quad W_m = \int_V w_m \, dV$$

so that

$$\frac{dW_e}{dt} = \int_V \boldsymbol{E} \cdot \frac{\partial \boldsymbol{D}}{\partial t} dV, \quad \frac{dW_m}{dt} = \int_V \boldsymbol{H} \cdot \frac{\partial \boldsymbol{B}}{\partial t} dV$$

Parameters Characterizing Linear Isotropic Media. For linear media the constitutive relations are simply

$$\begin{aligned} \boldsymbol{D} &= \varepsilon \boldsymbol{E} \quad (\varepsilon = \varepsilon_0(1+\chi_e)), \\ \boldsymbol{B} &= \mu \boldsymbol{H} \quad (\mu = \mu_0(1+\chi_m)), \\ \boldsymbol{J} &= \sigma \boldsymbol{E} \quad \text{(Ohm's Law)}. \end{aligned}$$

It is often necessary to take a Fourier transform with respect to time in which case the frequency parameter ω plays an important role in approximations. Thus the four parameters, $\omega, \sigma, \mu, \varepsilon$ describe the behavior of the constitutive relations at each point in space. From the point of view of wave propagation, it is usually more convenient to introduce a different set of parameters:

$$\begin{aligned} v &= \frac{1}{\sqrt{\mu\varepsilon}} && \text{(wave speed)} \\ \eta|_{\sigma=0} &= \sqrt{\frac{\mu}{\varepsilon}} && \text{(wave impedance in lossless media, i.e. } \sigma = 0) \\ \tau_e &= \frac{\varepsilon}{\sigma} && \text{(dielectric relaxation time)} \\ \delta &= \sqrt{\frac{2}{\omega\sigma\mu}} && \text{(skin depth)} \\ R_S &= \sqrt{\frac{\omega\mu}{2\sigma}} && \text{(surface resistivity)} \end{aligned}$$

(Note that $c = 1/\sqrt{\varepsilon_0\mu_0}$ is the speed of light in vacuum.) Although this set of parameters is unmotivated at this point, their interpretation and role in simplifying solutions will become clear when we consider electromagnetic waves.

The Standard Potentials for Maxwell's Equations. In this section we present the familiar potentials for the electromagnetic field as an efficient way to end up with "fewer equations in fewer unknowns." Recall Maxwell's equations in space-time $\mathbb{R}^3 \times \mathbb{R}$:

$$\operatorname{curl} \boldsymbol{H} = \boldsymbol{J} + \frac{\partial \boldsymbol{D}}{\partial t}, \tag{2–14}$$

$$\operatorname{div} \boldsymbol{D} = \rho, \tag{2–15}$$

$$\operatorname{curl} \boldsymbol{E} = -\frac{\partial \boldsymbol{B}}{\partial t}, \tag{2–16}$$

$$\operatorname{div} \boldsymbol{B} = 0. \tag{2–17}$$

It appears that we have eight scalar equations in twelve unknowns, the twelve unknowns being the components of $\boldsymbol{E}, \boldsymbol{B}, \boldsymbol{D}$, and $\boldsymbol{H}$. However, constitutive laws provide six more equations so that we should think of the above as eight equations in six unknowns. The standard potentials enable us to solve equations (2–16) and (2–17) explicitly so that we end up with four equations in four unknowns.

Since Equation (2–17) is valid throughout $\mathbb{R}^3$ and the second cohomology group of $\mathbb{R}^3$ is trivial, we have

$$\boldsymbol{B} = \operatorname{curl} \boldsymbol{A} \tag{2–18}$$

for some vector potential $\boldsymbol{A}$. Equations (2–16) and (2–18) can be combined to give

$$\operatorname{curl}\left(\boldsymbol{E} + \frac{\partial \boldsymbol{A}}{\partial t}\right) = 0.$$

Using the fact that this equation is valid throughout $\mathbb{R}^3$, one can conclude that for a scalar potential ϕ

$$\boldsymbol{E} + \frac{\partial \boldsymbol{A}}{\partial t} = -\operatorname{grad} \phi$$

or

$$\boldsymbol{E} = -\frac{\partial \boldsymbol{A}}{\partial t} - \operatorname{grad} \phi. \tag{2–19}$$

Equations (2–18) and (2–19) express the six vector components of $\boldsymbol{E}$ and $\boldsymbol{B}$ in terms of the four components of $\boldsymbol{A}$ and ϕ. Note that this is the most general solution to Equations (2–16) and (2–17). We have neglected to state conditions on the differentiability of $\boldsymbol{A}$ and ϕ. However it is straightforward to verify that the interface conditions associated with equations (2–16) and (2–17), namely

$$\boldsymbol{n} \times (\boldsymbol{E}^a - \boldsymbol{E}^b) = 0, \qquad \boldsymbol{n} \cdot (\boldsymbol{B}^a - \boldsymbol{B}^b) = 0,$$

are satisfied if $\boldsymbol{A}$ and ϕ are continuous, and we appeal to energy arguments to show that $\boldsymbol{A}$ and ϕ are differentiable if the constitutive laws are smooth in spatial variables. Another way of understanding this is by regarding $\boldsymbol{A}$ and ϕ to be solutions to the integral forms associated with (2–16) and (2–17).

The substitution of the potentials (2–18) and (2–19) and constitutive laws relating $\boldsymbol{D}$ and $\boldsymbol{H}$ to $\boldsymbol{E}$ and $\boldsymbol{B}$ into Equations (2–14) and (2–15) now yields four scalar equations in four unknowns. It turns out that although Maxwell's equations and constitutive laws specify $\boldsymbol{E}$ and $\boldsymbol{B}$ uniquely, they do not specify $\boldsymbol{A}$ and ϕ uniquely. This ambiguity in the potentials is a result of equations (2–14) and (2–15) being dependent and consistent only if the conservation of charge is satisfied. Hence, for a physically meaningful prescription of $\boldsymbol{J}$ and ρ, the potentials are nonunique.

The nonuniqueness of the potentials is summarized by gauge transformations which will now be described. Suppose one has two sets of potentials $(\boldsymbol{A}, \phi)$ and $(\boldsymbol{A}', \phi')$ related by the "gauge transformation"

$$\begin{aligned} \boldsymbol{A}' &= \boldsymbol{A} + \operatorname{grad} \chi \\ \phi' &= \phi - \frac{\partial \chi}{\partial t}. \end{aligned} \tag{2–20}$$

It is easy to verify that the gauge transformation leaves the field vectors invariant:

$$\boldsymbol{B}' = \operatorname{curl} \boldsymbol{A}' = \operatorname{curl} \boldsymbol{A} + \operatorname{curl} \operatorname{grad} \chi = \boldsymbol{B}$$

since curl grad = 0, and

$$\begin{aligned}\boldsymbol{E}' &= -\frac{\partial A'}{\partial t} - \operatorname{grad} \phi' = -\frac{\partial \boldsymbol{A}}{\partial t} - \frac{\partial}{\partial t}(\operatorname{grad} \chi) - \operatorname{grad} \phi + \operatorname{grad} \frac{\partial \chi}{\partial t} \\ &= -\frac{\partial \boldsymbol{A}}{\partial t} - \operatorname{grad} \phi = \boldsymbol{E}.\end{aligned}$$

In this way we see that the substitution of equations (2–18) and (2–19) into (2–14) and (2–15) yields three independent scalar equations for three independent scalar fields. We expect a solution if and only if the sources $\boldsymbol{J}$ and ρ are prescribed in a way consistent with the conservation of charge. The nonuniqueness of $\boldsymbol{A}$ and ϕ enables us to add additional constraints which have no physical significance but result in mathematical convenience through a process called gauge-fixing. Gauge-fixing usually consists of imposing an additional constraint in the form of a linear differential operator acting on $\boldsymbol{A}$ and ϕ. We consider two examples describing the Lorentz and Coulomb gauges.

Example 2.1 The Lorentz gauge. Suppose that

$$\operatorname{div} \boldsymbol{A} + \mu_0 \varepsilon_0 \frac{\partial \phi}{\partial t} = 0 \tag{2–21}$$

is imposed on $\boldsymbol{A}$ and ϕ as an additional constraint. This places a constraint on the gauge function χ which can be computed as follows:

$$0 = \left(\operatorname{div} \boldsymbol{A}' + \mu_0 \varepsilon_0 \frac{\partial \phi'}{\partial t}\right) - \left(\operatorname{div} \boldsymbol{A} + \mu_0 \varepsilon_0 \frac{\partial \phi}{\partial t}\right)$$

or

$$0 = \operatorname{div}(\boldsymbol{A}' - \boldsymbol{A}) + \mu_0 \varepsilon_0 \frac{\partial}{\partial t}(\phi' - \phi).$$

By Equation (2–20) this becomes

$$0 = \operatorname{div} \operatorname{grad} \chi - \mu_0 \varepsilon_0 \frac{\partial^2 \chi}{\partial t^2},$$

so we see that χ must satisfy a homogeneous wave equation. If suitable boundary and initial conditions are imposed on div and ϕ, χ would also be constrained by homogeneous boundary and initial conditions. Furthermore $\chi = 0$ by the uniqueness of solutions to the wave equation. Hence, the Lorentz condition (2–21) enables $\boldsymbol{A}$ and ϕ to be specified uniquely. □

Example 2.2 The Coulomb gauge. Suppose that

$$\operatorname{div} \boldsymbol{A} = 0. \tag{2–22}$$

Then, proceeding as before, by (2–20) we have

$$0 = \operatorname{div} \boldsymbol{A}' - \operatorname{div} \boldsymbol{A} = \operatorname{div}(\boldsymbol{A}' - \boldsymbol{A}) = \operatorname{div} \operatorname{grad} \chi.$$

χ is now constrained to be a harmonic function, and suitable boundary conditions on $\boldsymbol{A}$ force χ to be a constant, and we can once more specify $\boldsymbol{A}$ and ϕ uniquely by using the Coulomb gauge (2–22). □

Wave Equations and Superposition Integrals. As promised, we will write four scalar wave equations in the components of $\boldsymbol{A}$ and ϕ by substituting equations (2–18) and (2–19) into (2–14) and (2–15) and appealing to the constitutive relations. In order to avoid getting bogged down in modeling of material properties, we will assume the simplest constitutive laws, namely linear, homogeneous, isotropic media. This is accomplished by letting

$$\boldsymbol{B} = \mu_0 \boldsymbol{H}$$

and

$$\boldsymbol{D} = \varepsilon \boldsymbol{E} \quad (\varepsilon \geq \varepsilon_0),$$

where ε and μ are constant scalars when the vectors are written in terms of Cartesian coordinates. Hence equations (2–18), (2–19), (2–14), and (2–15) give

$$\begin{aligned} \boldsymbol{J} &= \operatorname{curl} \boldsymbol{H} - \frac{\partial \boldsymbol{D}}{\partial t} = \operatorname{curl} \frac{\boldsymbol{B}}{\mu} - \frac{\partial}{\partial t}(\varepsilon \boldsymbol{E}) \\ &= \operatorname{curl}\left(\frac{1}{\mu}\operatorname{curl} \boldsymbol{A}\right) + \frac{\partial}{\partial t}\left(\varepsilon \frac{\partial \boldsymbol{A}}{\partial t}\right) + \frac{\partial}{\partial t}(\varepsilon \operatorname{grad} \phi) \end{aligned} \tag{2–23}$$

and

$$\rho = \operatorname{div} \boldsymbol{D} = \operatorname{div}(\varepsilon \boldsymbol{E}) = -\operatorname{div}\left(\varepsilon \frac{\partial \boldsymbol{A}}{\partial t}\right) - \operatorname{div}(\varepsilon \operatorname{grad} \phi).$$

That is,

$$-\operatorname{curl}\operatorname{curl} \boldsymbol{A} - \mu\varepsilon \frac{\partial^2 \boldsymbol{A}}{\partial t^2} - \operatorname{grad}\left(\mu\varepsilon \frac{\partial \phi}{\partial t}\right) = -\mu \boldsymbol{J} \tag{2–24}$$

and

$$\operatorname{div}\operatorname{grad} \phi + \frac{\partial}{\partial t}(\operatorname{div} \boldsymbol{A}) = -\frac{\rho}{\varepsilon}. \tag{2–25}$$

Although this is the final product, it is possible to do much better. From the discussion of gauge invariance we know that we have four equations but, assuming that $\boldsymbol{J}$ and ρ are prescribed in a manner consistent with the conservation of charge, only three equations are independent and one is free to impose a gauge-fixing condition for convenience. In the case at hand, the Lorentz gauge (2–21) uncouples $\boldsymbol{A}$ from ϕ in Cartesian coordinates. Putting (2–21) into (2–24) and (2–25) results in the following uncoupled equations:

$$\operatorname{grad}(\operatorname{div} \boldsymbol{A}) - \operatorname{curl}(\operatorname{curl} \boldsymbol{A}) - \mu\varepsilon \frac{\partial^2 \boldsymbol{A}}{\partial t^2} = -\mu \boldsymbol{J} \tag{2–26}$$

and

$$\operatorname{div}(\operatorname{grad} \phi) - \mu\varepsilon \frac{\partial^2 \phi}{\partial t^2} = -\frac{\rho}{\varepsilon}. \tag{2–27}$$

It is possible to take one more step by delving into the arcana of Cartesian coordinates. If some vector field $\boldsymbol{C}$ has the form

$$\boldsymbol{C} = \sum_{i=1}^{3} C_i \hat{e}_i$$

in terms of the unit coordinate vectors $\hat{e}_i$, $1 \le i \le 3$, we have the identity

$$\text{grad div}\, \boldsymbol{C} - \text{curl curl}\, \boldsymbol{C} = \sum_{i=1}^{3} (\text{div grad}\, C_i)\hat{e}_i.$$

Applying the vector identity to the vector potential $\boldsymbol{A}$ and substituting into (2–26) gives:

$$\text{div grad}\, A_i - \mu\varepsilon \frac{\partial^2 A_i}{\partial t^2} = -\mu \boldsymbol{J}_i \quad (i = 1, 2, 3). \tag{2–28}$$

Thus, ϕ and the Cartesian components of $\boldsymbol{A}$ all satisfy a wave equation of the form

$$\text{div grad}\, \psi - \frac{1}{v^2}\frac{\partial^2 \psi}{\partial t^2} = f \tag{2–29}$$

where (Section 2A)

$$v = \frac{1}{\sqrt{\mu\varepsilon}} < c = \frac{1}{\sqrt{\mu_0\varepsilon_0}} = 3 \times 10^8 \text{ meters/second.}$$

As a historical note, it was Maxwell who made the fundamental discovery that adding the displacement current term to Ampère's Law enables one to show that electromagnetic disturbances can propagate and that they do so at the speed of light. Others had computed the value of $(\varepsilon_0\mu_0)^{-1/2}$ but attached no physical significance to the number. A further comment is that at this stage, the Lorentz gauge seems to work "like magic." To demystify it and give a forward pointer to the use of differential forms, we will simply say that the Lorentz gauge makes the four-component 1-form $(\boldsymbol{A}, \phi)$ a coclosed form on Minkowski space and that the Cartesian vector identity is a special case of the Weizenbrock identity relating the Laplace–Beltrami operator to the curvature tensor and covariant derivatives.

A more pedestrian view of the Lorentz gauge is that equations (2–27) and (2–28) can now be "solved" by appealing to the fundamental solution of equation (2–29). However, before doing so it is useful to reflect on the possibility of misinterpreting these equations by contrasting Equations (2–24) and (2–25) with (2–27) and (2–28). Equations (2–24) and (2–25) have a solution only if the conservation of charge is respected in the prescription of $\boldsymbol{J}$ and ρ. When this is the case, the potentials are not unique but related by gauge transformation (2–20). On the other hand, Equations (2–27) and (2–28) always have a unique, regardless of how $\boldsymbol{J}$ and ρ are prescribed. To reconcile these statements, it is possible to derive from equations (2–27) and (2–28) the identity

$$\text{div grad}\, f - \frac{1}{\mu_0\varepsilon_0}\frac{\partial^2 f}{\partial t^2} = \text{div}\, \boldsymbol{J} + \frac{\partial \rho}{\partial t}$$

where

$$f = \frac{1}{\mu_0}\left(\text{div}\, \boldsymbol{A} + \mu_0\varepsilon_0 \frac{\partial \phi}{\partial t}\right).$$

Thus, we see that if the conservation of charge is violated in the modeling of an electromagnetic device, the potentials cannot satisfy the Lorentz gauge and so Equations (2–27) and (2–28) are no longer equivalent to (2–24) and (2–25) which describe the electromagnetic field.

Quasistatics and Limitations of Superposition Integrals. It is important to interpret equations (2–27) and (2–28) in the context of quasistatics. In particular, the limits of μ and ε tending to zero give two ways to think of the limiting case of the speed of light $c = 1/\sqrt{\mu\varepsilon}$ becoming infinite. Electrical engineers call these two limits the electroquasistatic and magnetoquasistatic limits, respectively.

Electroquasistatics (EQS). For the moment let's ignore interface conditions and constitutive relations and naively set $\mu = 0$ in Equations (2–21), (2–27), and (2–28). This results in the following equations for the potentials:

$$\begin{aligned} \operatorname{div}\operatorname{grad}\phi &= -\frac{\rho}{\varepsilon}, \\ \operatorname{div}\operatorname{grad}A_i &= 0, \quad \text{for } i = 1, 2, 3, \\ \operatorname{div}\boldsymbol{A} &= 0. \end{aligned}$$

Thus, since a harmonic function vanishing at infinity is zero, we see that $\boldsymbol{A}$ vanishes. The electric field is described by the gradient of a scalar potential which can be time-varying. The advantage of the EQS approximation is that it enables us to work with a scalar potential while ignoring the magnetic field, and then determine the magnetic field as a perturbation by letting μ be a small parameter.

Magnetoquasistatics (MQS). If we now naively set $\varepsilon = 0$ in equations (2–21), (2–27), and (2–28) we see that the irrotational part of the electric field is undefined, and that

$$\begin{aligned} \operatorname{div}\operatorname{grad}A_i &= -\mu J_i \quad \text{for } i = 1, 2, 3, \\ \operatorname{div}\boldsymbol{A} &= 0. \end{aligned}$$

Here the Lorentz gauge reduces to the Coulomb gauge, and the magnetic field can be computed from a vector potential. The solenoidal part of the electric field is also deducible from the vector potential even though the irrotational part is undefined. If one reintroduces a small permittivity ε, and specifies a charge distribution ρ, the electric field can be modeled quite accurately.

To appreciate the limitations of these two extreme quasistatic limits, consider the "series RLC circuit" usually encountered in a first course on circuits (Figure 2.2). The second-order linear constant-coefficient differential equation which results from applying Kirchhoff's voltage law is a mathematical model of the circuit. The model is not valid if the electromagnetic field around the actual circuit is not quasistatic. Note that the converse of this statement is not generally true. For the circuit to be quasistatic, we require that the wavelengths of electromagnetic radiation at the Fourier components of $v(t)$ with nontrivial energy be large compared to the overall size of the circuit. Note however that even if the circuit can be described by quasistatics, it is neither MQS nor EQS since the magnetic field surrounding the inductor is not EQS and the electric field in the capacitor is not MQS. Furthermore, the circuit model can break down at

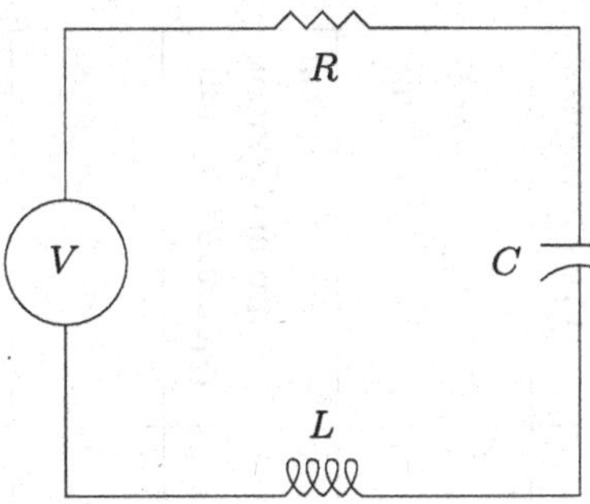

Figure 2.2. *RLC* circuit.

quasistatic frequencies where "parasitic capacitances" between windings in the inductor come into play.

In general engineering practice the aim is to break a quasistatic system into EQS and MQS subsystems and a quasistatic field analysis is performed on each of these subsystems. These subsystems (inductors, capacitors, and so on) are then given black box models where the terminal variables are related to periods of closed differential forms. Kirchhoff's laws then tie the topological information together in order to arrive at circuit equations. In this way electromagnetism and topology become the servants of circuit theory.

There is another perspective on quasistatics and engineering methodology. If one does not prescribe the current density $\boldsymbol{J}$, but rather invokes Ohm's law, then in a system which is rigid and can't do any mechanical work there are only three things that can happen with the energy in the system. It can be

(1) stored in the electric field,
(2) stored in the magnetic field, or
(3) dissipated as heat.

Hence, in the analysis of quasistatic systems, one identifies subsystems which are EQS, MQS, or characterized by the laws of steady current conduction. In the case of linear circuits, we end up with capacitors, inductors, and resistors. To clarify the process of extracting circuit parameters from EQS, MQS, and ohmic subsystems from the boundary value problems of computational electromagnetics, we devote one section of this chapter to each of these topics. The final section of the chapter then combines these three aspects of the electromagnetic field into a variational picture. However, before leaving the general picture of electromagnetics, we consider the fundamental solution for the scalar wave equation and see how it provides another framework for quasistatics, one which is not computationally useful but reconciles quasistatics with the theory of electromagnetic waves.

We return to Equation (2–29) as a means to investigating equations (2–27) and (2–28). Since we are trying to avoid issues of boundary values, interfaces, and topology in this section, we consider the wave equations in all of $\mathbb{R}^3$, and all

	EQS	MQS	Steady Currents	
	$(n=3,\ p=1)$	$(n=3,\ p=2)$	3-d $(n=3,\ p=1)$	2-d $(n=2,\ p=1)$
lumped parameter defined via relative period	$V=\int_c \boldsymbol{E}\cdot dl$	$\Phi=\int_S \boldsymbol{B}\cdot n\,dS$	$V=\int_c \boldsymbol{E}\cdot dl$	$V=\int_z \boldsymbol{E}\cdot dl$
relative groups	$[\boldsymbol{E}\cdot dl]\in H^1(\Omega,\partial\Omega)$ $[c]\in H_1(\Omega,\partial\Omega)$	$[\boldsymbol{B}\cdot dS]\in H^2(\Omega,\partial\Omega)$ $[S]\in H_2(\Omega,\partial\Omega)$	$[\boldsymbol{E}\cdot dl]\in H^1(\Omega,S_2)$ $[c]\in H_1(\Omega,S_2)$	$[\boldsymbol{E}\cdot dl]\in H^1(\Omega,S_2)$ $[z]\in H_1(\Omega,S_2)$
lumped parameter defined via dual period	$Q=\int_S \boldsymbol{D}\cdot\boldsymbol{n}\,dS$	$I=\int_c \boldsymbol{H}\cdot dl$	$I=\int_S \boldsymbol{J}\cdot\boldsymbol{n}\,dS$	$I=\int_c(\boldsymbol{K}\times\boldsymbol{n}')\cdot dl$
dual groups	$[\boldsymbol{D}\cdot dS]\in H^2(\Omega)$ $[S]\in H_2(\Omega)$	$[\boldsymbol{H}\cdot dl]\in H^1(\Omega)$ $[c]\in H_1(\Omega)$	$[\boldsymbol{J}\cdot dS]\in H^2(\Omega,S_1)$ $[S]\in H_2(\Omega,S_1)$	$[(\boldsymbol{K}\times\boldsymbol{n}')\cdot dl]\in H^1(\Omega,S_1)$ $[c]\in H_1(\Omega,S_1)$
quadratic form which descends to cohomology classes (energy)	$\int_\Omega \boldsymbol{E}\cdot\boldsymbol{D}\,dV=\int_{\partial\Omega}\phi\boldsymbol{D}\cdot dS=\sum_i V_iQ_i=\sum_{i,j}C_{ij}V_iV_j$	$\int_\Omega \boldsymbol{H}\cdot\boldsymbol{B}\,dV=\int_{\partial\Omega}(\boldsymbol{H}\times\boldsymbol{A})\cdot dS=\sum_i\phi_iI_i=\sum_{i,j}L_{ij}I_iI_j$	$\int \boldsymbol{E}\cdot\boldsymbol{J}dV=\int\phi\boldsymbol{J}\cdot dS=\sum_i V_iI_i=\sum_{i,j}V_iR_{ij}V_j$	$\int_\Omega \boldsymbol{E}\cdot\boldsymbol{K}dS$
topological obstructions to dual formulations (Lefschetz duality)	$\boldsymbol{D}=\operatorname{curl}\boldsymbol{C}$	$\boldsymbol{H}=\operatorname{grad}\psi$	$\boldsymbol{J}=\operatorname{curl}\boldsymbol{H}$	$\boldsymbol{K}=\boldsymbol{n}'\times\operatorname{grad}\chi$

Table 2.1. Lumped parameters and cohomology groups

time. In this case, the solution to Equation (2–29) takes the form

$$\psi(\boldsymbol{r},t) = -\int_{V'} \frac{f(\boldsymbol{r}',t-\frac{|\boldsymbol{r}-\boldsymbol{r}'|}{v})}{4\pi|\boldsymbol{r}-\boldsymbol{r}'|}dV' \tag{2–30}$$

so that the solutions to (2–27) and (2–28) are:

$$\begin{aligned} \boldsymbol{A}(\boldsymbol{r},t) &= -\int_{V'} \frac{\mu \boldsymbol{J}(\boldsymbol{r}',t-\frac{|\boldsymbol{r}-\boldsymbol{r}'|}{v})}{4\pi|\boldsymbol{r}-\boldsymbol{r}'|}dV', \\ \Phi(\boldsymbol{r},t) &= -\int_{V'} \frac{\rho(\boldsymbol{r}',t-\frac{|\boldsymbol{r}-\boldsymbol{r}'|}{v})}{4\pi\varepsilon|\boldsymbol{r}-\boldsymbol{r}'|}dV'. \end{aligned} \tag{2–31}$$

These solutions to the wave equations are called the *retarded potentials*, since they look like the solutions to the MQS and EQS problems, respectively, but the time variable in the sources is delayed by the time it takes for an electromagnetic disturbance to travel from the source to the point of observation. These integral formulae for the solutions also yield the correct insights into precise definition of quasistatics. Fourier transforms of (2–30) with respect to time look like the MQS and EQS integrals with an extra factor of

$$\exp^{i(2\pi f/v)|\boldsymbol{r}-\boldsymbol{r}'|}$$

in the integrand (where $i = \sqrt{-1}$). Hence in order to say that a system is quasistatic we need to define a highest frequency f_{max} above which we can agree that the Fourier components of $\boldsymbol{J}$ and ρ have negligible energy and a maximum distance $l_{\max}$ which is an upper bound on the distance between any observation point of interest and any point where the sources are nonzero. In terms of $f_{\max}$ and $l_{\max}$, the engineer's criterion for a system to be quasistatic is

$$f \le f_{\max} \ll \frac{v}{l_{\max}} < \frac{v}{|\boldsymbol{r}-\boldsymbol{r}'|}. \tag{2–32}$$

Here, f is any frequency of interest, and a factor of 10 or greater is usually used to interpret the middle inequality. Care must be taken in interpreting v. If $\varepsilon = \varepsilon_r\varepsilon_0$ is evaluated at the frequency of interest, then assuming $\mu = \mu_0$, we have (section 2A)

$$v = \frac{1}{\sqrt{\mu_0\varepsilon}} = \frac{1}{\sqrt{\varepsilon_r}\sqrt{\mu_0\varepsilon_0}} = \frac{c}{\sqrt{\varepsilon_r}}.$$

For example, if a computer circuit board has $\varepsilon_r \simeq 9$ when evaluated in the range of $10^8 < f < 10^9$, then we have that v is about one-third the speed of light or $v = 10^8$ m/s. Another way to understand the inequalities (2–32) is to rephrase them in terms of wavelengths. That is, if we define

$$\lambda = \frac{v}{f}, \qquad \lambda_{\min} = \frac{v}{f_{\max}},$$

then, in terms of wavelength, (2–32) becomes

$$\lambda \ge \lambda_{\min} \gg l_{\max} \ge |\boldsymbol{r}-\boldsymbol{r}'| \tag{2–33}$$

This inequality makes clear that there is no clearly defined frequency at which a generic electromagnetic system stops being quasistatic. Thus, a 60 Hertz power grid extending across a portion of North America, a 100 MHz circuit board in a computer, and a package for a 1 GHz microprocessor chip are all similar situations from the point of view of pushing the limits of quasistatic analysis. Furthermore, it is common to have systems which are not easily categorized. For example, the circuitry in a 1.8 GHz cell phone is designed using quasistatic assumptions while the antenna can't possibly be designed in this framework.

In summary we can state that Equations (2–31) enable us to articulate the validity of quasistatic analysis but do not help us analyze or design quasistatic systems. There are several reasons why (2–31) play a minor role in quasistatics:

(1) The superposition integrals (2–31) are, in a sense, a fortunate accident resulting from the ability to exploit the symmetries of Euclidean space in Fourier analysis. There is no analogous closed form solution in the case of inhomogeneous media or in the interface conditions associated with generic device geometries. This is a practical way of saying that (2–31) are derived under very restrictive assumptions on constitutive laws.
(2) In typical situations, the "sources" $\boldsymbol{J}$ and ρ are not known *a priori* but rather are part of the solution. For instance, in the case of Ohm's law $\boldsymbol{J}$ is a function of $\boldsymbol{E}$ so that, when restricted to a conducting body, Equations (2–31) become integral equations. The solution to these integral equations is seldom attempted when the electromagnetic field in a region of space is to be determined (as opposed to a single parameter such as antenna impedance or a scattering parameter), or when there are volumetric currents. For a survey of quasistatic applications of integral operators in the context of low frequency problems, see [Bos98].
(3) As an elaboration on the above point and as an introduction to this chapter as a whole, we should note that the "excitations" of a quasistatic system are nonlocal and topological in nature. Specifically, excitations come in the forms of voltages, currents, charges, and fluxes, which are periods of closed differential forms. Hence, we seek a formulation and language that brings us closer to these lumped parameters and associated capacitance, inductance, and resistance matrices (i.e. period matrices). This will be accomplished by identifying EQS and MQS substructures by focusing on the movement between different forms of energy. We shall see that (2–31) provide a framework for defining quasistatics, but the formalism of analytical mechanics gives us the power to make "simple" models of quasistatic systems.
(4) Although integral equations have their place in computational electromagnetics, and direct solution of differential equations has the limitation of not being useful for unbounded domains, there is no need for ideological debates. Computational electromagnetism has a brutally simple way of resolving these types of debates: once a given computational problem is defined, formulations and algorithms have been established, and a computer implementation is chosen, one "merely" looks at the accuracy of the computed solution for a given amount of computation and how this ratio scales with problem size.

As we go along, the above considerations will return to us time and time again. We will strive for formulations which are useful for nonlinear constitutive laws and inhomogeneous media, have a clear link between quasistatic fields and the topological constraints which generate them, propose finite element solution strategies, and for the more innovative aspects, we will develop techniques to estimate the amount of computational work for a given task so that there is a basis for comparing different methods.

2B. Variational Principles For Electroquasistatics

The Electroquasistatics Problem. The objective of this section is to understand how the three-step procedure for determining homology generators (page 38) comes into calculations involving the EQS coenergy principle. In particular, we will give a basic statement of the EQS problem and draw connections to the groups and the 3-step procedure outlined in Chapter 1. Then we will state the energy and coenergy functionals for electroquasistatics in the case of a linear constitutive law and see how these connect to Chapter 1. Finally, for the case of a linear constitutive law, we evaluate the coenergy functional at the extremum and see how the familiar notion of the capacitance matrix comes out of this process.

Consider a compact region Ω which contains no conducting bodies or free charges. The boundary $\partial\Omega = S_1 \cup S_2$ where $S_1 \cap S_2$ has no area and there are conditions

$$\begin{aligned} \boldsymbol{n} \times \boldsymbol{E} &= 0 \ \text{ on } S_1, \\ \boldsymbol{D} \cdot \boldsymbol{n} &= 0 \ \text{ on } S_2. \end{aligned}$$

The boundary condition on S_1 is associated with an interface to a conducting body or to a symmetry plane of the geometry while the boundary condition on S_2 can be associated with a symmetry plane. The boundary conditions associated with S_1 and S_2 can also arise on $\partial\Omega$ if there exists an interface where there is a sudden change in permittivity across $\partial\Omega$. Figure 2.3 shows S_1 and S_2 for the case of a charged loop after the geometry has been reduced by considering symmetries.

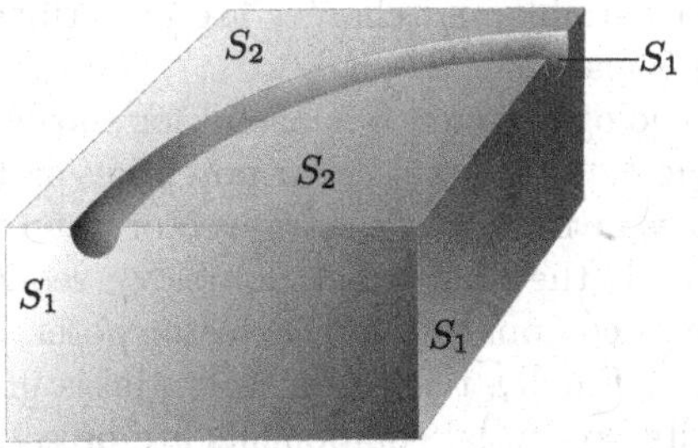

Figure 2.3. Geometry of a charged loop in a box after reduction by problem symmetry to an octant.

At this point it is possible to begin to employ the terminology of Chapter 1. Elements of $C^1_c(\Omega - S_1)$ are associated with vector fields whose components tangent to S_1 vanish. For the electric field intensity $\boldsymbol{E}$, since

$$\operatorname{curl} \boldsymbol{E} = 0 \ \text{ in } \Omega,$$
$$\boldsymbol{n} \times \boldsymbol{E} = 0 \ \text{ on } S_1,$$

it is possible to associate $\boldsymbol{E}$ with an element of $Z_c^1(\Omega - S_1)$. Similarly (and dually), elements of $C_c^2(\Omega - S_2)$ can be identified with vector fields whose component normal to S_2 vanishes. The electric field flux density $\boldsymbol{D}$, characterized by

$$\operatorname{div} \boldsymbol{D} = 0 \ \text{ in } \Omega,$$
$$\boldsymbol{D} \cdot \boldsymbol{n} = 0 \ \text{ on } S_2,$$

can therefore be identified with an element of $Z_c^2(\Omega - S_2)$.

Going to the level of homology, it can be stated that the nondegenerate bilinear pairings

$$\int : H_1(\Omega, S_1) \times H_c^1(\Omega - S_1) \to \mathbb{R},$$
$$\int : H_2(\Omega, S_2) \times H_c^2(\Omega - S_2) \to \mathbb{R},$$

induced on homology and cohomology classes by integration, are associated with potential differences and charges respectively. The fact that there are as many independent potential differences as there are independent charges suggests that

$$\beta_1(\Omega, S_1) = \beta_2(\Omega, S_2).$$

This result is discussed in Chapter 3. For now, by considering a few concrete situations the reader can convince himself that the periods of $\boldsymbol{E}$ on generators of $H_1(\Omega, S_1)$ are associated with prescribed electromotive forces while the periods of $\boldsymbol{D}$ on the generators of $H_2(\Omega, S_2)$ correspond to charges.

For variational principles where the electric field $\boldsymbol{E}$ is the independent variable and potential differences are prescribed, the variation of the extremal lies in the space $B_c^1(\Omega - S_1)$. Dually, for variational principles where the electric flux density $\boldsymbol{D}$ is the independent variable and charges are prescribed, the variation of the extremal occurs in the space $B_c^2(\Omega - S_2)$.

The long exact homology sequence is useful for showing the appropriateness of the variational principles involving scalar potentials and the limited usefulness of variational principles involving an electric vector potential. When the electric vector potential is used, the long exact homology sequence indicates how to prescribe the tangential components of the vector potential in terms of a scalar function defined on S_2. Finally, the vector potential is unique up to an element of $Z_c^1(\Omega - S_2)$ when its tangential components are prescribed on S_2.

Introduction of the Metric. It is useful to illustrate how the various spaces associated with the cochain complexes $C_c^*(\Omega - S_1)$ and $C_c^*(\Omega - S_2)$ arise in variational principles. Assume that there is a tensor constitutive relation

$$\boldsymbol{D} = \mathcal{D}(\boldsymbol{E}, \boldsymbol{r})$$

and an inverse transformation

$$\boldsymbol{E} = \mathcal{E}(\boldsymbol{D}, \boldsymbol{r})$$

such that

$$\mathcal{D}\left(\mathcal{E}(\boldsymbol{D}, \boldsymbol{r}), \boldsymbol{r}\right) = \boldsymbol{D}.$$

Furthermore, assuming the matrix

$$\frac{\partial \mathcal{D}_i}{\partial E_j}$$

is symmetric positive definite, then

$$\frac{\partial \mathcal{E}_i}{\partial D_j}$$

is symmetric positive definite.

The Capacitive Coenergy and Energy Principles. With the metric as described above, the principles of stationary capacitive energy and coenergy can be stated as follows [Mac70, pp. 332–333]. The stationary capacitive coenergy principle is

$$U(\boldsymbol{E}) = \inf_{\boldsymbol{E} \in Z_c^1(\Omega - S_1)} \int_\Omega \left(\int_0^{\boldsymbol{E}} \mathcal{D}(\xi, \boldsymbol{r}) \cdot d\xi \right) dV$$

subject to the constraint that on generators of $H_1(\Omega, S_1)$ periods are prescribed as follows

$$V_i = \int_{c_i} \boldsymbol{E} \cdot d\boldsymbol{l} \quad 1 \leq i \leq \beta_1(\Omega, S_1).$$

The stationary capacitive energy principle is

$$U'(\boldsymbol{D}) = \inf_{\boldsymbol{D} \in Z_c^2(\Omega - S_2)} \int_\Omega \left(\int_0^{\boldsymbol{D}} \mathcal{D}(\xi, \boldsymbol{r}) \cdot d\xi \right) dV$$

subject to the constraint that on generators of $H_2(\Omega, S_2)$ periods are prescribed as follows

$$Q_i = \int_{\Sigma_i} \boldsymbol{D} \cdot d\boldsymbol{l} \quad 1 \leq i \leq \beta_1(\Omega, S_2).$$

Note that in both of these variational principles the extremal is a relative cocycle and when the principal conditions are prescribed on the generators of a (co)homology group the variation of the extremal is constrained to be a relative coboundary. This is readily seen from the identities

$$Z_c^1(\Omega - S_1) \simeq H_c^1(\Omega - S_1) \oplus B_c^1(\Omega - S_1),$$
$$Z_c^2(\Omega - S_2) \simeq H_c^2(\Omega - S_2) \oplus B_c^2(\Omega - S_2).$$

In the case where the constitutive relations are linear, the coenergy and energy principles can be used to obtain upper bounds on capacitance and elastance lumped parameters respectively. This is achieved by expressing the minimum of the functional as a quadratic form in the prescribed periods and making the identifications

$$U(\boldsymbol{E}) = \frac{1}{2}\sum_{i,j=1}^{\beta_1(\Omega,S_1)} V_i C_{ij} V_j, \qquad U'(\boldsymbol{D}) = \frac{1}{2}\sum_{i,j=1}^{\beta_2(\Omega,S_2)} Q_i p_{ij} Q_j.$$

From the upper bound on elastance, a lower bound on capacitance can be obtained in the usual way. The estimation of partial capacitance can be obtained by leaving certain periods free so that their values can be determined as a by product of the minimization.

Not only is the above statement of stationary capacitive energy and coenergy principles succinct but it also gives a direct correspondence with the lumped parameter versions of the same principles. The derivation of the corresponding variational principles in terms of scalar and vector potentials is instructive since insight is gained into why the coenergy principle is naturally formulated in terms of a scalar potential while the formulation of the energy principle in terms of a vector potential requires topological constraints on the model in order for the principle to be valid. Let the coenergy principle in terms of a scalar potential be considered first.

Capacitance Via Direct Variational Methods with the Stationary Capacitive Coenergy Principle. When $\boldsymbol{E} = \operatorname{grad}\phi$ the coenergy principle can be restated as follows:

$$U(\operatorname{grad}\phi) = \inf_{\operatorname{grad}\phi\in Z^1_c(\Omega-S_1)} \int_\Omega \left(\int_0^\phi \mathcal{D}(\operatorname{grad}\eta, \boldsymbol{r})\cdot \operatorname{grad}(d\eta)\right) dV,$$

subject to the constraints

$$V_{(i+\dim\, \operatorname{im} \tilde{\jmath}_1)} = \phi(p_i) - \phi(p_0), \quad 1 \le i \le \dim\ker \tilde{\imath}_0.$$

Note that for this functional the space of admissible variations is still $B^1_c(\Omega - S_1)$, that is, ϕ can be varied by any scalar which vanishes on S_1.

Although the coenergy principle seems to be more natural when expressed in terms of a scalar potential, the only difference between the two principles is how they treat magnetic circuits in $\mathbb{R}^3 - \Omega$. The situation is quite different when one tries to express the energy principle in terms of a vector potential since, in general, the energy principle cannot be reformulated in terms of the vector potential alone.

We now elaborate on this scalar potential form of the coenergy in order to show how, when evaluated at the extremum, the capacitance matrix is a natural result of the variational principle. To begin, we write the coenergy functional as above:

$$U(\operatorname{grad}\phi^*) = \inf_{\operatorname{grad}\phi\in Z'_c(\Omega-S_1)} \int_\Omega \int_0^\phi \mathcal{D}(\operatorname{grad}\eta, \boldsymbol{r})\cdot \operatorname{grad}(d\eta)\, dV. \tag{2–34}$$

It helps to recall the meaning of $\operatorname{grad}\phi \in Z^1_c(\Omega - S_1)$. For this we need the long exact cohomology sequence

$$\cdots \leftarrow H^1(\Omega) \xleftarrow{\tilde{\jmath}^1} H^1(\Omega, S_1) \xleftarrow{\tilde{\delta}^1} H^0(S_1) \xleftarrow{i^0} H^0(\Omega) \leftarrow H^0(\Omega, S_1) \leftarrow \cdots$$

The connecting homomorphism $\tilde{\delta}^1$ describes fields for which $\boldsymbol{E} = -\operatorname{grad}\phi$, and $\operatorname{grad}\phi \times \boldsymbol{n} = 0$ on S_1 but $\phi \neq 0$ on S_1.

In electrostatics, where $\partial \boldsymbol{B}/\partial t = 0$, the periods of $\int \boldsymbol{E} \cdot dl$ in $H^1(\Omega)$ vanish so that $\tilde{j}^1$ is trivial. The nontrivial relative periods are then associated with

$$(\delta_1)^{-1}\left(\frac{H^0(S_1)}{(i^0)^{-1}(H^0(\Omega))}\right).$$

Let

$$S_1 = \bigcup_{i=0}^{\operatorname{Rank}\delta^1} S_{1i}$$

with S_{1i} connected and disjoint from S_{1j}, for $i \neq j$. If Ω is connected and ϕ is zero a single connected component S_{10} of S_1 (i.e., one component is "grounded"), then $\phi \in Z_c^1(\Omega - S_1)$ means that

$$\phi = \begin{cases} 0 & \text{on } S_{10}, \\ V_i, \ \text{for } 1 \le i \le \operatorname{Rank}\delta^1, & \text{on } S_{1i}, \end{cases}$$

where the V_i are constants.

Differentiating the integral in the space of fields, the variation of this functional becomes

$$U(\operatorname{grad}(\phi + \delta\phi)) - U(\operatorname{grad}\phi) = \int_\Omega \mathcal{D}(\operatorname{grad}\phi, \boldsymbol{r}) \cdot \operatorname{grad}\delta\phi \, dV + \mathcal{O}(\|\delta\phi\|^2),$$

where, since the V_i are constants,

$$\delta\phi = 0 \ \text{ on } S_{1i}, \ \ 0 \le i \le \operatorname{Rank}\delta^1. \tag{2–35}$$

The first variation δU is integrated by parts as follows. Note from the variation of the functional above that

$$\delta U = \int_\Omega \mathcal{D}(\operatorname{grad}\phi, \boldsymbol{r}) \cdot \operatorname{grad}(\delta\phi) \, dV.$$

Integrating by parts:

$$\delta U = \int_{\partial\Omega} \delta\phi \, \mathcal{D}(\operatorname{grad}\phi, \boldsymbol{r}) \cdot \boldsymbol{n} \, dS - \int_\Omega \delta\phi \, \operatorname{div}(\mathcal{D}(\operatorname{grad}\phi, \boldsymbol{r})) \, dV. \tag{2–36}$$

Since δU must vanish for all admissible variations $\delta\phi$, Equations (2–35) and (2–36) imply that

$$\begin{aligned} \operatorname{div}(\mathcal{D}(\operatorname{grad}\phi^*, \boldsymbol{r})) &= 0 \ \text{ in } \Omega, \\ \mathcal{D}(\operatorname{grad}\phi^*, \boldsymbol{r}) \cdot \boldsymbol{n} &= 0 \ \text{ on } \partial\Omega - S_1, \\ \phi^* &= V_i \ \text{ on } S_{1i}, \ \ 0 \le i \le \operatorname{Rank}\tilde{\delta}^1, \end{aligned} \tag{2–37}$$

where the conditions on ϕ^* are global constraints. This is the boundary value problem for ϕ^*.

Next, we would like to evaluate the coenergy functional (2–34) at the extremum using (2–37). Note that the relation

$$\frac{\partial \mathcal{D}_i}{\partial E_j} = \frac{\partial \mathcal{D}_j}{\partial E_i}$$

ensures that the line integral in the functional

$$U(\operatorname{grad}\phi^*) = \int_\Omega \left(\int_0^{\phi^*} \mathcal{D}(\operatorname{grad}\eta, \boldsymbol{r}) \cdot \operatorname{grad}(d\eta) \right) dV$$

is independent of path. This is how ϕ goes from the function that vanishes everywhere to ϕ^*. Now consider a family of functions $\phi_s^*(\boldsymbol{r})$, parametrized by s which are solutions to the following boundary value problem which, for $0 \le s \le 1$, is a variant of (2–37) and which interpolates between 0 and ϕ^*.

$$\begin{aligned} \operatorname{div}(\mathcal{D}(\operatorname{grad}\phi_s^*, \boldsymbol{r})) &= 0 \ \text{ in } \Omega \\ \mathcal{D}(\operatorname{grad}\phi_s^*, \boldsymbol{r}) \cdot \boldsymbol{n} &= 0 \ \text{ on } \partial\Omega - S_1 \\ \phi_s^* &= sV_i \ \text{ on } S_{1i},\ 0 \le i \le \operatorname{Rank}\delta^1. \end{aligned} \tag{2–38}$$

Substituting ϕ_s^* into (2–37) gives the following functional:

$$U(\operatorname{grad}\phi_s^*) = \int_\Omega \left(\int_0^1 \mathcal{D}(\operatorname{grad}\phi_s^*, \boldsymbol{r}) \cdot \operatorname{grad}\left(\frac{d\phi_s^*}{ds}\right) ds \right) dV$$

Interchanging the order of integration, integrating by parts, and using (2–38) gives

$$\begin{aligned} &U(\operatorname{grad}\phi_s^*) \\ &\quad = \int_0^1 \left(\int_\Omega \mathcal{D}(\operatorname{grad}\phi_s^*, \boldsymbol{r}) \cdot \operatorname{grad}\left(\frac{d\phi_s^*}{ds}\right) dV \right) ds \\ &\quad = \int_0^1 \left(\int_{\partial\Omega} \frac{d\phi_s^*}{ds} \mathcal{D}(\operatorname{grad}\phi_s^*, \boldsymbol{r}) \cdot \boldsymbol{n}\, dS - \int_\Omega \frac{d\phi_s^*}{ds} \operatorname{div}(\mathcal{D}(\operatorname{grad}\phi_s^*, \boldsymbol{r}))\, dV \right) ds, \end{aligned}$$

where the second term in the last expression is 0 because $\operatorname{div}\mathcal{D}(\operatorname{grad}\phi_s^*, \boldsymbol{r})$ vanishes in Ω. The remaining term can be split:

$$\begin{aligned} U(\operatorname{grad}\phi_s^*) = \int_0^1 \Bigg(\int_{S_1} &\frac{d\phi_s^*}{ds} \mathcal{D}(\operatorname{grad}\phi_s^*, \boldsymbol{r}) \cdot \boldsymbol{n}\, dS \\ &+ \int_{\partial\Omega - S_1} \frac{d\phi_s^*}{ds} \mathcal{D}(\operatorname{grad}\phi_s^*, \boldsymbol{r}) \cdot \boldsymbol{n}\, dS \Bigg) ds, \end{aligned}$$

where the second term vanishes because $\mathcal{D}(\operatorname{grad}\phi_s^*, \boldsymbol{r}) \cdot \boldsymbol{n} = 0$. Then

$$U(\operatorname{grad}\phi_s^*) = \int_0^1 \left(\sum_{i=1}^{\operatorname{Rank}\delta^1} \int_{S_{1i}} \frac{d\phi_s^*}{ds} \mathcal{D}(\operatorname{grad}\phi_s^*, \boldsymbol{r}) \cdot \boldsymbol{n}\, dS \right) ds.$$

From (2–38), on S_{1i}

$$\frac{d\phi_s^*}{ds} = \frac{d}{ds}(sV_i) = V_i.$$

which is a constant, so

$$U(\operatorname{grad}\phi^*) = \int_0^1 \left(\sum_{i=1}^{\operatorname{Rank}\delta^1} V_i \int_{S_{1i}} \mathcal{D}(\operatorname{grad}\phi_s^*, \boldsymbol{r}) \cdot \boldsymbol{n}\, dS \right) ds.$$

This can be rewritten as

$$U(\operatorname{grad}\phi^*) = \sum_{i=1}^{\operatorname{Rank}\delta^1} V_i \int_0^1 Q_i(s)\, ds, \tag{2–39}$$

where

$$Q_i(s) = \int_{S_{1i}} \mathcal{D}(\operatorname{grad}\phi_s^*, \boldsymbol{r}) \cdot \boldsymbol{n}\, dS.$$

Having a relatively neat expression for the stationary value of the functional, we turn to the case of a linear isotropic constitutive law and the notion of a capacitance matrix. In this case, the boundary value problem is linear and $Q_i(s) = sQ_i$ so that

$$\int_0^1 Q_i(s)\, ds = Q_i \int_0^1 s\, ds = \frac{Q_i}{2}$$

and Equation (2–39) becomes

$$U(\operatorname{grad}\phi^*) = \frac{1}{2} \sum_{i=1}^{\operatorname{Rank}\delta^1} V_i Q_i. \tag{2–40}$$

To develop a basic notion of capacitance matrix, we use the linearity of the present case. Let φ_i, $1 \le i \le \operatorname{Rank}\delta^1$ be the unique solutions to the boundary value problem

$$\begin{aligned} \operatorname{div}(\varepsilon(\boldsymbol{r}) \operatorname{grad}\varphi_i) &= 0 \text{ in } \Omega, \\ (\varepsilon(\boldsymbol{r}) \operatorname{grad}\varphi_i) \cdot \boldsymbol{n} &= 0 \text{ on } \partial\Omega - S_1, \\ \varphi_i &= \begin{cases} 1 & \text{on } S_{1i}, \\ 0 & \text{on } S_{1j} \text{ if } j \neq i. \end{cases} \end{aligned}$$

Using the functions defined above, we can express the solution to this linear isotropic version of (2–37) as

$$\phi^* = \sum_{j=1}^{\operatorname{Rank}\delta^1} V_j \varphi_j.$$

Note that φ_i is dimensionless but ϕ is not.

Furthermore, to evaluate (2–40), we note that

$$Q_i = \int_{S_{1i}} (\varepsilon(\boldsymbol{r}) \operatorname{grad}\varphi_j^*) \cdot \boldsymbol{n}\, dS = \int_{S_{1i}} \sum_{j=1}^{\operatorname{Rank}\delta^1} V_j (\varepsilon(\boldsymbol{r}) \operatorname{grad}\varphi_j) \cdot \boldsymbol{n}\, dS.$$

This can be written as

$$Q_i = \sum_{j=1}^{\operatorname{Rank}\delta^1} V_j C_{ij},$$

where the C_{ij} are the entries of the capacitance matrix:

$$C_{ij} = \int_{S_{1i}} (\varepsilon(\boldsymbol{r}) \operatorname{grad}\varphi_j) \cdot \boldsymbol{n}\, dS.$$

Combining the last two equations with (2–40), we finally have

$$U(\operatorname{grad}\phi^*) = \frac{1}{2}\sum_{i=1}^{\operatorname{Rank}\delta^1}\sum_{j=1}^{\operatorname{Rank}\delta^1} C_{ij}V_iV_j.$$

That is, the energy of the linear system is a quadratic form in the voltages, and the coefficients of the capacitance matrix are as given above. Physically, the C_{ij} represent the charge on S_{1i} due to a unit potential on S_{1j} and zero potential on S_{1k} when $k \neq j$. The capacitance matrix is symmetric and positive definite.

2C. Variational Principles for Magnetoquasistatics

The Magnetoquasistatics Problem. In order to make the connection between the long exact sequence in homology and the definition of lumped parameters in circuit theory, we will now look at boundary value problems involving the magnetic field. Consider a connected compact three-dimensional region Ω which contains no infinitely permeable or superconducting material, but whose boundary may contain an interface with an infinitely permeable or superconducting body. The boundary $\partial\Omega = S_1 \cup S_2$ where the two parts comprising the boundary are disjoint, that is $S_1 \cap S_2$ has no area. On each part of the boundary let there be the following boundary conditions:

$$\boldsymbol{B}\cdot\boldsymbol{n} = 0 \text{ on } S_1,$$
$$\boldsymbol{n}\times\boldsymbol{H} = 0 \text{ on } S_2.$$

The boundary condition on S_1 is associated with the boundary of a perfect conductor, a superconductor, or a symmetry plane. Alternatively, when Ω contains a very permeable body, part of whose boundary coincides with $\partial\Omega$ and it is known that no flux can escape through that part, the boundary condition for S_1 is also appropriate. This latter situation occurs if Ω is an ideal magnetic circuit. The boundary condition on S_2 is associated with boundaries of infinitely permeable bodies or symmetry planes.

Assume that no free currents flow in Ω. Since elements of $C_c^2(\Omega - S_1)$ can be identified with vector fields whose component normal to S_1 vanishes, the magnetic flux density $\boldsymbol{B}$ can be associated with an element of $Z_c^2(\Omega - S_1)$ because

$$\operatorname{div}\boldsymbol{B} = 0 \text{ in } \Omega,$$
$$\boldsymbol{B}\cdot\boldsymbol{n} = 0 \text{ on } S_1.$$

Similarly, elements of $C_c^1(\Omega - S_2)$ can be associated with vector fields whose components tangent to S_2 vanish. Since it is assumed that no free currents can flow, the magnetic field intensity $\boldsymbol{H}$ can be associated with an element of $Z_c^1(\Omega - S_2)$ because

$$\operatorname{curl}\boldsymbol{H} = 0 \text{ in } \Omega,$$
$$\boldsymbol{n}\times\boldsymbol{H} = 0 \text{ on } S_2.$$

The magnetic flux density $\boldsymbol{B}$ is associated with an element of $Z_c^2(\Omega - S_1)$ while the magnetic field intensity $\boldsymbol{H}$ is associated with an element of $Z_c^1(\Omega - S_2)$. The nondegenerate bilinear pairings

$$\int : H_2(\Omega, S_1) \times H_c^2(\Omega - S_1) \to \mathbb{R},$$
$$\int : H_1(\Omega, S_2) \times H_c^1(\Omega - S_2) \to \mathbb{R},$$

induced on homology and cohomology classes by integration, are associated with fluxes and magnetomotive forces respectively. In particular, these lumped variables can be associated with the generators of relative homology groups in the following sense. Let

$$\Sigma_i,\ 1 \le i \le \beta_2(\Omega, S_1)$$

is a set of surfaces associated with a basis of $H_2(\Omega, S_1)$ and

$$c_i,\ 1 \le i \le \beta_1(\Omega, S_2)$$

is a set of curves associated with a basis of $H_1(\Omega, S_2)$. Then the periods of $\boldsymbol{B}$ on the Σ_i,

$$\Phi_i = \int_{\Sigma_i} \boldsymbol{B} \cdot \boldsymbol{n}\, dS, \quad 1 \le i \le \beta_2(\Omega, S_1),$$

are associated with fluxes, and the periods of $\boldsymbol{H}$ on the c_i,

$$I_i = \int_{c_i} \boldsymbol{H} \cdot dl, \quad 1 \le i \le \beta_1(\Omega, S_2),$$

are associated with currents in $\mathbb{R}^3 - \Omega$.

For variational principles involving the magnetic flux density $\boldsymbol{B}$ where fluxes are prescribed, the variation of the extremal lies in the space $B_c^2(\Omega - S_1)$. Typically it is convenient to reformulate such variational principles in terms of a vector potential $\boldsymbol{A}$. When this is done the tangential components of $\boldsymbol{A}$ are prescribed on S_1 in order to specify fluxes corresponding to generators of $H_c^2(\Omega - S_1)$ and to ensure that the normal component of $\boldsymbol{B}$ vanishes on S_1. In such cases the vector potential which gives the functional its stationary value is unique to within an element of $Z_c^1(\Omega - S_1)$.

Dually, for variational principles where the magnetic field intensity $\boldsymbol{H}$ is the independent variable and magnetomotive forces are prescribed, the variation of the extremal is in the space $B_c^1(\Omega - S_2)$. Though it is not possible in general to reformulate such principles in terms of a continuous single-valued scalar potential, it is possible to find a scalar potential formulation if one introduces suitable barriers, called cut surfaces, into Ω, prescribing jumps to the scalar potential as the surfaces are crossed and fixing the scalar potential to be a different fixed constant on each connected component of S_2.

Whether the variational principles are formulated in terms of vector or scalar potentials, the long exact homology sequence plays a crucial role in understanding the topological implications of the formulations. In addition it sheds light on prescribing boundary conditions for the vector potential formulation.

Introduction of Metric. In order to illustrate the role of cochain complexes in the statement of variational principles, a constitutive relation must be introduced. Let

$$\boldsymbol{H} = \mathcal{H}(\boldsymbol{B}, \boldsymbol{r})$$

be a a tensor constitutive relation and let

$$\boldsymbol{B} = \mathcal{B}(\boldsymbol{H}, \boldsymbol{r})$$

be the inverse transformation satisfying

$$\mathcal{B}\left(\mathcal{H}(\boldsymbol{B}, \boldsymbol{r}), \boldsymbol{r}\right) = \boldsymbol{B}.$$

Furthermore, assume that the matrix with entries

$$\frac{\partial \mathcal{H}_i}{\partial B_j}$$

is symmetric positive definite. Consequently,

$$\frac{\partial \mathcal{B}_i}{\partial H_j}$$

is symmetric positive definite.

Stationary Inductive Coenergy and Energy Principles. As stated before for electroquasistatics, we now write the inductive energy principles. With the metric as stated, the principles of stationary inductive coenergy and energy are the following (see also [Mac70, pp. 330–332]) The coenergy principle is

$$T'(\boldsymbol{H}) = \inf_{\boldsymbol{H} \in Z_c^1(\Omega - S_2)} \int_\Omega \int_0^{\boldsymbol{H}} \mathcal{B}(\xi, \boldsymbol{r}) \cdot d\xi \, dV,$$

subject to the constraints that prescribe periods of $\boldsymbol{H}$ on generators of $H_1(\Omega, S_2)$:

$$I_i = \int_{c_i} \boldsymbol{H} \cdot d\boldsymbol{l}, \quad 1 \le i \le \beta_1(\Omega, S_2).$$

The stationary inductive energy principle is

$$T(\boldsymbol{B}) = \inf_{\boldsymbol{B} \in Z_c^2(\Omega - S_1)} \int_\Omega \int_0^{\boldsymbol{B}} \mathcal{H}(\xi, \boldsymbol{r}) \cdot d\xi \, dV,$$

subject to the constraints which prescribe periods of $\boldsymbol{B}$ on generators of $H_2(\Omega, S_1)$

$$\Phi_j = \int_{\Sigma_i} \boldsymbol{B} \cdot \boldsymbol{n} \, dS \quad 1 \le i \le \beta_2(\Omega, S_1).$$

As in the case of electroquasistatics, the extremals in both variational principles are constrained to be relative cocycles. When principal conditions are prescribed on the generators of a (co)homology group, the variation of the extremal is constrained to be a relative coboundary. This is readily seen from the identities

$$Z_c^1(\Omega - S_2) \simeq H_c^1(\Omega - S_2) \oplus B_c^1(\Omega - S_2),$$
$$Z_c^2(\Omega - S_1) \simeq H_c^2(\Omega - S_1) \oplus B_c^1(\Omega - S_1),$$

and the fact that the following relative de Rham isomorphisms have been assumed:

$$H_c^1(\Omega - S_2) \simeq H_1(\Omega, S_2) \qquad H_c^2(\Omega - S_1) \simeq H_2(\Omega, S_1).$$

For linear constitutive relations the coenergy principle gives an upper bound for inductance while the energy principle gives an upper bound for inverse inductance. This is done by expressing the minimum of the functional as a quadratic form in the prescribed periods and making the identifications

$$T'(\boldsymbol{H}) = \frac{1}{2} \sum_{i,j=1}^{\beta_1(\Omega,S_2)} I_i L_{ij} I_i, \qquad T(\boldsymbol{B}) = \frac{1}{2} \sum_{i,j=1}^{\beta_2(\Omega,S_1)} \Phi_i \Gamma_{ij} \Phi_j.$$

From the upper bound on inverse inductance a lower bound on inductance can be found in the usual way. The estimation of partial inductances can be obtained by leaving some of the periods free in a given variational principle so that their values can be determined by the minimization.

These variational principles are interesting since they provide a direct link with lumped parameters and show how the various subspaces of the complexes $C_c^*(\Omega - S_1)$ and $C_c^*(\Omega - S_2)$ play a role. As in the case of electrostatics it is useful to further investigate the relative (co)homology groups of concern in order to see how the above variational principles can be rephrased in terms of vector and scalar potentials and to know the topological restrictions which may arise.

The Inductive Energy Principle with a Vector Potential. While we will briefly state the stationary inductive energy principle in terms of a vector potential, we will not use it for calculations, preferring to concentrate on calculations with the coenergy principle and scalar potential. The stationary inductive energy principle with $\boldsymbol{B} = \operatorname{curl} \boldsymbol{A}$ is

$$T(\operatorname{curl} \boldsymbol{A}) = \inf_{\boldsymbol{A}} \int_\Omega \int^{\boldsymbol{A}} \mathcal{H}(\operatorname{curl} \xi, \boldsymbol{r}) \cdot \operatorname{curl}(d\xi)\, dV,$$

subject to the principal boundary condition

$$\boldsymbol{n} \times \boldsymbol{A} = \operatorname{curl} \psi \quad \text{on } S_1,$$

where $[\psi]_{d_j}$ is prescribed on generators of $H_1(S_1, \partial S_1)$ but ψ is otherwise an arbitrary single-valued function. Note that in this formulation of the energy principle the extremal $\boldsymbol{A}$ is unique to within an element of $Z_c^1(\Omega - S_1)$. Since

$$Z_c^1(\Omega - S_1) = H_c^1(\Omega - S_1) \oplus B_c^1(\Omega - S_1),$$

the nonuniqueness can be overcome by specifying the periods of $\boldsymbol{A}$ on the generators of $H_1(\Omega, S_1)$ and, in analogy with the uniqueness considerations of the electric vector potential, specifying the divergence of $\boldsymbol{A}$ and its normal component on S_2 eliminates the ambiguity in $B_c^1(\Omega - S_1)$. More detailed explorations of topological aspects of the vector potential in magnetostatics occur in Examples 1.15 and 3.4.

The Inductive Coenergy Principle with a Scalar Potential. As in the case of the capacitive coenergy principle, the three-step procedure from page 38 which came about from the long exact homology sequence plays a role in the inductive coenergy principle. The added twist is that in order to formulate the principle in terms of a scalar potential, cuts must be introduced in order to make the scalar potential single-valued. The introduction of cuts requires a certain amount of additional bookkeeping, but the method is essentially the same as before. There are questions of what cuts are and how they might be computed. These matters are treated more fully in later chapters.

In terms of a scalar potential, the inductive coenergy functional is

$$T'(\operatorname{grad}\psi) = \inf_{\operatorname{grad}\psi \in Z^1_c(\Omega - S_2)} \int_{\Omega^-} \int_0^{\psi} \mathcal{B}(\xi, \boldsymbol{r}) \cdot d\xi \, dV,$$

where $\Omega^- = \Omega - \cup \Sigma_i'$, $1 \le i \le \dim \operatorname{im} \tilde{\jmath}_1$ denotes the region Ω with cuts removed.

Consider the following portion of the long exact homology sequence associated with the pair (Ω, S_2):

$$\cdots \xrightarrow{\tilde{\delta}_2} H_1(S_2) \xrightarrow{\tilde{\imath}_1} H_1(\Omega) \xrightarrow{\tilde{\jmath}_1} H_1(\Omega, S_2) \xrightarrow{\tilde{\delta}_1} H_0(S_2) \xrightarrow{\tilde{\imath}_0} H_0(\Omega) \xrightarrow{\tilde{\jmath}_0} \cdots$$

According to the prescription given by Equation (1–3) (page 38), the generators of $H_1(\Omega, S_2)$ can be constructed from the three-step procedure for

$$H_1(\Omega, S_2) \simeq \tilde{\delta}_1^{-1}(\ker \tilde{\imath}_0) \oplus \tilde{\jmath}_1 \left(\frac{H_1(\Omega)}{\tilde{\imath}_1 (H_1(S_2))} \right).$$

Let c_i, for $1 \le i \le \beta_1(\Omega, S_2)$, be a set of curves which are associated with the generators of $H_1(\Omega, S_2)$. The generators can be classified into two groups which lend some insight into the periods of $\boldsymbol{H}$.

The first group consists of $\dim \operatorname{im} \tilde{\jmath}_1$ generators of $H_1(\Omega, S_2)$ which are homologous in the absolute sense to generators of $H_1(\Omega)$. These generators can be associated with closed curves $c_i, 1 \le i \le \dim \operatorname{im} \tilde{\jmath}_1$. In this case, the period

$$\int_{c_i} \boldsymbol{H} \cdot dl = I_i$$

is equal to the current in $\mathbb{R}^3 - \Omega$ which links the generator of $H_1(\Omega, S_2)$ associated with c_i.

The second group consists of $\dim \ker \tilde{\imath}_0$ remaining generators of $H_1(\Omega, S_2)$ which are associated with simple open curves whose end points lie in distinct connected components of S_2. That is, in each connected component of Ω, one can find curves $c_{i+\dim \operatorname{im} \tilde{\jmath}_1}$, for $1 \le i \le \dim \ker \tilde{\imath}_0$, such that

$$\partial c_{i+\dim \operatorname{im} \tilde{\jmath}_1} = p_i - p_0,$$

where p_0 is a datum node lying in some connected component of S_2 and each p_i lies in some other distinct connected component of S_2. In this case, the period

$$\int_{c_j} \boldsymbol{H} \cdot dl = I_j, \qquad \dim \operatorname{im} \tilde{\jmath}_1 < j \le \beta_1(\Omega, S_2)$$

is associated with a magnetomotive force.

It is often convenient to describe the magnetic field intensity $\boldsymbol{H}$ in terms of a scalar potential ψ since, from a practical point of view, it is much easier to work with a scalar function than three components of a vector field. However, when the periods of $\boldsymbol{H}$ do not vanish on the 1-cycles in the first group, then it is not possible to make ψ continuous and single-valued since this would imply

$$I_i = \int_{c_i} \boldsymbol{H} \cdot dl = \int_{c_i} \operatorname{grad} \psi \cdot dl = 0$$

since

$$\partial c_i = 0, \quad 1 \le i \le \dim \operatorname{im} \tilde{\jmath}_1.$$

In order to overcome this difficulty one can perform an analog of the procedure used to prescribe the tangential components of the vector potential. There exist cut surfaces

$$\Sigma'_i, \quad 1 \le i \le \dim \operatorname{im} \tilde{\jmath}_1,$$

such that the Σ'_i are associated with subspaces of $H_2(\Omega, S_1)$ of dimension equal to $\dim \operatorname{im} \tilde{\jmath}_1$. Let $\operatorname{Int}(c_i, \Sigma'_j)$ denote the number of oriented intersections of the 1-chain c_i with the relative 2-chain Σ'_j and let $\mathcal{I}$ be the the $\dim \operatorname{im} \tilde{\jmath}_1 \times \dim \operatorname{im} \tilde{\jmath}_1$ matrix such that

$$\mathcal{I}_{ij} = \operatorname{Int}(c_i, \Sigma'_j).$$

The intersection matrix $\mathcal{I}$ is nonsingular as will be apparent when duality theorems are considered. It turns out that one can make the scalar potential single-valued on

$$\Omega^- = \Omega - \bigcup_{i=1}^{\dim \operatorname{im} \tilde{\jmath}_1} \Sigma'_i.$$

Assuming that the Σ'_i have been chosen properly, the periods of $\boldsymbol{H}$ on the generators of $H_1(\Omega, S_2)$ which lie in the first group of generators are easily expressed in terms of the jumps in ψ as the Σ'_i are traversed. That is, if $[\psi]_{\Sigma'_i}$ are these jumps, then for $1 \le i \le \dim \operatorname{im} \tilde{\jmath}_1$ one has

$$I_i = \int_{c_i} \boldsymbol{H} \cdot \boldsymbol{dl} = \int_{c_i} \operatorname{grad} \psi \cdot \boldsymbol{dl} = \sum_{j=1}^{\dim \operatorname{im} \tilde{\jmath}_1} \mathcal{I}_{ij} [\psi]_{\Sigma'_j}.$$

This forms a set of linear equations which can be inverted to yield

$$[\psi]_{\Sigma'_i} = \sum_{j=1}^{\dim \operatorname{im} \tilde{\jmath}_1} \mathcal{I}^{-1}_{ij} I_j.$$

Thus, once the cuts have been selected, one has an explicit way of prescribing the first $\dim \operatorname{im} \tilde{\jmath}_1$ periods of $\boldsymbol{H}$ in terms of the jumps in ψ. In order to specify the remaining $\dim \ker \tilde{\imath}_1$ periods of $\boldsymbol{H}$ in terms of the scalar function ψ, one defines a $\dim \ker \tilde{\imath}_0 \times \dim \operatorname{im} \tilde{\jmath}_1$ intersection matrix $\mathcal{I}'_{ij}$ whose entries are the number of oriented intersections of curves $c_{i+\dim \operatorname{im} \tilde{\jmath}_1}$ with surfaces Σ'_i. Hence the periods $\boldsymbol{H}$ on the remaining generators can be expressed as follows:

$$\begin{aligned} I_{i+\dim \operatorname{im} \tilde{\jmath}_1} &= \int_{c_{i+\dim \operatorname{im} \tilde{\jmath}_1}} \boldsymbol{H} \cdot d\boldsymbol{l} = \int_{c_{i+\dim \operatorname{im} \tilde{\jmath}_1}} \operatorname{grad} \psi \cdot d\boldsymbol{l} \\ &= \psi(p_i) - \psi(p_0) + \sum_{j=1}^{\dim \operatorname{im} \tilde{\jmath}_1} \mathcal{I}'_{ij} [\psi]_{\Sigma'_j}. \end{aligned}$$

Hence, if $\psi(p_0)$ is chosen arbitrarily, we have

$$\begin{aligned} \psi(p_i) &= \psi(p_0) + I_{i+\dim \operatorname{im} \tilde{\jmath}_1} - \sum_{j=1}^{\dim \operatorname{im} \tilde{\jmath}_1} \mathcal{I}'_{ij} [\psi]_{\Sigma'_j} \\ &= \psi(p_0) + I_{i+\dim \operatorname{im} \tilde{\jmath}_1} - \sum_{l,j=1}^{\dim \operatorname{im} \tilde{\jmath}_1} \mathcal{I}'_{ij} (\mathcal{I}_{jl})^{-1} I_l, \end{aligned}$$

which is an explicit formula giving the value of $\psi(p_i)$ in terms of the remaining periods to be prescribed.

Note that on the ith connected component of S_2

$$\psi = \psi(p_i), \quad 0 \le i \le \dim \operatorname{im} \delta_1$$

since the tangential components of $\boldsymbol{H}$ vanish on S_2.

Inductance Parameters from the Inductive Coenergy Principle. With the foregoing discussion in mind, we will show the calculations which lead to inductance parameters. Once cuts have been introduced, the calculations are, in almost every respect, parallel to the ones previously done for capacitance parameters. We start with the inductive coenergy functional

$$T'(\operatorname{grad} \psi) = \inf_{\operatorname{grad} \psi \in Z^1_c(\Omega - S_2)} \int_{\Omega^-} \int_0^{\psi} \mathcal{B}(\xi, \boldsymbol{r}) \cdot d\xi \, dV, \tag{2–41}$$

where

$$\Omega^- = \Omega - \bigcup_{1 \le i \le \dim \operatorname{im} \tilde{\jmath}_1} \Sigma'_i$$

denotes the region Ω with cuts removed. The functional is subject to the following constraints:

$$\begin{aligned} [\psi]_{\Sigma'_i} &\ne 0 && \text{on cuts } \Sigma'_i,\ 1 \le i \le \dim \operatorname{im} \tilde{\jmath}_1, \\ \boldsymbol{n} \cdot \mathcal{B}(\operatorname{grad} \psi, \boldsymbol{r}) &= 0 && \text{on } S_1, \\ \boldsymbol{n} \times \operatorname{grad} \psi &= 0 && \text{on } S_2, \\ \psi &= \psi(p_i) && \text{on } S_{2i}. \end{aligned}$$

The last constraint specifies a given constant on each connected component of S_2. In particular, for the last constraint let

$$S_2 = \bigcup_{i=0}^{\operatorname{Rank} \ker \tilde{\imath}_0 - 1} S_{2i},$$

with S_{2i} connected and disjoint from S_{2j} when $i \neq j$. If Ω is connected and ψ is zero on S_{20} (i.e., one component is grounded), then $\psi \in Z_c^1(\Omega - S_2)$ means that

$$\psi = \begin{cases} 0, & \text{on } S_{20}, \\ \psi(p_i),\ 1 \le i \le \operatorname{Rank}\ker \tilde{\imath}_0 - 1) & \text{on } S_{2i}, \end{cases}$$

where the $\psi(p_i)$ are constants.

Differentiating functional (2–41) in the space of fields and taking the variation, the functional becomes

(2–42)

$$T'(\operatorname{grad}\psi + \delta\psi) - T'(\operatorname{grad}\psi) = \int_{\Omega^-} \mathcal{B}(\operatorname{grad}\psi, \boldsymbol{r}) \cdot \operatorname{grad}(\delta\psi)\, dV + \mathcal{O}(\|\delta\psi\|^2).$$

Since the $\psi(p_i)$ are constants, $\delta\psi = 0$ on S_{2i}.

The first variation of the functional is the part of (2–42) which is linear in $\delta\psi$. Taking the first variation and integrating by parts,

$$\delta T' = \int_{\partial\Omega^-} \delta\psi\, \mathcal{B}(\operatorname{grad}\psi, \boldsymbol{r})\, dS - \int_{\Omega^-} \delta\psi\, \operatorname{div}\mathcal{B}(\operatorname{grad}\psi, \boldsymbol{r})\, dV.$$

Since $\delta T'$ must vanish for all admissible variations of T', this equation implies that the following boundary value problem is satisfied at the extremum ψ^*:

$$\begin{aligned} \operatorname{div}\mathcal{B}(\operatorname{grad}\psi^*, \boldsymbol{r}) &= 0 && \text{in } \Omega^-, \\ \mathcal{B}(\operatorname{grad}\psi^*, \boldsymbol{r}) \cdot \boldsymbol{n}0 & && \text{on } S_1, \\ \psi^* &= \psi(p_i) && \text{on } S_{2i}, \end{aligned} \tag{2–43}$$

with $[\psi^*]_{\Sigma_i'}$ prescribed on Σ_i' for $1 \le i \le \dim \operatorname{im} \tilde{\jmath}_1$.

Recall that the constitutive law satisfies the relation

$$\frac{\partial \mathcal{B}_i}{\partial \mathcal{H}_j} = \frac{\partial \mathcal{H}_j}{\partial \mathcal{B}_j},$$

so that the line integral in the functional

$$T'(\operatorname{grad}\psi^*) = \int_{\Omega^-} \left(\int_0^{\psi^*} \mathcal{B}(\operatorname{grad}\eta, \boldsymbol{r}) \cdot \operatorname{grad}(d\eta) \right) dV$$

is independent of path. We now take a family of functions $\psi_s^*(\boldsymbol{r})$, parametrized by s, with $0 \le s \le 1$, which are solutions to the boundary value problem which is a variant of (2–43) and provides a linear interpolation between 0 and ψ^*. Putting ψ_s^* into the preceding,

$$T'(\operatorname{grad}\psi_s^*) = \int_{\Omega^-} \left(\int_0^1 \mathcal{B}(\operatorname{grad}\psi_s^*, \boldsymbol{r}) \cdot \operatorname{grad}\left(\frac{d\psi_s^*}{ds}\right) ds \right) dV.$$

Now interchange the order of integration and integrate by parts:

$$\begin{aligned}
T'(\operatorname{grad}\psi_s^*)
&= \int_0^1 \left(\int_{\Omega^-} \mathcal{B}(\operatorname{grad}\psi_s^*, \boldsymbol{r}) \cdot \operatorname{grad}\left(\frac{d\psi_s^*}{ds}\right) dV \right) ds \\
&= \int_0^1 \left(\int_{\partial\Omega^-} \frac{d\psi_s^*}{ds} \mathcal{B}(\operatorname{grad}\psi_s^*, \boldsymbol{r}) \cdot \boldsymbol{n}\, dS - \int_{\Omega^-} \frac{d\psi_s^*}{ds} \operatorname{div}(\mathcal{B}(\operatorname{grad}\psi_s^*, \boldsymbol{r}))\, dV \right) ds \\
&+ \int_0^1 \left(\sum_{i=1}^{\dim \operatorname{im} \tilde{\jmath}_1} \int_{\Sigma_i'} \left[\frac{\partial \psi_s^*}{\partial s}\right]_{\Sigma_i'} \mathcal{B}(\operatorname{grad}\psi_s^*, \boldsymbol{r}) \cdot \boldsymbol{n}\, dS \right) ds.
\end{aligned}$$

The second term in the last expression vanishes because $\operatorname{div}(\mathcal{B}(\operatorname{grad}\psi_s^*, \boldsymbol{r})) = 0$ and the first term can be split so that

$$\begin{aligned}
T'(\operatorname{grad}\psi_s^*) = \int_0^1 \Bigg(\int_{S_2} &\frac{d\psi_s^*}{ds} \mathcal{B}(\operatorname{grad}\psi_s^*, \boldsymbol{r}) \cdot \boldsymbol{n}\, dS \\
&+ \int_{\partial\Omega^- - S_2} \frac{d\psi_s^*}{ds} \mathcal{B}(\operatorname{grad}\psi_s^*, \boldsymbol{r}) \cdot \boldsymbol{n}\, dS \Bigg) ds \\
&+ \int_0^1 \left(\sum_{i=1}^{\dim \operatorname{im} \tilde{\jmath}_1} \int_{\Sigma_i'} \left[\frac{\partial \psi_s^*}{\partial s}\right]_{\Sigma_i'} \mathcal{B}(\operatorname{grad}\psi_s^*, \boldsymbol{r}) \cdot \boldsymbol{n}\, dS \right) ds,
\end{aligned}$$

where the second term vanishes because $\mathcal{B}(\operatorname{grad}\psi_s^*, \boldsymbol{r}) \cdot \boldsymbol{n} = 0$ on S_2. Then

$$\begin{aligned}
T'(\operatorname{grad}\psi_s^*) = \int_0^1 &\left(\sum_{i=1}^{\dim \ker \tilde{\imath}_0 - 1} \int_{S_{2i}} \frac{d\psi_s^*}{ds} \mathcal{B}(\operatorname{grad}\psi_s^*, \boldsymbol{r}) \cdot \boldsymbol{n}\, dS \right) ds \\
&+ \int_0^1 \left(\sum_{i=1}^{\dim \operatorname{im} \tilde{\jmath}_1} \int_{\Sigma_i'} \left[\frac{\partial \psi_s^*}{\partial s}\right]_{\Sigma_i'} \mathcal{B}(\operatorname{grad}\psi_s^*, \boldsymbol{r}) \cdot \boldsymbol{n}\, dS \right) ds.
\end{aligned}$$

Note that

$$\frac{d\psi_s^*}{ds} = \frac{d}{ds}(sI_i) = I_i$$

is a constant on S_{2i} while a similar relation exists on Σ_i'. Hence,

$$\begin{aligned}
T'(\operatorname{grad}\psi_s^*) = \int_0^1 &\left(\sum_{i=1}^{\dim \ker \tilde{\imath}_0 - 1} I_i \int_{S_{2i}} \mathcal{B}(\operatorname{grad}\psi_s^*, \boldsymbol{r}) \cdot \boldsymbol{n}\, dS \right) ds \\
&+ \int_0^1 \left(\sum_{i=1}^{\dim \operatorname{im} \tilde{\jmath}_1} \int_{\Sigma_i'} \left[\frac{\partial \psi_s^*}{\partial s}\right]_{\Sigma_i'} \mathcal{B}(\operatorname{grad}\psi_s^*, \boldsymbol{r}) \cdot \boldsymbol{n}\, dS \right) ds.
\end{aligned}$$

This can be rewritten as

$$\begin{aligned}
\text{(2–44)} \quad T'(\operatorname{grad}\psi_s^*) = &\sum_{i=1}^{\dim \ker \tilde{\imath}_0 - 1} I_i \int_0^1 \Phi_i(s)\, ds \\
&+ \sum_{i=1}^{\dim \operatorname{im} \tilde{\jmath}_1} [\psi]_{\Sigma_i'} \int_0^1 \left(\int_{\Sigma_i'} \mathcal{B}(\operatorname{grad}\psi_s^*, \boldsymbol{r}) \cdot \boldsymbol{n}\, dS \right) ds,
\end{aligned}$$

where

$$\Phi_i(s) = \int_{S_{2i}} \mathcal{B}(\operatorname{grad} \psi_s^*, \boldsymbol{r}) \cdot \boldsymbol{n}\, dS,$$

for $1 \le i \le \dim \ker \tilde{\imath}_0 - 1$. On the cuts we have

$$\Phi_{\dim \ker \tilde{\imath}_0 - 1 + i} = \int_{\Sigma_i'} \mathcal{B}(\operatorname{grad} \psi_s^*, \boldsymbol{r}) \cdot \boldsymbol{n}\, dS$$

for $\dim \ker \tilde{\imath}_0 < i \le \beta_1(\Omega, S_2)$.

Finally, we develop the expression for the energy of a linear system of currents which comes out of the preceding variational process. This expression comes about by a calculation similar to the one made for electroquasistatics and contains the notion of inductance parameters. We take the case of a linear isotropic constitutive law so that the boundary value problem is linear and $\Phi_i(s) = s\Phi_i$. Then

$$\int_0^1 \Phi_i(s)\, ds = \Phi_i \int_0^1 s\, ds = \frac{\Phi_i}{2}$$

and (2–44) becomes

$$T'(\operatorname{grad} \psi^*) = \frac{1}{2} \sum_{i=1}^{\dim \ker \tilde{\imath}_0 - 1} I_i \Phi_i + \sum_{j=1}^{\dim \operatorname{im} \tilde{\jmath}_1} [\psi^*]_{\Sigma_j'} \Phi_{\dim \ker \tilde{\imath}_0 - 1 + j}. \tag{2–45}$$

Let ψ_i, $1 \le i \le \dim \ker \tilde{\imath}_0 - 1$ be the unique solutions to the boundary value problem

$$\begin{aligned} [\psi_i] &= 0 \quad \text{on } \Sigma_k' \text{ for all } k, \\ \operatorname{div}(\mu(\boldsymbol{r}) \operatorname{grad} \psi_i) &= 0 \quad \text{in } \Omega \\ (\mu(\boldsymbol{r}) \operatorname{grad} \psi_i) \cdot \boldsymbol{n} &= 0 \quad \text{on } \partial\Omega - S_2 \\ \psi_i &= \begin{cases} 1 & \text{on } S_{2i}, \\ 0 & \text{on } S_{2j} \text{ if } j \ne i. \end{cases} \end{aligned}$$

Note that ψ_i is dimensionless but ψ is not. Similarly, for $\dim \operatorname{im} \tilde{\jmath}_1 \le i \le \beta_1(\Omega, S_2)$, we let

$$\begin{aligned} [\psi_i]_{\Sigma_j'} &= \begin{cases} 1 & \text{if } i = \dim \ker \tilde{\imath}_0 + j, \\ 0 & \text{otherwise,} \end{cases} \\ \operatorname{div}(\mu(\boldsymbol{r}) \operatorname{grad} \psi_i) &= 0 \quad \text{in } \Omega, \\ \mu(\boldsymbol{r}) \operatorname{grad} \psi_i \cdot \boldsymbol{n} &= 0 \quad \text{on } \partial\Omega - S_2, \\ \psi_i &= 0 \quad \text{on } S_{2k} \text{ for all } k. \end{aligned}$$

Using the functions defined above, we can express the solution to this linear isotropic version of (2–43) as

$$\psi^* = \sum_{j=1}^{\dim \ker \tilde{\imath}_0 - 1} I_j \psi_j + \sum_{j=1}^{\dim \operatorname{im} \tilde{\jmath}_1} [\psi]_{\Sigma_j'} \Phi_{\dim \ker \tilde{\imath}_0 - 1 + j}.$$

To evaluate (2–45), note that

$$\begin{aligned}\Phi_i &= \int_{S_{2i}} (\mu(\boldsymbol{r})\,\mathrm{grad}\,\psi^*)\cdot \boldsymbol{n}\,dS \\ &= \int_{S_{2i}} \sum_{j=1}^{\dim\ker \tilde{\imath}_0 - 1} I_j(\mu(\boldsymbol{r})\,\mathrm{grad}\,\psi_j)\cdot \boldsymbol{n}\,dS \\ &\qquad + \int_{S_{2i}} \sum_{j=1}^{\dim \mathrm{im}\,\tilde{\jmath}_1} [\psi]_{\Sigma'_j}(\mu(\boldsymbol{r})\,\mathrm{grad}\,\psi_{\dim\ker\tilde{\imath}_0-1+j})\cdot \boldsymbol{n}\,dS.\end{aligned}$$

In principle this can be written as

$$\Phi_i = \sum_{j=1}^{\beta_1(\Omega,S_2)} I_j L_{ij},$$

where the L_{ij} are the entries of the inductance matrix.This step necessarily involves rewriting the $[\psi^*]_{\Sigma'_j}$ in terms of exterior currents by means of linking information. Combining the last equation with (2–45), we have

$$T'(\mathrm{grad}\,\psi^*) = \frac{1}{2}\sum_{i=1}^{\beta_1(\Omega,S_2)}\sum_{i=1}^{\beta_1(\Omega,S_2)} L_{ij} I_i I_j.$$

The energy of the linear system is a quadratic form in the currents, and the coefficients of the inductance matrix are as given above. Physically, the entries of the first block of L_{ij} represent the flux linking surface S_{2i} due to a unit current on S_{2j} and zero current on S_{2k} when $k \neq j$. The second block on the diagonal of the inductance matrix relates to the currents circulating on the boundaries of the cuts. The inductance matrix is symmetric and positive definite.

2D. Steady Current Flow

We will soon see that in the modeling of quasistatic electromagnetic systems where no mechanical work is being performed, it is important to track the electric and magnetic energy, as well as the energy dissipated as heat. Having considered electrostatics and magnetostatics in the previous two sections, we now turn to steady current conduction. This sets the stage for our discussion of electroquasistatic modeling in the next section.

Example 2.3 Steady Current conduction in three dimensions: $n = 3$, $p = 2$. Consider a connected compact region Ω of finite, nonzero conductivity and whose boundary may contain interfaces with nonconducting or perfectly conducting bodies. Let $\partial\Omega = S_1 \cup S_2$, where $S_1 \cap S_2$ has no area, and suppose

$$\begin{aligned}\mathrm{div}\,\boldsymbol{J} &= 0 \ \text{ in } \Omega,\\ \boldsymbol{J}\cdot\boldsymbol{n} &= 0 \ \text{ on } S_1,\\ \mathrm{curl}\,\boldsymbol{E} &= 0 \ \text{ in } \Omega,\\ \boldsymbol{n}\times\boldsymbol{E} &= 0 \ \text{ on } S_2.\end{aligned}$$

It is readily seen that the transformation

$$\boldsymbol{J} \to \boldsymbol{B}, \qquad \boldsymbol{E} \to \boldsymbol{H},$$

makes this problem formally equivalent to example 1.15. It is clear that current density $\boldsymbol{J}$ can be associated with an element of $Z_c^2(\Omega - S_1)$. The electric field intensity $\boldsymbol{E}$ can be associated with an element of $Z_c^1(\Omega - S_2)$. Note that the boundary condition on S_1 can be associated with a symmetry plane or interface with a nonconducting body while the boundary condition on S_2 can be associated with another type of symmetry plane or the interface of a perfectly conducting body. If

$$\Sigma_i, \quad 1 \le i \le \beta_2(\Omega, S_1),$$

is a set of surfaces associated with a basis of $H_2(\Omega, S_1)$ then the periods of $\boldsymbol{J}$ on the Σ_i,

$$I_i = \int_{\Sigma_i} \boldsymbol{J} \cdot \boldsymbol{n}\, dS, \quad 1 \le i \le \beta_2(\Omega, S_1),$$

are associated with currents. In addition, if

$$c_i, \quad 1 \le i \le \beta_1(\Omega, S_2),$$

is a set of curves associated with a basis of $H_1(\Omega, S_2)$, then the periods of $\boldsymbol{E}$ on the c_i

$$V_i = \int_{c_i} \boldsymbol{E} \cdot d\boldsymbol{l}, \quad 1 \le i \le \beta_1(\Omega, S_2),$$

are associated with voltages, or electromotive forces.

To obtain a variational formulation of the problem, consider a constitutive relation

$$\boldsymbol{E} = \mathcal{E}(\boldsymbol{J}, \boldsymbol{r})$$

and an inverse constitutive relation

$$\boldsymbol{J} = \mathcal{J}(\boldsymbol{E}, \boldsymbol{r})$$

which satisfies

$$\mathcal{E}\left(\mathcal{J}(\boldsymbol{E}, \boldsymbol{r}), \boldsymbol{r}\right) = \boldsymbol{E}.$$

Furthermore, assume that the two matrices with elements

$$\frac{\partial \mathcal{E}_i}{\partial J_j}, \qquad \frac{\partial \mathcal{J}_i}{\partial E_j}$$

are symmetric and positive definite. In this case the principles of stationary content and cocontent can be stated as follows (see [Mac70, pp. 329–330]).

Stationary Content Principle.

$$G(\boldsymbol{J}) = \inf_{\boldsymbol{J} \in Z_c^2(\Omega - S_1)} \int_\Omega \left(\int_0^{\boldsymbol{J}} \mathcal{E}(\xi, \boldsymbol{r}) \cdot d\xi \right) dV,$$

subject to the constraints which prescribe the periods of $\boldsymbol{J}$ on generators of $H_2(\Omega, S_1)$

$$I_i = \int_{\Sigma_i} \boldsymbol{J} \cdot \boldsymbol{n}\, dS, \quad 1 \le i \le \beta_2(\Omega, S_1)$$

Stationary Cocontent Principle.

$$G'(\boldsymbol{E}) = \inf_{\boldsymbol{E}\in Z_c^1(\Omega - S_2)} \int_\Omega \left(\int_0^{\boldsymbol{E}} \mathcal{J}(\xi, \boldsymbol{r}) \cdot d\xi \right) dV$$

subject to the constraints which prescribe the periods of $\boldsymbol{E}$ on generators of $H_1(\Omega, S_2)$

$$V_i = \int_{c_i} \boldsymbol{E} \cdot \boldsymbol{dl}, \quad 1 \leq i \leq \beta_1(\Omega, S_2).$$

As in the previous two examples, the two variational principles stated above constrain the extremal to be a relative cocycle and when additional constraints are prescribed on the generators of a (co)homology group, the variation of the extremal is constrained to be a relative coboundary.

For linear constitutive relations the content principle gives an upper bound for resistance while the cocontent principle gives an upper bound on conductance. As usual the upper bounds are obtained by expressing the minimum of the functional as a quadratic form in the prescribed periods and making the identifications

$$G(\boldsymbol{J}) = \sum_{i,j=1}^{\beta_2(\Omega,S_1)} I_i R_{ij} I_j, \qquad G(\boldsymbol{E}) = \sum_{i,j=1}^{\beta_1(\Omega,S_2)} V_i G_{ij} V_j.$$

From the upper bound on conductance, a lower bound on resistance can be found in the usual way.

By attempting to express the variational principles in terms of vector and scalar potentials one will find, as in previous examples, many topological subtleties. Noticing the mathematical equivalence between this example involving steady currents and the previous one involving magnetostatics, one may form a transformation of variables

$$\boldsymbol{J} \to \boldsymbol{B}, \qquad \boldsymbol{E} \to \boldsymbol{H}, \qquad \boldsymbol{T} \to \boldsymbol{A}, \qquad \phi \to \psi,$$

as soon as one tries to define potentials ϕ and $\boldsymbol{T}$ such that

$$\boldsymbol{J} = \operatorname{curl} \boldsymbol{T}, \qquad \boldsymbol{E} = \operatorname{grad} \phi.$$

Summarizing the results of exploiting the mathematical analogy, one can say the following about the cocontent principle in terms of a vector potential $\boldsymbol{T}$. As in the previous example, one can consider the long exact homology sequence for the pair (Ω, S_1) and obtain

$$H_2(\Omega, S_1) = \delta_2^{-1}(\ker \tilde{\imath}_1) \oplus \tilde{\jmath}_2 \left(\frac{H_2(\Omega)}{\tilde{\imath}_2 (H_2(S_1))} \right)$$

where the relevant portion of the long exact homology sequence is

$$\cdots \xrightarrow{\delta_3} H_2(S_1) \xrightarrow{\tilde{\imath}_2} H_2(\Omega) \xrightarrow{\tilde{\jmath}_2} H_2(\Omega, S_1) \xrightarrow{\delta_2} H_1(S_1) \xrightarrow{\tilde{\imath}_1} H_1(\Omega) \xrightarrow{\tilde{\jmath}_1} \cdots$$

As in the case of the vector potential $\boldsymbol{A}$ in Example 1.15, the surfaces $\Sigma_i, 1 \leq i \leq \beta_2(\Omega, S_1)$ can be split up into two groups where the first group is associated with a basis of the second term in the direct sum and the periods vanish on this first group. That is: the set

$$\Sigma_i, \quad 1 \leq i \leq \dim \operatorname{im} \tilde{\jmath}_2,$$

is related to a basis for $\operatorname{im} \tilde{\jmath}_2$ and

$$I_i = \int_{\Sigma_i} \boldsymbol{J} \cdot \boldsymbol{n} \, dS = \int_{\partial \Sigma_i} \boldsymbol{T} \cdot \boldsymbol{dl} = 0$$

since

$$\partial \Sigma_i = 0, \quad 1 \leq i \leq \dim \operatorname{im} \tilde{\jmath}_2.$$

The second group is associated with a basis for the first term in the direct sum. The periods of the current density $\boldsymbol{J}$ is easily expressed in terms of the vector potential in this case. Let

$$\partial \Sigma_{i+\dim \operatorname{im} \tilde{\jmath}_2}, \quad 1 \leq i \leq \dim \operatorname{im} \delta_2,$$

be associated with a basis of $\operatorname{im} \delta_2$:

$$I_{i+\dim \operatorname{im} \tilde{\jmath}_2} = \int_{\Sigma_{\dim \operatorname{im}(\tilde{\jmath}_2)+i}} \boldsymbol{J} \cdot \boldsymbol{n} \, dS = \int_{\partial \Sigma_{i+\dim \operatorname{im} \tilde{\jmath}_2}} \boldsymbol{T} \cdot \boldsymbol{dl}.$$

The fact that the periods of $\boldsymbol{J}$ vanish on the first $\dim \operatorname{im} \tilde{\jmath}_2$ generators of $H_2(\Omega, S_1)$ in order for a vector potential to exist imposes no real constraint on the problem since these periods represent the rate of change of net charge in some connected component of $\mathbb{R}^3 - \Omega$. Since the problem is assumed to be static, these periods are taken to be zero. The next problem which arises is the prescription of the tangential components of the vector potential on S_1 so that the normal component of $\boldsymbol{J}$ vanishes there and the periods of $\boldsymbol{J}$ on the $\delta_2^{-1}(\ker \tilde{\imath}_1)$ remaining generators of $H_2(\Omega, S_1)$ can be prescribed in terms of the vector potential. This problem can be overcome by using exactly the same technique as in Example 1.15. That is, let

$$\boldsymbol{n} \times \boldsymbol{T} = \overline{\operatorname{curl}} \, \psi \text{ on } S_1$$

where the jumps

$$[\psi]_{d_j}, \quad 1 \leq j \leq \beta_1(S_1, \partial S_1),$$

are prescribed on the curves d_i which are associated with a basis of $H_1(S_1, \partial S_1)$. As in that previous example, selecting a set of curves z_i, $1 \leq i \leq \beta_1(S_1)$, associated with a basis of $H_1(S_1)$ where

$$\partial \Sigma_{i+\dim \operatorname{im} \tilde{\jmath}_2} = z_i, \quad 1 \leq i \leq \dim \operatorname{im} \delta_2,$$

one can explicitly describe the jumps $[\psi]_{d_j}$ in terms of the periods

$$I_{i+\dim \operatorname{im} \tilde{\jmath}_2}, \quad 1 \leq i \leq \dim \operatorname{im} \delta_2.$$

In this way it is possible to restate the stationary content principle as follows:

Stationary Content Principle ($\boldsymbol{J} = \operatorname{curl}\boldsymbol{T}$).

$$G(\operatorname{curl}\boldsymbol{T}) = \inf_{\boldsymbol{T}} \int_\Omega \int^{\boldsymbol{T}} \mathcal{E}(\operatorname{curl}\xi, \boldsymbol{r}) \cdot \operatorname{curl}(d\xi)\, dV,$$

subject to the principal boundary condition

$$\boldsymbol{n} \times \boldsymbol{T} = \overline{\operatorname{curl}\psi} \quad \text{on } S_1,$$

where $[\psi]_{d_j}$ are prescribed on curves representing generators of $H_1(S_1, \partial S_1)$ and ψ is otherwise an arbitrary single-valued function.

Turning to the other variational principle, and using the mathematical analogy between this example and the previous example, one sees that the stationary cocontent principle cannot in general be expressed in terms of a continuous single-valued scalar potential. To see why this is so, one considers the following portion of the long exact homology sequence of the pair (Ω, S_2):

$$\cdots \xrightarrow{\delta_2} H_1(S_2) \xrightarrow{\tilde{\imath}_1} H_1(\Omega) \xrightarrow{\tilde{\jmath}_1} H_1(\Omega, S_2) \xrightarrow{\delta_1} H_0(S_2) \xrightarrow{\tilde{\imath}_0} H_0(\Omega) \xrightarrow{\tilde{\jmath}_0} \cdots$$

Let

$$c_i, \quad 1 \le i \le \dim \operatorname{im} \tilde{\jmath}_1,$$

be a set of curves associated with a basis of $\operatorname{im}\tilde{\jmath}_1$ in $H_1(\Omega, S_2)$. The periods of the electric field intensity $\boldsymbol{E}$ on these curves are in general nonzero, but would be zero if $\boldsymbol{E}$ is the gradient of a continuous single-valued scalar potential. As in Example 1.15, this problem can be overcome by letting

$$\Sigma_i', \quad 1 \le i \le \dim \operatorname{im} \tilde{\jmath}_1$$

be a set of surfaces associated with generators of $H_2(\Omega, S_1)$ which act like barriers which cause the scalar potential to be single-valued on

$$\Omega^- = \Omega - \bigcup_{i=1}^{\dim \operatorname{im} \tilde{\jmath}_1} \Sigma_i'.$$

Furthermore, the periods

$$V_i = \int_{c_i} \boldsymbol{E} \cdot d\boldsymbol{l}, \quad 1 \le i \le \dim \operatorname{im} \tilde{\jmath}_1,$$

can be prescribed in terms of the jumps $[\psi]_{\Sigma_i'}$, and when this is done the remaining periods

$$V_{i+\dim \operatorname{im} \tilde{\jmath}_1} = \int_{c_{i+\dim \operatorname{im} \tilde{\jmath}_1}} \boldsymbol{E} \cdot d\boldsymbol{l}$$

can be expressed in terms of the scalar potential, which is constant-valued on each connected component of S_2. That is, if

$$\partial c_{i+\dim \operatorname{im} \tilde{\jmath}_1} = p_i - p_0, \quad 1 \le i \le \dim \operatorname{im} \delta_1,$$

then

$$\phi = \phi(p_i)$$

on the ith connected component of S_2, and if p_0 is some datum node then the last $\dim \operatorname{im} \delta_1$ periods of $\boldsymbol{E}$ can be prescribed by specifying the potential differences

$$\phi(p_i) - \phi(p_0), \quad 1 \le i \le \dim \operatorname{im} \delta_1.$$

When this is done the stationary cocontent principle can be rephrased in terms of a scalar potential as follows.

Stationary Cocontent Principle ($\boldsymbol{E} = \operatorname{grad}\phi$).

$$G'(\operatorname{grad}\phi) = \inf_{\phi} \int_{\Omega} \int^{\phi} \mathcal{J}(\operatorname{grad}\xi, \boldsymbol{r}) \cdot \operatorname{grad}(d\xi)\, dV$$

subject to the constraints

$$[\phi]_{\Sigma_i}, \text{ prescribed on cut surfaces } \Sigma_i'$$

and

$$\phi = \phi(p_i), \text{ on the } i\text{th connected component of } S_2.$$

In summary, the problem of calculating steady current distributions in conducting bodies and the problem of three-dimensional magnetostatics are equivalent under the change of variables

$$\boldsymbol{A} \leftrightarrow \boldsymbol{T}, \qquad \boldsymbol{B} \leftrightarrow \boldsymbol{J} \qquad \psi \leftrightarrow \phi, \qquad \boldsymbol{H} \leftrightarrow \boldsymbol{E},$$

hence the mathematical considerations in using vector or scalar potentials are the same in both problems. Thus it is necessary to summarize only the physical interpretations of the periods and potentials. In this example vector fields $\boldsymbol{J}$ and $\boldsymbol{E}$ were associated with elements of $Z_c^2(\Omega - S_1)$ and $Z_c^1(\Omega - S_2)$ respectively and the nondegenerate bilinear pairings

$$\int : H_2(\Omega, S_1) \times H_c^2(\Omega - S_1) \to \mathbb{R},$$

$$\int : H_1(\Omega, S_2) \times H_c^1(\Omega - S_2) \to \mathbb{R},$$

induced in homology and cohomology by integration, are associated with currents and electromotive forces respectively. The formulas

$$Z_c^2(\Omega - S_1) \simeq H_c^2(\Omega - S_1) \oplus B_c^2(\Omega - S_1),$$
$$Z_c^1(\Omega - S_2) \simeq H_c^1(\Omega - S_2) \oplus B_c^1(\Omega - S_2),$$

show that when there is a variational principle where either $\boldsymbol{J}$ or $\boldsymbol{E}$ are independent variables, conditions fixing the periods of these relative cocycles restrict the variation of the extremal to be a relative coboundary. Alternatively, when the variational principles are formulated in terms of potentials, the potentials are unique to within an element of

$$Z_c^1(\Omega - S_1) \quad \text{for } \boldsymbol{T},$$
$$Z_c^0(\Omega - S_2) \quad \text{for } \phi,$$

and techniques of the previous example show how to eliminate this nonuniqueness in the case of the vector potential. □

Example 2.4 Currents on conducting surfaces: $n = 2$, $p = 1$. Consider again the two dimensional surface of Example 1.5, which is homeomorphic to a sphere with n handles and k holes, and suppose that the component of the magnetic field normal to the surface is negligible and that the frequency of excitation is low enough to make displacement currents negligible. Hence let $\partial\Omega = S_1 \cup S_2$, where $S_1 \cap S_2$ has no length, and assume

$$\begin{aligned} \operatorname{div} \boldsymbol{J} &= 0 \ \text{ on } \Omega, \\ J_n &= 0 \ \text{ on } S_1, \\ \operatorname{curl} \boldsymbol{E} &= 0 \ \text{ on } \Omega, \\ E_t &= 0 \ \text{ on } S_2. \end{aligned}$$

Thus S_1 is associated with the edge of the plate which does not touch any other conducting body and S_2 is associated with the interface of a perfect conductor. Alternatively S_1 or S_2 can be identified with symmetry planes. In this case the electric field $\boldsymbol{E}$ is associated with an element of $Z_c^1(\Omega - S_2)$.

Let $\boldsymbol{n}'$ be the vector normal to the sheet. Then $\boldsymbol{n}' \times \boldsymbol{J}$ can be associated with an element of $Z_c^1(\Omega - S_1)$. That is,

$$\begin{aligned} \operatorname{curl}(\boldsymbol{n}' \times \boldsymbol{J}) \cdot \boldsymbol{n}' &= 0 \ \text{ on } \Omega, \\ (\boldsymbol{n}' \times \boldsymbol{J})_t &= 0 \ \text{ on } S_1. \end{aligned}$$

Let c_i, for $1 \le i \le \beta_1(\Omega, S_1)$, be a set of curves associated with generators of $H_1(\Omega, S_1)$ and let z_j, $1 \le j \le \beta_1(\Omega, S_2)$, be another set of curves associated with generators of $H_1(\Omega, S_2)$. Let these sets of curves be arranged in intersecting pairs as in Example 1.5. That is, if $\operatorname{Int}(c_i, z_j)$ is the number of oriented intersections of c_i with z_j, then

$$\operatorname{Int}(c_i, z_j) = \delta_{ij} \text{ (Kronecker delta).}$$

Furthermore let the periods of the two cocycles on these sets of cycles be denoted by

$$\begin{aligned} I_i &= \int_{c_i} (\boldsymbol{J} \times \boldsymbol{n}') \cdot d\boldsymbol{l}, \qquad 1 \le i \le \beta_1(\Omega, S_1), \\ V_j &= \int_{z_j} \boldsymbol{E} \cdot d\boldsymbol{l}, \qquad\qquad\ 1 \le j \le \beta_1(\Omega, S_2). \end{aligned}$$

If d is the thickness of the plate and σ the conductivity of the material, the stationary content and cocontent principles can be restated as follows (note that $\boldsymbol{J}$ has units of current per length in this problem).

Stationary Content Principle.

$$G(\boldsymbol{J}) = \inf_{\boldsymbol{J} \times \boldsymbol{n}' \in Z_c^1(\Omega - S_1)} \int_\Omega \frac{|\boldsymbol{J}|^2}{\sigma d}\, dS,$$

subject to the constraints prescribing the periods of $\boldsymbol{J} \times \boldsymbol{n}'$ on generators of $H_1(\Omega, S_1)$:

$$I_i = \int_{c_i} (\boldsymbol{J} \times \boldsymbol{n}') \cdot d\boldsymbol{l}, \quad 1 \le i \le \beta_1(\Omega, S_1).$$

Stationary Cocontent Principle.

$$G'(\boldsymbol{E}) = \inf_{\boldsymbol{E}\in Z_c^1(\Omega-S_2)} \int_\Omega \frac{\sigma d\,|\boldsymbol{E}|^2}{2}\, dS$$

subject to the constraint prescribing the periods of $\boldsymbol{E}$ on generators of $H_1(\Omega, S_2)$:

$$V_j = \int_{z_j} \boldsymbol{E}\cdot d\boldsymbol{l}, \quad 1 \leq j \leq \beta_1(\Omega, S_2).$$

As in Examples 1.14, 1.15 and 2.3, the extremals are constrained to be relative cocycles, and when the periods on generators of relative homology groups are specified, the space of admissible variations of the extremal is the space of coboundaries. This follows from the identities

$$Z_c^1(\Omega - S_1) \simeq H_c^1(\Omega - S_1) \oplus B_c^1(\Omega - S_1),$$
$$Z_c^1(\Omega - S_2) \simeq H_c^1(\Omega - S_2) \oplus B_c^1(\Omega - S_2).$$

The relationship of these variational principles to the lumped parameters of resistance and conductance is the same as in example 2.3 and hence will not be discussed here. Instead, it is instructive to reformulate the above variational principles in terms of scalar potentials. By interpreting the c_i and the z_i as cuts, one can set

$$\boldsymbol{J} = \overline{\operatorname{curl}}\,\chi \;\text{ on } \Omega - \bigcup_{i=1}^{\beta_1(\Omega,S_2)} z_i, \qquad \chi = 0 \;\text{ on } S_1,$$

and let $[\chi]_{z_i}$ denote the jump of χ as z_i is traversed. Similarly one can set

$$\boldsymbol{E} = \operatorname{grad}\phi \;\text{ on } \Omega - \bigcup_{i=1}^{\beta_1(\Omega,S_1)} c_i, \qquad \phi = 0 \;\text{ on } S_2,$$

where the jumps of ϕ on the c_i are denoted by $[\phi]_{c_i}$.

It is quite natural to associate $\boldsymbol{J}$ with an element of $Z_c^1(\Omega - S_1)$ having nonzero periods. These periods result naturally from a nonzero current being forced by time varying magnetic fields. On the other hand the assumption that the electric field is irrotational in the plate seems to preclude the possibility of the electric field having nonzero periods, but this is not necessarily the case.

The periods of the cocycles in terms of the jumps in the corresponding potentials are easily calculated. Here the duality

$$H_1(\Omega, S_1) \simeq H_1(\Omega, S_2)$$

comes in nicely, for if the entries of the intersection matrix satisfy

$$\mathcal{I}_{ij} = \operatorname{Int}(c_i, z_j) = \delta_{ij},$$

then

$$\begin{aligned}
I_i &= \int_{c_i} (\boldsymbol{J} \times \boldsymbol{n}') \cdot d\boldsymbol{l} = \int_{c_i} ((\overline{\operatorname{curl}} \chi) \times \boldsymbol{n}') \cdot d\boldsymbol{l} = \int_{c_i} \operatorname{grad} \chi \cdot d\boldsymbol{l} \\
&= \sum_{i=1}^{\beta_1(\Omega, S_2)} \mathcal{I}_{ij} [\chi]_{z_j} \quad (\text{since } \partial c_i \in C_0(S_1)) \\
&= [\chi]_{z_i}
\end{aligned}$$

for $1 \leq i \leq \beta_1(\Omega, S_1)$, and similarly

$$\begin{aligned}
V_j &= \int_{z_j} \boldsymbol{E} \cdot d\boldsymbol{l} = \int_{z_j} \operatorname{grad} \phi \cdot d\boldsymbol{l} \\
&= \sum_{i=1}^{\beta_1(\Omega, S_1)} \mathcal{I}_{ij} [\phi]_{c_i} \quad \text{since } \partial z_i \in C_0(S_2) \\
&= [\phi]_{c_i}.
\end{aligned}$$

Thus when the bases of $H_1(\Omega, S_1)$ and $H_1(\Omega, S_2)$ are arranged so that the intersection matrix is the unit matrix, the stationary content and cocontent principles can be restated as:

Stationary Content Principle ($\boldsymbol{J} = \operatorname{curl} \chi$).

$$G(\operatorname{curl} \chi) = \inf_{\chi} \int_{\Omega} \frac{|\operatorname{curl} \chi|^2}{2\sigma d} \, dS$$

subject to

$$\chi = 0 \text{ on } S_1$$

and the constraints prescribing the periods of $\boldsymbol{J} \times \boldsymbol{n}'$ on generators of $H_1(\Omega, S_1)$:

$$I_i = [\chi]_{z_i} \quad 1 \leq i \leq \beta_1(\Omega, S_1).$$

Stationary Cocontent Principle ($\boldsymbol{E} = \operatorname{grad} \phi$).

$$G'(\operatorname{grad} \phi) = \inf_{\phi} \int_{\Omega} \frac{\sigma d \, |\operatorname{grad} \phi|^2}{2} \, dS$$

subject to $\phi = 0$ on S_2 and the constraints prescribing the periods of $\boldsymbol{E}$ on generators of $H_1(\Omega, S_2)$:

$$V_i = [\phi]_{c_i}.$$

Thus, by playing down the role implicitly played by the long exact homology sequence, the role of duality theorems in handling topological aspects has become more apparent in this example. Duality theorems will be explored in greater detail in Chapter 3. □

The previous examples show that homology groups arise naturally in boundary value problems of electromagnetics. It is beyond the intended scope of these introductory chapters to go beyond a heuristic account of axiomatic homology theory in the context of the boundary value problems being considered because additional mathematical machinery such as categories, functors, and homotopies are required to explain the axioms which underlie the theory (see [HY61, Section

7.7] for an explanation of the axioms). It is sufficient to say that the existence of a long exact homology sequence is only one of the seven axioms of a homology theory. The other six axioms of a homology theory, once understood, are "intuitively obvious" in the present context and have been used implicitly in many of the previous examples.

2E. The Electromagnetic Lagrangian and Rayleigh Dissipation Functions

Introduction. In the beginning of this chapter, we defined quasistatics as follows. Given an electromagnetic system with overall dimension l, and highest frequency of interest $f_{\max}$, the system is quasistatic if the corresponding wavelength, $\lambda_{\min} = c/f_{\max}$, is considerably larger than l. Unfortunately this criterion does not give us any tools for modeling quasistatic systems. The tools for extracting circuit models from quasistatic systems are based on tracking energy and identifying circuit variables with the periods of the closed differential forms which are responsible for the low frequency modes of energy storage and power dissipation. If the system is rigid, there is no possibility of doing mechanical work, and energy can only be stored in the electric or magnetic field, or dissipated as heat. These three aspects of the electromagnetic field where introduced in sections 2B, 2C, 2D, respectively. If we interpret the magnetic energy as the energy of moving charges and the electric energy as the energy of stationary charges, then we can naively view the Lagrangian for the electromagnetic field as the difference between magnetic and electric energies. This is indeed the case and is the 19th century view of the Lagrangian for the electromagnetic field. In this section we will add flesh to this bare bones point of view. To do so, we will first review the Lagrangian mechanics of dissipative systems and the calculus of variations for multiple integral problems. This provides the basis for building our electromagnetic Lagrangian and associated Rayleigh dissipation function

Lagrangian Mechanics and Rayleigh Dissipation Function. Lagrangian mechanics [Lan70] is one of the most beautiful chapters in the history of mathematical physics. In this section we introduce some key aspects which are needed to develop our picture of electroquasistatics. In order to get started and to avoid an overly detailed treatment, we will focus on three terms which describe the simplest systems: dissipationless, holonomic, and scleronomic.

Dissipationless systems are systems whose internal energy may change from on form to another in a reversible way, but is never dissipated as heat. Holonomic systems are systems whose energy can be described by a set independent coordinates. Motorists may recognize that a car with tires constrained to roll on the pavement without slipping is an example of a nonholonomic system. If the position and orientation of the car are described by three coordinates (x, y, θ), a car cannot move sideways by using only one of the two variables which can be changed, namely the position of the steering wheel and the ability to move forward and backward. Furthermore, the no-slip condition provides a relation between the differentials dx, dy, and $d\theta$. However the car cannot be described by two coordinates alone since, by moving forward, turning wheels, moving backward, turning wheels in opposite direction, and repeating these four steps, the car can be moved sideways as in parallel parking. In effect, one can place and orient the car to coincide with any chosen values of x, y, and θ, but these coordinates are dependent and no two coordinates suffice. This characterizes a nonholonomic system.

Since the analysis of nonholonomic systems tends to be involved, and the subject is often avoided in elementary treatments, we will consider just one more example before moving on:

Example 2.5 A billiard ball on a flat table as a nonholonomic system. Consider Cartesian coordinates (x, y, z) in Euclidean space and a table top whose surface is given by the equation $z = 0$. Suppose also that a billiard ball whose surface described by spherical coordinates θ and ϕ is lying on the table. The orientation of the ball is described by three coordinates: θ, ϕ fix the point of the ball touching the table, and ψ describes a rotation about this fixed point. Adding the constraint that the ball must always remain in contact with the table, the five coordinates $(x, y, \theta, \phi, \psi)$ describe the position of the ball. If the motion is constrained by requiring that the ball roll without slipping, the coordinates are not independent since the no-slip condition involves the differentials of all coordinates. The subtlety is that no four coordinates describe the ball completely, since there always exists a way of starting at any given $(x_0, y_0, \theta_0, \phi_0, \psi_0)$ and rolling the ball on the table without slipping to end up at another given $(x_0, y_0, \theta_1, \phi_1, \psi_1)$. (Try it!) Hence five coordinates subject to a differential relation describe the ball but no four coordinates suffice, and we see that the system is nonholonomic. □

We will not consider detailed examples of nonholonomic systems arising in electromechanical energy conversion, but rather be satisfied with the terms dissipationless and holonomic. For a dissipationless holonomic system, we can consider the energy E which depends on some number n of independent coordinates $\{q_i\}_{i=1}^n$, their time derivatives and time. The space (or manifold) described by these local coordinates is called the configuration space. For nonrelativistic systems, we write

$$E = T + V$$

where V is the potential energy which is a function of the coordinates $\{q_i\}_{i=1}^n$ and time, and T is the kinetic energy which is a symmetric positive definite quadratic form in the velocities $\{\dot{q}_i\}_{i=1}^n$, whose coefficients can be functions of $\{q_i\}_{i=1}^n$ and time. Thus at a given time, the energy is a function of $2n$ variables.

The Lagrangian of a dissipationless holonomic system is defined as the difference between the kinetic and potential energies, $L = T - V$, and the principle of least action states that the system evolves in time such that the functional

$$\int L(q_1, \ldots, q_n, \dot{q}_1, \ldots, \dot{q}_n)\, dt$$

is stationary with respect to variations in the functions $\{q_i(t)\}_{i=1}^n$.

The significance of the Lagrangian is appreciated when one solves complicated problems in mechanics. We shall soon see that the techniques of the calculus of variations show that the above functional is stationary if the functions $\{q_i(t)\}_{i=1}^n$ satisfy the Euler–Lagrange equations:

$$\frac{d}{dt}\left(\frac{\partial L}{\partial \dot{q}_i}\right) - \frac{\partial L}{\partial q_i} = 0, \quad 1 \le i \le n.$$

For nontrivial systems such as tops, coupled pendulums and spherical pendulums, one quickly learns that it is far simpler to form the Lagrangian and try solve a system of Euler–Lagrange equations than to attempt to write down the equations of motion using Newton's equations. This is the simplest vindication of the Lagrangian approach.

Often the Lagrangian of a system does explicitly depend on time. Such systems are called scleronomic systems. This class of systems is of secondary importance to us since we want to consider systems which are subject to external excitations.

It must be stressed that quasistatic electromagnetic systems may or may not be dissipationless, holonomic, or scleronomic. Given the elements of Lagrangian mechanics introduced so far and our goal of modeling electromagnetic systems, it is imperative that we consider dissipative systems. Although dissipation is a result of processes described by huge numbers of molecular degrees of freedom, on a macroscopic scale we are limited to models which use only $\{q_i, \dot{q}_i\}$ as variables. To see how one attempts to model dissipation, we consider the variation of the Euler–Lagrange equations when external forces are present:

$$\frac{d}{dt}\left(\frac{\partial L}{\partial \dot{q}_i}\right) - \frac{\partial L}{\partial q_i} = F_i, \quad 1 \le i \le n.$$

Here the F_i are external forces which for the moment we will assume model dissipative processes. The simplest model involves assuming frictional forces so that each F_i is a linear combination of the velocities $\{\dot{q}_j\}$. The Rayleigh dissipation function arises when one assumes that the heat generated is a well-defined function of the velocities. For a linear friction law we then have a quadratic form

$$F = \frac{1}{2}\sum_{i,j=i}^{n} c_{ij}\dot{q}_i\dot{q}_j \quad \text{such that} \quad F_i = \frac{\partial F}{\partial \dot{q}_i}.$$

Thus, a holonomic system whose dissipation is a result of linear friction laws is characterized by T, V, and F. The equations of motion are given by

$$\frac{d}{dt}\left(\frac{\partial L}{\partial \dot{q}_i}\right) - \frac{\partial L}{\partial q_i} = \frac{\partial F}{\partial \dot{q}_i}, \quad 1 \le i \le n, \tag{2–46}$$

where $L = T - V$. This recipe works for continuum systems but before applying the formalism to electroquasistatic systems where T, V, and F correspond to magnetic energy, electric energy, and Joule dissipation, we need to invest in the techniques of the calculus of variations.

The Calculus of Variations for Multiple Integral Problems. To extend the formalism of the Lagrangian mechanics of dissipative systems to electromagnetic systems, we need the calculus of variations. To get started in the multidimensional case, we need some notation for differential operators. Consider an n-dimensional region R, and two vector spaces V_1 and V_2 of dimension r_1 and r_2, respectively. Let $\mathcal{V}_1$ and $\mathcal{V}_2$ be the spaces of smooth functions from the region R to V_1 and V_2 respectively. A differential operator is a map

$$D : \mathcal{V}_1 \to \mathcal{V}_2$$

of a special form between two such spaces. Concretely, D is represented by an $r_1 \times r_2$ matrix whose entries are variable coefficient differential operators of some specified degree. We consider a few examples:

(1) d^p/dt^p is a degree p differential operator. Here R is $\mathbb{R}$ and $r_1 = r_2 = 1$.
(2) grad, div, and curl are first-degree differential operators. Here R is a three-dimensional subset of $\mathbb{R}^3$ and (r_1, r_2) are $(1,3)$, $(3,1)$, and $(3,3)$ respectively.
(3) The Laplacian operators $-\operatorname{div}\operatorname{grad}$ and $\operatorname{curl}\operatorname{curl} - \operatorname{grad}\operatorname{div}$ are second order differential operators defined on subsets of $\mathbb{R}^3$. Here (r_1, r_2) are $(1,1)$ and $(3,3)$, respectively.
(4) The wave operators

$$-\operatorname{div}\operatorname{grad} + \frac{1}{c^2}\frac{\partial^2}{\partial t^2}$$

and

$$-\operatorname{grad}\operatorname{div} + \operatorname{curl}\operatorname{curl} + \frac{1}{c^2}\frac{\partial^2}{\partial t^2}$$

are second order differential operators defined on subsets of $\mathbb{R}^4$ (space-time). Here again (r_1, r_2) are $(1,1)$ and $(3,3)$ respectively.

In order to talk about the multidimensional version of an Euler–Lagrange equation, we need some acquaintance with the notion of an adjoint operator. To get started, suppose V_1 and V_2 are inner product spaces with inner products $\langle \cdot, \cdot \rangle_{V_1}$ and $\langle \cdot, \cdot \rangle_{V_2}$ respectively. More generally we can replace the inner products with nondegenerate symmetric bilinear forms. We can then define inner products on $\mathcal{V}_1$ and $\mathcal{V}_2$ by

$$\langle f_1, f_2 \rangle_{\mathcal{V}_i} = \int_R \langle f_1, f_2 \rangle_{V_i} \, dV.$$

Formally, we can define a transpose operator D^T by the formula

$$\langle Du, v \rangle_{\mathcal{V}_2} = \langle u, D^T v \rangle_{\mathcal{V}_1},$$

where $D^T : \mathcal{V}_2 \to \mathcal{V}_1$, and if we are not dealing with inner product spaces, we should define D^T as a map between dual spaces. In practice, the operators D and D^T are supplemented by boundary conditions, and given D, D^T is found via integration by parts. Returning to our examples, we have, ignoring boundary conditions:

(1) If $D = d/dt$ then $D^T = -d/dt$ on $\mathbb{R}^1$, $r_1 = r_2 = 1$.

(2) For $R \in \mathbb{R}^3$, we have from standard vector identities

$$D = \text{grad} \Longrightarrow D^T = -\text{div} \quad (r_1, r_2) = (1,3),$$
$$D = \text{curl} \Longrightarrow D^T = -\text{curl} \quad (r_1, r_2) = (3,3),$$
$$D = \text{div} \Longrightarrow D^T = -\text{grad} \quad (r_1, r_2) = (3,1).$$

(3) For $R \in \mathbb{R}^3$ we can iterate the above identities to get

$$D = -\text{div grad} \Longrightarrow D^T = D \text{ modulo boundary conditions},$$
$$D = -\text{grad div} + \text{curl curl} \Longrightarrow D^T = D \text{ modulo boundary conditions}.$$

(4) If D is a wave operator as above, then $D^T = D$ modulo boundary and initial conditions. This follows from examples 1 and 3 above.

Refusing to slow down for a careful specification of $\mathcal{V}_1$ and $\mathcal{V}_2$, or to specify the boundary conditions picked up when integrating by parts, then we are ready for a naive introduction to multiple integral problems in the calculus of variations. Suppose now that u is a map from the region R to a vector space V, $u : R \to V$ which is, in some sense, smooth. If V is $\mathbb{R}$, then u is a function and if V is $\mathbb{R}^n$ then we can naively think of u as a vector or tensor field. Next, consider a set of inner product spaces $\{V_i\}_{i=1}^m$ and spaces of maps

$$\mathcal{V}_i = \text{Maps}(R, V_i), \quad 1 \le i \le m.$$

If each V_i is an inner product space with inner product $\langle \cdot, \cdot \rangle_{V_i}$, then

$$\langle f_1, f_2 \rangle_{\mathcal{V}_i} = \int_R \langle f_1, f_2 \rangle_{V_i} \, dV$$

is an inner product on $\mathcal{V}_i$. Suppose further that we have differential operators

$$D_i : \mathcal{V} \to \mathcal{V}_i$$

and associated adjoint operators defined via

$$\langle D_i u, v \rangle_{\mathcal{V}_i} = \langle u, D_i^T v \rangle_{\mathcal{V}_i}.$$

In this context, consider the following functional defined in terms of an integral over the n-dimensional region R:

$$\mathcal{F}[u] = \int_R F(u, D_1 u, \ldots, D_m u) \, dV, \tag{2–47}$$

where F is a smooth function

$$F : V \oplus (\oplus_{i=1}^m V_i) \longrightarrow \mathbb{R}.$$

To derive an Euler equation for such a functional, we can compute the variation of the functional, take the linear part if the functional is differentiable, and set it equal to zero. Although this is a tall order, the form of the functional allows us to reduce the problem to finding the Taylor series of a function, and integration by parts.

Consider a family of functions $\tilde{u}_\varepsilon$ parametrized by $\varepsilon \in [0, 1]$:

$$\tilde{u}_\varepsilon : R \times [0,1] \longrightarrow V.$$

If we assume that $\tilde{u}_\varepsilon$ is analytic in the parameter ε, the $\mathcal{F}[\tilde{u}_\varepsilon]$ becomes a function of ε when $\tilde{u}_\varepsilon$ is given. We call this function $f(\varepsilon)$. That is ,

$$f(\varepsilon) = \mathcal{F}[\tilde{u}_\varepsilon],$$

and $f'(\varepsilon)|_{\varepsilon=0}$ is the variation of $\mathcal{F}$ in the direction of $\tilde{u}_0'$. Here we can think of $\tilde{u}_\varepsilon$ as a curve in the space of functions and $\tilde{u}_0'$ as the tangent vector to the endpoint $\tilde{u}_0$. Using the chain rule we have

$$f'(0) = \int_R \left\langle \frac{\partial F}{\partial u}, \tilde{u}_0' \right\rangle + \sum_{k=1}^{m} \left\langle \frac{\partial F}{\partial (D_k u)}, D_k \tilde{u}_0' \right\rangle_{V_i} dV.$$

Here we have taken the liberty of identifying the finite-dimensional vector spaces with their duals. By the definition of the adjoint operators we have

(2–48)
$$\left\langle \frac{\delta F}{\delta u}, \tilde{u}_0' \right\rangle_{\mathcal{V}} = \int_R \left\langle \frac{\partial F}{\partial u} + \sum_{k=1}^{m} D_k^T \left(\frac{\partial F}{\partial (D_k u)} \right), \tilde{u}_0' \right\rangle_V dV + \text{Boundary Terms.}$$

Since we must have the first variation of the functional vanish for all possible "admissible variations" $\tilde{u}_0'$, we conclude that

$$\frac{\partial F}{\partial u} + \sum_{k=1}^{m} D_k^T \left(\frac{\partial F}{\partial (D_k u)} \right) = 0 \quad \text{in } R \tag{2–49}$$

is the Euler equation for the functional (2–47). Boundary conditions and global topological aspects emerge when the boundary terms arising from integration by parts are scrutinized.

At this point we see that we have achieved our immediate goal of generalizing the Euler–Lagrange equation of classical mechanics

$$\frac{\partial L}{\partial q_i} - \frac{d}{dt} \left(\frac{\partial L}{\partial (dq_i/dt)} \right) = 0.$$

Before reconciling quasistatic electromagnetic systems with the Lagrangian mechanics of dissipative systems, some of the more subtle aspects of the above development should be mentioned. They help motivate some ideas which will be developed in later chapters.

(1) The details of how one performs the integrations by parts in order to deduce the transpose operators and boundary conditions can be a huge chore. Luckily, all of the differential operators encountered in electromagnetism are special cases of the exterior derivative. This is because they can all be related to Maxwell's equations in integral form. As we shall see in the next chapter, Stokes' theorem on manifolds will make all of these conditions simple to compute. Furthermore, our investment in cohomology theory will pay off when it comes time to extract lumped parameter information from the boundary integrals.

(2) The argument that the functional is stationary when its first derivative vanishes requires that the functional be differentiable and that the remainder term in the functional Taylor series be bounded in some neighborhood of the

stationary point. A careful treatment requires defining a space of "admissible variations". This is a delicate issue which must be handled case by case. In most cases, the physics of the situation dictates a suitable "energy norm" to resolve these issues.

(3) We have considered nonquadratic functionals. We now need to consider how one reduces a minimization problem to something amenable to computer solution. Loosely speaking there are three basic steps involved. Although they are elaborated in the remainder of the book, it is fun to have a sneak preview:
   (a) If we restrict ourselves to convex functionals, we are guaranteed a unique minimum. If the convex functional is nonquadratic, one can use the Newton–Kantorovich method to get a sequence of linear boundary value problems which inherit a variational structure. In this way, minimizing a convex, nonquadratic functional is reduced to minimizing a sequence of quadratic functionals.
   (b) The second step in reducing a quasistatic problem to something amenable to computer solution is to consider elliptic problems. Technically speaking, elliptic problems are those for which the characteristics of the system of equations are trivial. This opens the door to elliptic regularity theory, that is, tools for proving that the solution is smooth. This in turn guarantees that the solution can be approximated quite well in some finite-dimensional vector space. In other words, elliptic equations, unlike hyperbolic equations where singularities can propagate, have smooth solutions and this is a key step in reliably computing solutions. We will elaborate on the notion of ellipticity in the context of quadratic functionals below.
   (c) Finite element theory provides a way of discretizing the solution space of an elliptic differential operator and approximating the solution by a function of a finite number of degrees of freedom. In this way, the problem of finding a solution reduces to finding the minimum of a matrix quadratic form. The resulting matrix equation involves a large, sparse, symmetric, positive definite matrix — a great playground for high-performance computing. Furthermore, as we shall see in following chapters, our investment in the formalism of chains and cochains points to "Whitney forms" as an interpolation scheme for finite element discretizations of problems involving vector fields. A leisurely but accurate survey of these ideas can be found in the thesis of Tarhasaari [Tar02].

Before we refocus on modeling quasistatic electromagnetic systems by building a continuum analog of Lagrangian mechanics, we consider the special case of Equations (2–48) and (2–49) where the functional is quadratic. If $D_i : \mathcal{V} \to \mathcal{V}_i$, $1 \leq i \leq n$ as before, and

$$\mathcal{F}[u] = \langle u, u \rangle_{\mathcal{V}} + \sum_{i=1}^{k} \langle D_i u, D_i u \rangle_{\mathcal{V}_i} + \langle u, f \rangle$$

In this case, the Euler–Lagrange equation has the form

$$u + \sum_{i=1}^{n} D_i^T D_i u = -f \quad \text{in } R,$$

and before we consider a few concrete examples we will elaborate on the notion of ellipticity. Suppose each D_i is a first order differential operator. Then the operator in the equation above is a matrix of second order differential operators. If we replace the partial derivatives by Fourier variables, we obtain the "symbol" of the differential operator. The principal symbol, denoted by σ_p is a $r \times r$ matrix of homogeneous second-order polynomials in the Fourier variables. The equation is elliptic if

$$\det\left(\sigma_p\left(\sum_{i=1}^{n} D_i^T D_i\right)\right) \neq 0$$

for any choice of Fourier variables restricted to the unit sphere in the Fourier domain. This can be clarified by means of a few examples.

Example 2.6 Ellipticity of the scalar Laplacian. If

$$-\operatorname{div}\operatorname{grad} = \frac{\partial^2}{\partial x^2} + \frac{\partial^2}{\partial y^2} + \frac{\partial^2}{\partial z^2}$$

and

$$\left(\frac{\partial}{\partial x}, \frac{\partial}{\partial y}, \frac{\partial}{\partial z}\right) \to (i\xi, i\eta, i\zeta)$$

in Fourier space, then $\sigma_p(-\nabla^2) = +(\xi^2 + \eta^2 + \zeta^2)$ is a matrix and

$$\det(\sigma_p(-\nabla^2))\big|_{\xi^2+\eta^2+\zeta^2=1} = 1,$$

so that $-\nabla^2$ is elliptic. □

Example 2.7 Ellipticity of the vector Laplacian. curl curl − grad div $= -I_{3\times 3}\nabla^2$ in Cartesian coordinates. Using the same coordinates as before,

$$\det(\sigma_p(-I_{3\times 3}\nabla^2))|_{\xi^2+\eta^2+\zeta^2=1} = 1,$$

so that $-I_{3\times 3}\nabla^2$ is elliptic. □

Example 2.8 Nonellipticity of the wave equation. If $-\operatorname{div}\operatorname{grad} + \dfrac{1}{c^2}\dfrac{\partial^2}{\partial t^2}$ is the wave operator and

$$\left(\frac{\partial}{\partial x}, \frac{\partial}{\partial y}, \frac{\partial}{\partial z}, \frac{\partial}{\partial t}\right) \to (i\xi, i\eta, i\zeta, i\tau),$$

then $\sigma_p = \xi^2 + \eta^2 + \zeta^2 - \tau^2/c^2$. Finally,

$$\det(\sigma_p)|_{\xi^2+\eta^2+\zeta^2+\tau^2=1} = 0$$

when

$$\xi^2 + \eta^2 + \zeta^2 = \frac{\tau^2}{c^2}.$$

Hence, the wave operator is not elliptic and there can be discontinuous solutions on the light cone. □

Note that in the case of Lagrangian mechanics there is no need to fuss about ellipticity. This is because any ordinary differential operator whose highest order derivative has a nonvanishing coefficient is automatically elliptic. However, once we know where to be careful, we can return to finding the continuum analogs of $L = T - V$ and F for an electroquasistatic system.

We now return to the process of reconciling the Lagrangian mechanics of dissipative systems, with the analysis of quasistatic electromagnetic systems. It turns out that the Lagrangian for Maxwell's equations subject to the constitutive laws of free space and prescribed current and charge distributions, is most simply expressed in terms of differential forms. Since Maxwell's equations in four dimensions are considered in Example 7.10 (page 229), and the Hodge star is presented in Section MA-L, we have the basic ingredients of the electromagnetic Lagrangian at our fingertips. However, we will now see that in the analysis of quasistatic systems, there is no need to assemble the Lagrangian. There are practical reasons for not doing so:

(1) The underlying geometry in space-time is not Riemannian, but Lorentzian and the Hodge star operator is easily modified in order to account for the signature of the metric. This essential detail, however, has profound consequences; the resulting Euler–Lagrange equations are wave equations and not elliptic equations like Poisson's equation. Hence they are much more difficult to deal with since the regularity results which apply to elliptic equations are no longer applicable.

(2) If the Lagrangian is expressed as $L = T - V$, where T and V are positive semi-definite, then T and V can be individually tied to systems of elliptic equations. The data required to ensure uniqueness of solution is specified by the topological degrees of freedom of circuit theory, as well as boundary and initial conditions. Hence, in the case of quasistatics, the primary coupling between the T and V parts of the Lagrangian, are via degrees of freedom given by cohomology groups.

(3) By forming T, V and the Rayleigh dissipation function F, and minimizing each of these functions subject to boundary and initial data, the time evolution of the system is given by the time evolution of the topological degrees of freedom. That is, a system of ordinary differential equations involving the periods of closed forms and period matrices. These equations have the same structure as those of analytical mechanics.

This is the big picture of how ellipticity rules in the realm of quasistatics. It is somewhat vague until we make the identifications of T, V, and F, which lead to both Table 2.1, and the equations of motion described in Section 2E (Equation (2–46)). First, the kinetic energy, T, is the magnetic energy. Depending on context, it is given by the functionals presented in sections 2C or 3B. The stationary points of these functionals are parametrized by topological degrees of freedom, and are given by the quadratic form involving currents and the inductance matrix.This is a natural and intuitive identification since magnetic fields arise from the motion of charges. Next, the potential energy V, is the electrostatic energy. This functional was considered in Section 2B. The stationary points of this functional are again parametrized by topological degrees of freedom, this time by the quadratic form involving charges and the capacitance matrix. The electromagnetic Lagrangian, $L = T - V$, is quite remarkable in that it is both an elegant expression in four dimensions, and intuitive in terms of a clear connection with classical mechanics. The latter interpretation was the one, which made sense to Maxwell in the days before his theory of electromagnetic radiation was verified by Hertz.

The final identification is that of the Rayleigh dissipation function F, with the Joule heating of steady currents. This functional was considered in Section 2D. Once again, the stationary points of this functional are parametrized by topological degrees of freedom, this time by the quadratic form involving currents and a resistance matrix.

Having made the identifications which are crucial to the modeling process, we can step back and see how the remainder of the book is organized around some details of how the engineer's modeling process works in a computational setting. The first chapter introduced homology and cohomology theory as a bridge between the languages of circuit theory and electromagnetics. The present chapter introduced the notion of quasistatic electromagnetic fields and outlined how the formalism of circuit theory is tied to quasistatic fields via the formalism of analytical mechanics. The next chapter relates the quadratic forms arising in this chapter to the bilinear forms underpinning duality theorems in cohomology theory. This solidifies the intuitions we sought to develop in the first chapter. The bridge to data structures and finite element theory is finally laid down in Chapter 4. Chapter 5 applies the modeling strategy developed in this section to the problem of modeling eddy currents on sheets. This problem is chosen because it is of practical importance in the context of eddy current nondestructive testing, we can give it a relatively complete treatment, and because of the richness of the topological aspects. Chapter 6 fills in the main gap left in Chapter 5: the computation of cuts for magnetic scalar potentials in the context of the finite element method. Finally, all of the variational principles we encounter are put into a common framework in chapter seven. A transition to the formalism of differential forms is required to pull this off, but the benefit of formulating such a paradigm problem yields a recognition that both cohomology theory is the formalism to articulate all of the topological aspects which relate to circuit theory, and that Whitney form interpolation is precisely the tool that captures the topological properties in the discrete setting of finite element analysis.

"The cyclomatic number of a closed surface is twice that of either of the regions it bounds... The space outside the region has the same cyclomatic number as the region itself."

James Clerk Maxwell, A Treatise on Electricity and Magnetism, 1891

3

Duality Theorems for Manifolds With Boundary

3A. Duality Theorems

The next topic from homology theory which sheds light on the topological aspects of boundary value problems is that of duality theorems . Duality theorems serve three functions, namely to show

(1) a duality between certain sets of lumped electromagnetic parameters which are conjugate in the sense of the Legendre transformation;
(2) the relationship between the generators of the pth homology group of an n-dimensional manifold and the $(n-p)$-dimensional barriers which must be inserted into the manifold in order to make the pth homology group of the base manifold trivial;
(3) a global duality between compatibility conditions on the sources in a boundary value problem and the gauge transformation or nonuniqueness of a potential.

In order to simplify ideas, the discussion is restricted to manifolds and homology calculated with coefficients in the field $\mathbb{R}$. The duality theorems of interest to us are formulated for orientable n-dimensional manifolds M and have the form

$$H_c^p(\Omega_1) \simeq H_{n-p}(\Omega_2),$$

where Ω_1 and Ω_2 are manifolds having some geometric relation. In general, the geometric relationship of interest to us is that of a manifold and its boundary or the manifold and its complement. A complete development of duality theorems requires the calculus of differential forms, but we will merely state the relevant duality theorems without proof and give examples to illustrate their application. These duality theorems are a result of a nondegenerate bilinear pairing in cohomology classes and integration. The Mathematical Appendix covers details of the exterior product which leads to the necessary bilinear pairing, but for now we note that we note that the exterior product of a p-form and an $n-p$-form gives an n-form. For orientable manifolds an n-form can be constructed so that integration over the manifold behaves like a nondegenerate bilinear pairing:

$$\int_M : C_c^p(M) \times C^{n-p}(M) \to \mathbb{R}.$$

The pairing induces a nondegenerate bilinear pairing on cohomology

$$\int_M : H_c^p(M) \times H^{n-p}(M) \to \mathbb{R},$$

where the product on forms induces a multiplication on homology classes which is called the cup product . In summary, duality theorems are a consequence of identifying a nondegenerate bilinear pairing associated with integration just as the de Rham theorem comes about as a result of a nondegenerate bilinear pairing between chains and cochains. See [Mas80, Chapter 9] or [GH81, Part 3] for derivations of the most useful duality theorems which do not depend on the formalism of differential forms.

The oldest of these duality theorems is the Poincaré duality theorem, which says that for an orientable n-dimensional manifold M without boundary,

$$H_c^p(M) \simeq H_{n-p}(M).$$

For compact closed manifolds, an intuitive geometric sense of the meaning of this duality can be gained by writing

$$H_p(M) \simeq H_{n-p}(M)$$

and verifying the theorem for some 1- and 2-dimensional manifolds.

In electromagnetics boundary value problems require duality theorems which apply to manifolds with boundary. The classical prototype of this type of theorem is the Lefschetz duality theorem, which states that for a compact n-dimensional region Ω with boundary

$$H_c^{n-p}(\Omega) \simeq H_p(\Omega, \partial\Omega).$$

By de Rham's Theorem this implies

$$H_{n-p}(\Omega) \simeq H_c^p(\Omega - \partial\Omega).$$

Hence

$$\beta_{n-p}(\Omega) = \beta_p(\Omega, \partial\Omega),$$

and again by de Rham's Theorem

$$H_{n-p}(\Omega) \simeq H_p(\Omega, \partial\Omega).$$

To appreciate this duality theorem, intuitively consider the following examples.

3B. Examples of Duality Theorems in Electromagnetism

The Poincaré–Lefschetz duality theorem will turn out to be an immensely useful tool in later chapters. In particular, it is worth illustrating in the context of how it relates $H^1(R;\mathbb{Z})$ to $H_2(R,\partial R;\mathbb{Z})$ for three-dimensional manifolds. At first, we define spaces so that a special case of Poincaré–Lefschetz duality can be written in terms of vector fields. Later it will be stated generally in terms of differential forms to illustrate that this construction is general. Let F and G be vector fields and set

$$\mathcal{F} = \{F \mid \operatorname{curl} F = 0 \text{ in } R \text{ such that } \oint_c F \cdot dl \in \mathbb{Z}\}$$

for a closed path c satisfying $\partial c = 0$ (nonbounding 1-cycle). Also let

$$\mathcal{G} = \{G \mid \operatorname{div} G = 0 \text{ in } R,\ G \cdot \boldsymbol{n}{=}0 \text{ on } \partial R\}.$$

Given any surface S used to calculate flux linkage, Poincaré–Lefschetz duality says that there exists an $F_S \in \mathcal{F}$ dual to S such that

$$\int_R F_S \cdot G\, dV = \int_S G \cdot \boldsymbol{n}\, dS \tag{3–1}$$

for all $G \in \mathcal{G}$. This illustrates the fact that $H^1(R;\mathbb{Z})$ and $H_2(R,\partial R;\mathbb{Z})$ are dual spaces of the space of vector fields G subject to the equivalence relation

$$G \sim G' \ \text{ if } G' = G + \operatorname{curl} A$$

for some A where $\boldsymbol{n} \times A = 0$ on ∂R.

We digress briefly to review some of the structures defined in sections 1A–1C and to note the link between Chapter 1 and the theory of differential forms. Here as in Chapter 1, $H^1(R)$ has been articulated in terms of vector fields for an intuitive understanding of the structure. However, differential forms as described in the Mathematical Appendix are needed for proper development. Loosely speaking, differential forms are the objects appearing under integral signs. They are integrands of p-fold integrals in an n-dimensional manifold where $0 \le p \le n$. For an n-dimensional manifold M^n, the set of all p-dimensional regions c over which p-fold integrals are evaluated is denoted by $C_p(M^n)$ while the set of all p-forms ω are denoted by $C^p(M^n)$. Since $C_p(M^n)$ and $C^p(M^n)$ are spaces with some algebraic structure, integration $\int_c \omega$ can be regarded as a bilinear map which is a nondegenerate bilinear pairing between the spaces:

$$\int : C_p(M^n) \times C^p(M^n) \longrightarrow \mathbb{R}.$$

In this context, the fundamental theorem of calculus, Gauss' and Stokes' theorems of multivariable calculus, and Green's theorem in the plane are generalized by the Stokes theorem on manifolds:

$$\int_c d\omega = \int_{\partial c} \omega \tag{3–2}$$

where $\partial c \in C_{p-1}(M^n)$ is the boundary of c and the exterior derivative operator d takes p-forms to $p+1$-forms so as to satisfy Stokes' theorem. Finally, there is a bilinear, associative, graded commutative product of forms, called the exterior product, which takes a p-form ω and a q-form η and gives a $p+q$-form $\omega \wedge \eta$.

The introduction to homology groups referred to the fact that $\partial_{p-1}\partial_p c = 0$. By a simple calculation, this combines with Stokes' theorem to show that p-forms satisfy $d^{p+1}d^p\omega = 0$ and motivates the following definition. A form μ is said to be *closed* if $d\mu = 0$. In vector calculus, this amounts to saying that $\operatorname{curl}\operatorname{grad}\phi = 0$ and $\operatorname{div}\operatorname{curl} A = 0$. Moreover, the existence of nonbounding cycles and Stokes' theorem also leads to the notion of exact forms: if $\mu = d\omega$ for some ω, μ is said to be exact. While all exact p-forms are closed, not all closed forms are exact, so H^p is the quotient group of closed p-forms modulo exact p-forms. Formally, the coboundary operator is a map $d^p : C^p(M) \to C^{p+1}(M)$ and

$$H^p(M;\mathbb{R}) = \ker d^p / \operatorname{im} d^{p-1}. \tag{3–3}$$

In general, Poincaré–Lefschetz duality says that for a compact, orientable n-dimensional manifold M^n with boundary,

$$H^k(M^n;\mathbb{Z}) \simeq H_{n-k}(M^n, \partial M^n;\mathbb{Z}) \quad \text{for } 0 \le k \le n. \tag{3–4}$$

It holds for any abelian coefficient group, but for us it is enough to deal with integer coefficients. In the case of $n = 3$, duality establishes a one-to-one correspondence between classes in $H^1(R;\mathbb{Z})$ and classes in $H_2(R, \partial R;\mathbb{Z})$. In the context of magnetoquasistatics, these are, respectively, equivalence classes of magnetic fields and equivalence classes of surfaces for flux linkage calculations.

Example 3.1 Poincaré–Lefschetz dual of a submanifold. The notion of the Poincaré–Lefschetz dual of a submanifold is needed in order to develop a formal definition of a cut. Given an n-dimensional compact, oriented manifold M^n with boundary, a closed oriented $n-k$-dimensional submanifold S of M^n, and a closed $n-k$-form ω whose restriction to ∂M is zero, the Poincaré–Lefschetz dual of S is a closed k-form η_S such that [BT82]

$$\int_{M^n} \omega \wedge \eta_S = \int_S \omega.$$

This is a generalized statement of Equation (3–1). When subjected to the (co)homology equivalence relations, the bilinear pairings on both sides of the preceding equation become nondegenerate bilinear pairings between $H^k(M^n)$ and $H^{n-k}(M^n, \partial M^n)$ on the left, and $H_{n-k}(M^n, \partial M^n)$ and $H^{n-k}(M, \partial M)$ on the right. Thus equation (3–4) arises since $H^{n-k}(M^n, \partial M^n)$ is the dual space to both spaces. In other words, for the homology class $[S] \in H_{n-k}(M^n, \partial M^n)$ associated with a submanifold $S \in M^n$, there is an associated unique cohomology class $[\eta_S] \in H^k(M^n)$. As a consequence of Poincaré–Lefschetz duality, cutsfor the magnetic scalar potential can be defined as representatives of classes in $H_2(R, \partial R;\mathbb{Z})$. □

Example 3.2 Lefschetz duality in 3-d electrostatics: $n = 3$, $p = 1$. Consider a nonconducting dielectric region Ω whose boundary $\partial\Omega$ is an interface to conducting bodies. Each connected component of $\partial\Omega$ is associated with an equipotential, and the generators of $H_1(\Omega, \partial\Omega)$ can be associated with curves c_i, for $1 \le i \le \beta_1(\Omega, \partial\Omega)$. The endpoints of the curves c_i are used to specify the $\beta_1(\Omega, \partial\Omega)$ independent potential differences in the problem. Dually, the generators of $H_2(\Omega)$ can be associated with closed surfaces Σ_j, for $1 \le j \le \beta_2(\Omega)$, which can be used to specify the net electrical flux of each of the $\beta_2(\Omega)$ independent charge distributions of the problem. That is, if $\boldsymbol{E}$ denotes the electric field, $\boldsymbol{D}$ the electric field density, V a potential difference, and Q a net charge,

$$\int_{c_i} \boldsymbol{E} \cdot \boldsymbol{t}\, dl = V_i, \quad \text{for } 1 \le j \le \beta_1(\Omega, \partial\Omega)$$

and

$$\int_{\Sigma_j} \boldsymbol{D} \cdot \boldsymbol{n}\, dS = Q_j, \quad \text{for } 1 \le j \le \beta_2(\Omega).$$

Furthermore, the fact that there are just as many independent potential differences as there are independent charges or charge distributions in the problem is expressed in the relationship between the ranks of homology groups which comes from Lefschetz duality:

$$\beta_1(\Omega, \partial\Omega) = \beta_2(\Omega).$$

Another interpretation of the Lefschetz duality theorem is obtained by constructing a matrix of intersection numbers $\mathcal{I}$ where $\mathcal{I}_{ij}$ is the number of oriented intersections of H_1 generators c_i with H_2 generators Σ_j. The Lefschetz theorem then asserts that this matrix is nonsingular simply because generators in classes can be paired. Hence the c_i can be interpreted as a minimal set of curves which, when considered as barriers make

$$H_2\left(\Omega - \bigcup_{i=1}^{\beta_1(\Omega,\partial\Omega)} c_i\right) \simeq 0.$$

In Maxwell's terminology, the c_i eliminate the periphraxity of the region Ω. If the c_i are replaced by tubular neighborhoods of the c_i, then it is always the case that

$$\boldsymbol{D} = \operatorname{curl} \boldsymbol{C} \text{ in } \Omega - \bigcup_{i=1}^{\beta_1(\Omega,\partial\Omega)} c_i$$

whenever $\operatorname{div} \boldsymbol{D} = 0$ in Ω, regardless of how charge is distributed in the exterior of the region and on the boundary. □

Example 3.3 Lefschetz duality in 3-d magnetostatics: $n = 3$, $p = 2$. Consider a nonconducting region Ω whose boundary $\partial\Omega$ is an interface to perfect conductors. In this case the generators of $H_2(\Omega, \partial\Omega)$ can be associated with open surfaces Σ_i, for $1 \le i \le \beta_2(\Omega, \partial\Omega)$, which can be used to compute the $\beta_2(\Omega, \partial\Omega)$ independent magnetic fluxes in the problem. Dually, the generators of $H_1(\Omega)$ are associated with closed curves c_j, for $1 \le j \le \beta_1(\Omega)$, which can be used to specify the number of independent currents in the problem. The magnetic flux density is denoted by $\boldsymbol{B}$, magnetic field intensity by $\boldsymbol{H}$, net magnetic flux by

ϕ, the magnetic vector potential by $\boldsymbol{A}$, and current by I. In these terms, the generators of homology are related to periods as follows:

$$\Phi_i = \int_{\Sigma_i} \boldsymbol{B} \cdot \boldsymbol{n}\, dS = \int_{\partial \Sigma_i} \boldsymbol{A} \cdot \boldsymbol{t}\, dl, \quad \text{for } 1 \leq i \leq \beta_2(\Omega, \partial\Omega)$$

and Ampère's law,

$$I_j = \int_{c_j} \boldsymbol{H} \cdot \boldsymbol{t}\, dl, \quad \text{for } 1 \leq j \leq \beta_1(\Omega).$$

As before, the fact that there are as many independent fluxes as there are currents comes out of the Lefschetz theorem: $\beta_2(\Omega, \partial\Omega) = \beta_1(\Omega)$.

As noted in Example 3.2, another interpretation of the Lefschetz duality theorem can be obtained by constructing the intersection matrix $\mathcal{I}$, where $\mathcal{I}_{ij}$ is the number of oriented intersections of Σ_i with c_j. The Lefschetz theorem asserts that this matrix is nonsingular. Hence the Σ_i can be interpreted as a set of barriers such that

$$\boldsymbol{H} = \operatorname{grad} \zeta \text{ in } \Omega - \bigcup_{i=1}^{\beta_2(\Omega,\partial\Omega)} \Sigma_i$$

whenever

$$\operatorname{curl} \boldsymbol{H} = 0 \text{ in } \Omega$$

regardless of how currents flow in the exterior of the region or on the boundary. □

Example 3.4 Duality and variational principles for magnetostatics. This example connects variational principles for magnetostatics problems with current sources to duality theorems for orientable manifolds with boundary. The reader who is in a hurry may wish to return to this example at a later time. Consider a magnetostatics problem in a compact region Ω with $\partial\Omega = S_1 \cup S_2$, where $S_1 \cap S_2$ has no 2-dimensional area, and satisfying

$$\begin{aligned} \operatorname{div} \boldsymbol{B} &= 0 \text{ in } \Omega, \\ \boldsymbol{B} \cdot \boldsymbol{n} &= 0 \text{ on } S_1, \\ \operatorname{curl} \boldsymbol{H} &= \boldsymbol{J} \text{ in } \Omega, \\ \boldsymbol{H} \times \boldsymbol{n} &= 0 \text{ on } S_2. \end{aligned}$$

As before, S_1 is the interface to a superconductor or a symmetry plane while S_2 is an interface to infinitely permeable bodies. As in Example 1.15, the magnetic flux density vector $\boldsymbol{B}$ can be associated with an element of $Z_c^2(\Omega - S_1)$ and in general it is not a relative coboundary. However, considering the long exact homology sequence for the pair (Ω, S_1) one has

$$H_2(\Omega, S_1) \simeq \delta_2^{-1}(\ker \tilde{\imath}_1) \oplus \tilde{\jmath}_2 \left(\frac{H_2(\Omega)}{\tilde{\imath}_2(H_2(S_1))} \right),$$

where the relevant portion of the long exact sequence is

$$\cdots \xrightarrow{\delta_3} H_2(S_1) \xrightarrow{\tilde{\imath}_2} H_2(\Omega) \xrightarrow{\tilde{\jmath}_2} H_2(\Omega, S_1) \xrightarrow{\delta_2} H_1(S_1) \xrightarrow{\tilde{\imath}_1} H_1(\Omega) \xrightarrow{\tilde{\jmath}_1} \cdots$$

Since the periods of $\boldsymbol{B}$ on the basis of $\operatorname{im}\tilde{\jmath}_2$ correspond to distributions of magnetic poles in $\mathbb{R}^3 - \Omega$, it is natural to set them to zero. When this is done the periods of $\boldsymbol{B}$ on the generators of $H_2(\Omega)$ vanish and $\boldsymbol{B}$ can be written as

$$\boldsymbol{B} = \operatorname{curl}\boldsymbol{A} \text{ in } \Omega.$$

However, as in Example 1.15, one cannot insist that the components of the vector potential tangent to S_1 vanish unless the periods of $\boldsymbol{B}$ on the basis of $\delta_2^{-1}(\ker\tilde{\imath}_1)$ vanish. Since this is not true in general, one lets

$$\boldsymbol{n}\times\boldsymbol{A} = \overline{\operatorname{curl}}\,\chi \quad \text{on } S_1 - \bigcup_{i=1}^{\beta_1(S_1,\partial S_1)} d_i$$

as in Example 1.15, where the d_i, for $1 \le i \le \beta_1(S_1, \partial S_1)$, form a basis for $H_1(S_1, \partial S_1)$. As in Example 1.15, one can express the periods of $\boldsymbol{B}$ on the basis of $\delta_2^{-1}(\ker\tilde{\imath}_1)$ by prescribing the jumps of χ on the d_i.

Assuming the constitutive relation

$$\boldsymbol{H} = \mathcal{H}(\boldsymbol{B}, \boldsymbol{r})$$

defined in Example 1.15, one can rewrite the variational formulation for magnetostatics as follows (see [Kot82], Section 1.21).

Variational Principle ($\boldsymbol{B} = \operatorname{curl}\boldsymbol{A}$).

$$F(\boldsymbol{A}) = \operatorname{ext}_{\boldsymbol{A}} \int_\Omega \int^{\operatorname{curl}\boldsymbol{A}} \mathcal{H}(\xi, \boldsymbol{r})\cdot d\xi - \boldsymbol{A}\cdot\boldsymbol{J}\,dV,$$

subject to the principal boundary condition

$$\boldsymbol{n}\times\boldsymbol{A} = \overline{\operatorname{curl}}\,\chi \quad \text{on } S_1 - \bigcup_{i=1}^{\beta_1(S_1,\partial S_1)} d_i$$

where the $[\chi]_{d_i}$, for $1 \le i \le \beta_1(S_1, \partial S_1)$, are chosen so that the periods of $\boldsymbol{B}$ on $\delta_2^{-1}(\ker\tilde{\imath}_1)$ have their desired values, and χ is otherwise arbitrarily chosen.

The above functional has a nonunique extremal whenever an extremal exists. As in the energy formulation of Example 1.15, the nonuniqueness of the extremal corresponds to an element of $Z_c^1(\Omega - S_1)$. That is, if $\boldsymbol{A}$ and $\boldsymbol{A}'$ correspond to two vector potentials which give the functional its stationary value, then

$$\boldsymbol{A} - \boldsymbol{A}' \in Z_c^1(\Omega - S_1).$$

As noted in Example 1.15, one can write

$$Z_c^1(\Omega - S_1) \simeq H_c^1(\Omega - S_1) \oplus B_c^1(\Omega - S_1)$$

and the nonuniqueness of $\boldsymbol{A}$ can be eliminated by specifying the periods of $\boldsymbol{A}$ on generators of $H_1(\Omega, S_1)$ as well as the divergence of $\boldsymbol{A}$ in Ω and the normal component of $\boldsymbol{A}$ on S_2. This can be done by either making these conditions principal conditions on the above functional or, as in [Kot82, Chapter 5], by constructing another variational formulation for which these conditions are a consequence of extremizing the functional. The question of alternate variational formulations for this problem is taken up in full generality in Chapter 3.

At this point one can expose the interplay between the nonuniqueness of $\boldsymbol{A}$ (the gauge transformation) and the conditions on the solvability of the associated boundary value problem (the conditions for the functional to have an extremum). As was noted in [Kot82], certain convexity conditions on the constitutive relation are sufficient to ensure that questions of solvability can be answered by a "Fredholm alternative" type of argument which implies that the problem has a solution if and only if

$$0 = F(\boldsymbol{A}) - F(\boldsymbol{A}') = \int_{\Omega} (\boldsymbol{A} - \boldsymbol{A}') \cdot \boldsymbol{J}\, dV \text{ for all } \boldsymbol{A} - \boldsymbol{A}' \in Z_c^1(\Omega - S_1).$$

By brute force calculation [Kot82, Chapter 4, Theorem 4.3], the preceding orthogonality condition can be restated entirely in terms of the current density vector $\boldsymbol{J}$. In the present case of homogeneous boundary conditions on S_2, assuming that Σ_i, with $1 \leq i \leq \beta_2(\Omega, S_2)$, is a set of generators of $H_2(\Omega, S_2)$, the conditions for the solvability of the equations for the extremum of the functional are as follows. The local conditions on the current density vector $\boldsymbol{J}$ are

$$\begin{aligned} &\operatorname{div} \boldsymbol{J} = 0 \quad \text{in } \Omega, \\ &\boldsymbol{J} \cdot \boldsymbol{n} \qquad \text{continuous across interfaces}, \\ &\boldsymbol{J} \cdot \boldsymbol{n} = 0 \quad \text{on } S_2. \end{aligned}$$

Meanwhile, the global conditions on $\boldsymbol{J}$ are

$$\int_{\Sigma_i} \boldsymbol{J} \cdot \boldsymbol{n}\, dS = 0, \quad 1 \leq i \leq \beta_2(\Omega, S_2),$$

where the global constraints are verified by using the three-step procedure and the long exact homology sequence for the pair (Ω, S_2) in the usual way. The local conditions in this set of solvability conditions merely state that $\boldsymbol{J}$ can be associated with an element of $Z_c^2(\Omega - S_2)$ while the global conditions ensure that the projection of this cocycle into $H_c^2(\Omega - S_2)$ is zero. Thus the solvability conditions merely state that $\boldsymbol{J}$ can be associated with a relative coboundary in $B_c^2(\Omega - S_2)$. But this is exactly what one requires of $\boldsymbol{J}$ in order to write

$$\boldsymbol{J} = \operatorname{curl} \boldsymbol{H} \quad \text{in } \Omega,$$

with

$$\boldsymbol{n} \times \boldsymbol{H} = 0 \quad \text{on } S_2.$$

Returning to the duality theorems, in Example 1.15 the duality between lumped variables was expressed by

$$H_2(\Omega, S_1) \simeq H_{3-2}(\Omega, S_2) = H_1(\Omega, S_2).$$

Here, in contrast, when sources are added the duality

$$H_1(\Omega, S_1) \simeq H_{3-1}(\Omega, S_2) = H_2(\Omega, S_2)$$

expresses a duality between global degrees of freedom in the gauge transformation and compatibility conditions on the prescription of the current density vector $\boldsymbol{J}$.

One final remark is in order. The global ambiguity of the gauge of the vector potential is associated with unspecified fluxes through "handles" of Ω or unspecified time integrals of electromotive forces between connected components of S_1, while the compatibility conditions on $\boldsymbol{J}$ ensure that Ampère's law can be used without contradiction. The intersection matrix for the generators of the groups $H_1(\Omega, S_1)$ and $H_2(\Omega, S_2)$ can be used to help see how a magnetostatics problem is improperly posed by checking to see which degrees of freedom in the gauge transformation do not leave the value of the functional invariant. □

Example 3.5 Lefschetz duality and currents on orientable surfaces: $n = 2$, $p = 1$. Consider a conducting sheet Ω which is homeomorphic to a sphere with n handles and k holes as in Example 1.5. Suppose that slowly-varying magnetic fields are inducing currents on Ω and that the boundary of Ω does not touch any other conducting body. If $\boldsymbol{J}$ denotes the current density on Ω, the physics requires that $\operatorname{div} \boldsymbol{J} = 0$ on Ω and $\boldsymbol{J} \cdot \boldsymbol{n} = 0$ on $\partial\Omega$. This problem is dual to the one considered in Example 1.8 in the sense that current flow normal to the boundary of the plate must vanish. Define a surface current density vector $\boldsymbol{K}$ by the relation

$$\boldsymbol{J} \times \boldsymbol{n}' = \boldsymbol{K}.$$

The book's Mathematical Appendix considers differential forms and phrases this operation in terms of the Hodge star operator. Readers familiar with differential forms will note that the operation $\boldsymbol{n}' \times$ corresponds to the Hodge star operator on 1-forms in two dimensions. In any case, locally,

$$\boldsymbol{J} = \overline{\operatorname{curl}}\, \psi = \boldsymbol{n}' \times \operatorname{grad} \psi,$$

so that

$$\boldsymbol{K} = (\boldsymbol{n}' \times \operatorname{grad} \psi) \times \boldsymbol{n}' = (\boldsymbol{n}' \cdot \boldsymbol{n}') \operatorname{grad} \psi - (\boldsymbol{n}' \cdot \operatorname{grad} \psi)\boldsymbol{n}' = \operatorname{grad} \psi.$$

In other words, the surface current can be expressed in terms of a potential ψ called a stream function. Since the current density $\boldsymbol{J}$ is tangent to the boundary, the vector field $\boldsymbol{K}$ has vanishing tangential components at the boundary and can be associated with an element of

$$Z_c^1(\Omega - \partial\Omega).$$

Thus if $\boldsymbol{K}$ can be described by the stream function ψ then ψ is a constant on each connected component of $\partial\Omega$.

Let c_i, for $1 \le i \le \beta_1(\Omega, \partial\Omega)$, be a set of curves which are associated with generators of $H_1(\Omega, \partial\Omega)$ and let z_i, for $1 \le i \le \beta_1(\Omega)$, be a set of curves associated with generators of $H_1(\Omega)$ as in Example 1.8. Dual to the situation in Examples 1.8 and 1.13, the z_i act like cuts which enable ψ to be single-valued on

$$\Omega - \bigcup_{i=1}^{\beta_1(\Omega)} z_i$$

while $\psi = 0$ on $\partial\Omega$. Furthermore, discontinuities $[\psi]_{z_i}$ in ψ across the z_i are given by calculating the periods

$$I_i = \int_{c_i} \boldsymbol{J} \times \boldsymbol{n}' \cdot \boldsymbol{t}\, dl = \int_{c_i} \boldsymbol{K} \cdot \boldsymbol{t}\, dl = \int_{c_i} \operatorname{grad} \psi \cdot \boldsymbol{t}\, dl = \sum_{j=1}^{\beta_1(\Omega)} \mathcal{I}_{ij} [\psi]_{z_j},$$

where $\mathcal{I}_{ij} = \operatorname{Int}(c_i, z_j)$ is the number of oriented intersections of c_i with z_j. $\mathcal{I}$ is square since by the Lefschetz duality theorem

$$\beta_1(\Omega) = \beta_1(\Omega, \partial\Omega).$$

This matrix is nonsingular if the c_i and z_j actually correspond to bases of $H_1(\Omega, \partial\Omega)$ and $H_1(\Omega)$ respectively. The matrix can be inverted to yield

$$[\psi]_{z_j} = \sum_{i=1}^{\beta_1(\Omega,\partial\Omega)} \mathcal{I}_{ij}^{-1} I_i.$$

Hence the duality between the homology groups is useful in prescribing periods of vector fields in terms of jumps in the scalar potential on curves associated with a dual group. □

Example 3.6 Lefschetz duality and stream functions on orientable surfaces: $n = 2$, $p = 1$. When considering the current density vector $\boldsymbol{J}$ on a sheet as in Example 1.8 or when prescribing the components of a vector tangent to a surface as in Examples 1.14, 1.15, 2.3, the following situation occurs. Given a two-dimensional surface S, suppose one of the following is true on S:

$$\begin{aligned} (\operatorname{curl} \boldsymbol{C}) \cdot \boldsymbol{n}' &= 0, \\ (\operatorname{curl} \boldsymbol{A}) \cdot \boldsymbol{n}' &= 0, \\ \text{or } (\operatorname{curl} \boldsymbol{T}) \cdot \boldsymbol{n}' &= 0. \end{aligned}$$

Alternatively, if $\boldsymbol{J}$ is a vector field defined in the surface, suppose $\operatorname{div} \boldsymbol{J} = 0$ on S and no boundary conditions are prescribed on ∂S. In these cases it is useful to set

$$\left.\begin{array}{r} \boldsymbol{n}' \times \boldsymbol{C} \\ \boldsymbol{n}' \times \boldsymbol{A} \\ \boldsymbol{n}' \times \boldsymbol{T} \\ \boldsymbol{J} \end{array}\right\} = \overline{\operatorname{curl}}\, \psi \quad \text{locally on } S.$$

Next let z_i, for $1 \le i \le \beta_1(S)$, be a set of curves associated with a basis of $H_1(S)$, and c_i, for $1 \le i \le \beta_1(S, \partial S)$, be a set of curves associated with a basis of $H_1(S, \partial S)$. In this case

$$H_1\left(S - \bigcup_{i=1}^{\beta_1(S,\partial S)} c_i\right) \simeq 0,$$

so that the stream function can be made continuous and single-valued when the curves c_i are regarded as cuts. Furthermore the discontinuities $[\psi]_{c_i}$ across cuts c_i can be used to prescribe the periods

$$p_i = \int_{z_i} \left\{ \begin{array}{c} -\boldsymbol{C} \\ -\boldsymbol{A} \\ -\boldsymbol{T} \\ \boldsymbol{J} \times \boldsymbol{n}' \end{array} \right\} \cdot \boldsymbol{t}\, dl,$$

since

$$p_i = \int_{z_i} (\overline{\text{curl}}\, \psi \times \boldsymbol{n}') \cdot \boldsymbol{t}\, dl = \int_{z_i} \text{grad}\, \psi \cdot \boldsymbol{t}\, dl = \sum_{j=i}^{\beta_1(S,\partial S)} \mathcal{I}_{ij} [\psi]_{c_i},$$

where $\mathcal{I}_{ij}$ is the number of oriented intersections of c_i with z_j. The Lefschetz duality theorem ensures that the above matrix equation is uniquely soluble for the $[\psi]_{c_i}$ if the c_i and z_i are actually associated with a full set of generators for $H_1(S, \partial S)$ and $H_1(S)$ respectively. □

Example 3.7 Lefschetz duality theorem and nonorientable surfaces. It has been stated without proof that the duality theorems are true for orientable manifolds. In order to see that the Lefschetz duality theorem is not true in the case of a nonorientable manifold, we consider the case when Ω is Möbius strip of Example 1.12 and Figure 1.14. Recall that $\beta_1(\Omega, S) = 0$ while $\beta_1(\Omega) = 1$. Clearly the Lefschetz duality theorem does not apply in this case.

Now, let $\text{div}\, \boldsymbol{J} = 0$ on Ω and $\boldsymbol{J} \cdot \boldsymbol{n} = 0$ on the Möbius band S. Since S is connected, one may attempt to set $J = \overline{\text{curl}}\, \psi$ on Ω and ψ equal to a constant on S. However, if the current flowing around the band is nonzero, it is not possible to set the stream function ψ to a constant on S, even if S is connected. That is, if I is the current which flows around the band and ψ is single-valued, then

$$I = \int_{\tilde{z}} \left((\overline{\text{curl}}\, \psi) \times \boldsymbol{n}' \right) \cdot \boldsymbol{t}\, dl = \psi_s - \psi_s = 0,$$

where ψ_s is the value of ψ on S. Furthermore, since $\beta_1(\Omega, S) = 0$, there is no way to take a curve associated with a generator of $H_1(\Omega, S)$ and use it to specify the current flowing around the loop. Thus the method of considering curves associated with generators of $H_1(\Omega)$ to specify periods of the vector field and generators of $H_1(\Omega, S)$ as cuts, is critically dependent on the Lefschetz duality theorem.

To see how the current flow can be described in terms of a stream function, it is best to look at the problem in terms of $H_1(\Omega, S; \mathbb{Z})$ where $\tilde{z} \sim z$. The reader may convince himself that this generator of the torsion subgroup of $H_1(\Omega, S; \mathbb{Z})$ can be used as a cut in Ω that enables one to describe the current density $\boldsymbol{J}$ in terms of a single-valued stream function. Considering the diagram, there are two obvious ways of doing this:

(1) Take z as the cut, impose the condition $\psi \to -\psi$ across the cut, and set

$$\psi_s = \pm \frac{I}{2},$$

where the sign is chosen depending on the sense of the current.

(2) Take $\tilde{z}$ as the cut, and note that $S - \tilde{z}$ has two connected components which shall be called S' and S''. In order to describe the current flow in terms of the stream function, let $\psi \to -\psi$ across the cut and let

$$\psi_{s'} = -\psi_{s''} = \pm \frac{I}{2},$$

where the sign is chosen depending on the sense of the current.

Techniques making implicit use of duality theorems do not necessarily work in situations where the hypotheses underlying the duality theorems are not satisfied. □

It is noteworthy that, for an n-dimensional manifold, the interpretation of the Lefschetz duality theorem in terms of oriented intersections of p and $(n-p)$-dimensional submanifolds makes the duality intuitive when $n < 4$. For a proper account of this interpretation see [GH81, Chapter 31]. A leisurely but rigorous development of intersections is in [GP74, Chapters 2 and 3].

The boundary value problems considered in Examples 1.14, 1.15, and 2.3 show that the Lefschetz duality theorem is inadequate for dealing with complicated problems where different boundary conditions are imposed on different connected components of $\partial\Omega$ or when symmetry is used to reduce the size of a given problem as commonly done for efficiency in numerical problems. In other words the Lefschetz duality theorem is inadequate for many problems formulated for numerical computation. For cases like Examples 1.14, 1.15, and 2.3, the following duality theorems apply:

$$H_c^p(\Omega - S_1) \simeq H_{n-p}(\Omega, S_2),$$
$$H_c^p(\Omega - S_2) \simeq H_{n-p}(\Omega, S_1),$$

where $\partial\Omega = S_1 \cup S_2$ and $S_1 \cap S_2$ has no $(n-1)$-dimensional volume. Here it is understood that the connected components of S_1 and S_2 correspond to intersections of planes of symmetry with some original problem or to connected components of the boundary of the original problem which was reduced by identifying symmetries. This duality was first observed by Conner [Con54] for the case where S_1 and S_2 are the union of connected components of $\partial\Omega$. The proof of the theorem in this case is outlined in [Vic94, Section 5.25]. The more general version which is assumed in this book can be obtained from the version known to Conner by the usual method of doubling (see [Duf52] or [Fri55]).

The duality theorem stated above implies that $\beta_p(\Omega, S_1) = \beta_{n-p}(\Omega, S_2)$ and $H_p(\Omega, S_1) \simeq H_{n-p}(\Omega, S_2)$. The isomorphism between these two homology groups can be interpreted as asserting that there is a nondegenerate bilinear pairing between the two groups which can be represented by a square nonsingular matrix whose entries count the number of oriented intersections of p and $(n-p)$-dimensional submanifolds associated with the generators of both groups. The special cases of this theorem for the case $n = 3$ can be found in Examples 1.14, 1.15 and 2.3 (where $p = 1, 2, 3$ respectively).

Finally, we consider the Alexander duality theorem. Although this theorem might not be considered as visual or easy as Lefschetz or Poincaré dualities, various special cases of this general theorem were known to Maxwell. In its most general form the Alexander duality theorem states that for an n-dimensional manifold M and a closed subset Ω,

$$H^p(\Omega) = H_{n-p}(M, M - \Omega).$$

There is a question of limits which we are sidestepping here (see Greenberg and Harper [GH81, p. 233] for an exact statement). Skipping over these details is justified since one can say that the exceptions are pathological (see [Mas80, Chapter 9, §6] for an example). For most applications, M is taken to be $\mathbb{R}^3$ in which case the theorem says

$$H^p(\Omega) = H_{n-p}(\mathbb{R}^3, \mathbb{R}^3 - \Omega).$$

There is a classical version of the Alexander duality theorem that can be obtained as a corollary of the one above by the following simplified argument (compare [GH81, §27.9]). Consider the long exact homology sequence for the pair $(\mathbb{R}^3, \mathbb{R}^3 - \Omega)$:

$$\begin{array}{l} 0 \longrightarrow H_3(\mathbb{R}^3 - \Omega) \xrightarrow{\tilde{\imath}_3} H_3(\mathbb{R}^3) \xrightarrow{\tilde{\jmath}_3} H_3(\mathbb{R}^3, \mathbb{R}^3 - \Omega) \longrightarrow \\ \xrightarrow{\delta_3} H_2(\mathbb{R}^3 - \Omega) \xrightarrow{\tilde{\imath}_2} H_2(\mathbb{R}^3) \xrightarrow{\tilde{\jmath}_2} H_2(\mathbb{R}^3, \mathbb{R}^3 - \Omega) \longrightarrow \\ \xrightarrow{\delta_2} H_1(\mathbb{R}^3 - \Omega) \xrightarrow{\tilde{\imath}_1} H_1(\mathbb{R}^3) \xrightarrow{\tilde{\jmath}_1} H_1(\mathbb{R}^3, \mathbb{R}^3 - \Omega) \longrightarrow \\ \xrightarrow{\delta_1} H_0(\mathbb{R}^3 - \Omega) \xrightarrow{\tilde{\imath}_0} H_0(\mathbb{R}^3) \xrightarrow{\tilde{\jmath}_0} H_0(\mathbb{R}^3, \mathbb{R}^3 - \Omega) \longrightarrow 0. \end{array}$$

Since

$$H_p(\mathbb{R}^3) \simeq \begin{cases} 0 & \text{if } p \neq 0, \\ \mathbb{R} & \text{if } p = 0, \end{cases}$$

the long exact sequence tells us that, if Ω is not the empty set,

$$\begin{aligned} H_{3-p}(\mathbb{R}^3, \mathbb{R}^3 - \Omega) &\simeq H_{2-p}(\mathbb{R}^3 - \Omega) \quad \text{if } p \neq 2, \\ \mathbb{R} \oplus H_{3-p}(\mathbb{R}^3, \mathbb{R}^3 - \Omega) &\simeq H_{2-p}(\mathbb{R}^3 - \Omega) \quad \text{if } p = 2. \end{aligned}$$

Combining this with the Alexander duality theorem yields

$$\begin{aligned} H^p(\Omega) &\simeq H_{2-p}(\mathbb{R}^3 - \Omega) \ \text{ if } p \neq 2, \\ \mathbb{R} \oplus H^2(\Omega) &\simeq H_0(\mathbb{R}^3 - \Omega), \end{aligned}$$

or again

$$\begin{aligned} \beta_p(\Omega) &= \beta_{2-p}(\mathbb{R}^3 - \Omega) \ \text{ if } p \neq 2, \\ 1 + \beta_2(\Omega) &= \beta_0(\mathbb{R}^3 - \Omega). \end{aligned}$$

These are the classical versions of the Alexander duality theorem. The case of $p = 1$ was known to Maxwell [Max91, Art. 18] in the following form: "The space outside the region has the same cyclomatic number as the region itself." And again: "The cyclomatic number of a closed surface is twice that of either of the regions it bounds."

The reader may turn back to Example 1.7 to see how the classical version of the Alexander duality theorem was used in the case where $p = 2$, and Example 1.9 for the case $p = 1$. In general, the classical version of the Alexander duality is very useful when one wants to consider how global aspects of gauge transformations, solvability conditions, or potential formulations for a problem defined in a region Ω are a result of sources in $\mathbb{R}^3 - \Omega$.

In summary, there are three types of duality theorems which are invaluable when considering electromagnetic boundary value problems in complicated domains. They are:

(1) Lefschetz duality theorem (Ω n-dimensional):
$$H_p(\Omega) \simeq H_{n-p}(\Omega, \partial\Omega).$$
(2) When Ω is n-dimensional and $\partial\Omega = S_1 \cup S_2$ where S_1 and S_2 are two regions whose intersection does not have any $(n-1)$-dimensional volume and which are associated with dual boundary conditions on symmetry planes and interfaces, then
$$H_p(\Omega, S_1) \simeq H_{n-p}(\Omega, S_2).$$
(3) Alexander duality theorem:
$$H_p(\Omega) \simeq H_{2-p}(\mathbb{R}^3 - \Omega) \quad \text{for } p \neq 2,$$
$$\mathbb{R} \oplus H_2(\Omega) \simeq H_0(\mathbb{R}^3 - \Omega).$$

The first two duality theorems can be interpreted in terms of an intersection matrix.

Note that in the foregoing discussion of duality theorems, the torsion subgroups of the integer homology groups are assumed to vanish. Although this is not the case in general, it is the case for three dimensional manifolds with boundary embedded in $\mathbb{R}^3$. Section 3D is devoted to a proof of this fact.

3C. Linking Numbers, Solid Angle, and Cuts

The classical version of Alexander Duality can be interpreted through the notion of a linking number [Fla89, §6.4]. This is the approach taken in this section.

Solid Angle. In order to simplify the following discussion on linking numbers, we will think of current flowing on a set of thin wires or curves. Such systems have an infinite amount of energy [PP62], however wires can be regarded as tubular neighborhoods of curves. This does not limit the generality of the arguments. In fact, the results hold for surface and volume distributions of current [Kot88], and for general constitutive laws.

Consider two nonintersecting curves c and c' as shown in Figure 3.1. Curve c is the boundary of a surface S' ($c = \partial S'$), and a current I flows on curve c' which transversely intersects S'. For magnetoquasistatics, displacement current

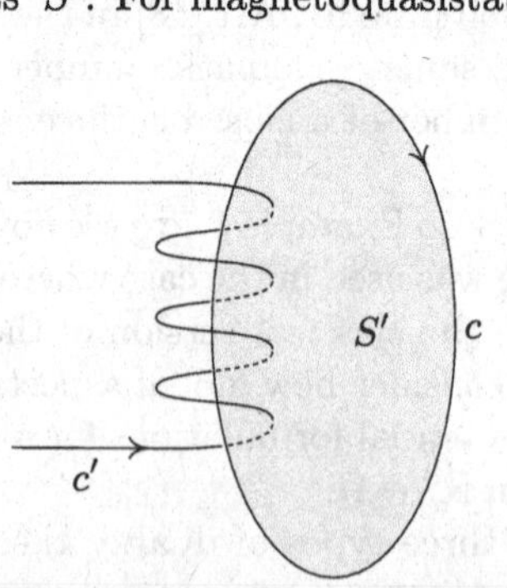

Figure 3.1. Current on curve c' tranversely crossing surface S'. Here $\text{Link}(c, c') = 4$.

is assumed to be negligible such that Ampère's Law is

$$\oint_c \boldsymbol{H} \cdot d\boldsymbol{r} = \int_{S'} \boldsymbol{J} \cdot \boldsymbol{n}\, ds = I\,\mathrm{Link}(c, c') \tag{3–5}$$

where $\mathrm{Link}(c, c')$ is the oriented linking number of curves c and c', $\boldsymbol{H}$ is the magnetic field intensity, and $\boldsymbol{J}$ is the conduction current density. The physical details of Ampère's Law are not so important here. The main point to note is that the periods of the field $\boldsymbol{H}$ are related with the linking number to the current on c'. Note that

$$\int_{S'} \boldsymbol{J} \cdot \boldsymbol{n}\, ds = I \quad \text{if } \mathrm{Link}(c, c') = 1.$$

In a region where $\boldsymbol{J} = 0$ the vector field $\boldsymbol{H}$ is irrotational ($\mathrm{curl}\,\boldsymbol{H} = 0$) and $\boldsymbol{H}$ may be expressed as the gradient of a magnetic scalar potential,

$$\boldsymbol{H}(\boldsymbol{r}) = -\nabla\psi,$$

so that

$$\psi(p) - \psi(p_0) = -\int_{p_0}^{p} \boldsymbol{H} \cdot d\boldsymbol{r}. \tag{3–6}$$

In general, the region in question is multiply connected so that a closed integration path c may link a current I. Hence the scalar potential is multivalued, picking up integral multiples of I depending on c. If c links the current n times, the value of the scalar potential at a point has an added quantity nI after traversing c (Figure 3.1).

The magnetic flux $\boldsymbol{B}$ can be expressed in terms of a vector potential $\boldsymbol{A}$, so that in linear, isotropic, homogeneous media we have

$$\nabla \times \boldsymbol{H} = \nabla \times \left(\frac{1}{\mu}\nabla \times \boldsymbol{A}\right) = \boldsymbol{J}. \tag{3–7}$$

However, since we are primarily concerned with topological issues the choice of constitutive law is not very important. When the Coulomb gauge ($\mathrm{div}\,\boldsymbol{A} = 0$) is applied in the magnetoquasistatic case, and $\boldsymbol{A}$ is expressed in terms of Cartesian coordinates, the components of $\boldsymbol{A}$ satisfy

$$\nabla^2 A_i = -\mu J_i \quad \text{for } i = 1, 2, 3.$$

For a vanishing vector potential as $\boldsymbol{r} \to \infty$ there is a Green's function solution

$$\boldsymbol{A}(\boldsymbol{r}) = \frac{\mu}{4\pi}\int_{V'} \frac{\boldsymbol{J}(\boldsymbol{r}')}{|\boldsymbol{r} - \boldsymbol{r}'|}\, dV', \tag{3–8}$$

where $\boldsymbol{r}'$ is a source point and the integral is over the conducting region. Noting that

$$\nabla_{\boldsymbol{r}} \times \frac{\boldsymbol{J}(\boldsymbol{r}')}{|\boldsymbol{r} - \boldsymbol{r}'|} = \frac{\boldsymbol{J}(\boldsymbol{r}') \times (\boldsymbol{r} - \boldsymbol{r}')}{|\boldsymbol{r} - \boldsymbol{r}'|^3}, \tag{3–9}$$

where the $\boldsymbol{r}$ in $\nabla_{\boldsymbol{r}}$ refers to differentiation with respect to unprimed variables, Equations (3–7), (3–8), and (3–9) give

$$\text{(3–10)} \qquad \boldsymbol{H}(\boldsymbol{r}) = \frac{1}{4\pi} \int_{V'} \frac{\boldsymbol{J}(\boldsymbol{r}') \times (\boldsymbol{r} - \boldsymbol{r}')}{|\boldsymbol{r} - \boldsymbol{r}'|^3} \, dV'.$$

Now consider c' contained in a tubular neighborhood as shown in Figure 3.2 such that $V' = c' \times D$ where D is a disc transverse to c'. Assume, for a point $\boldsymbol{r}'$

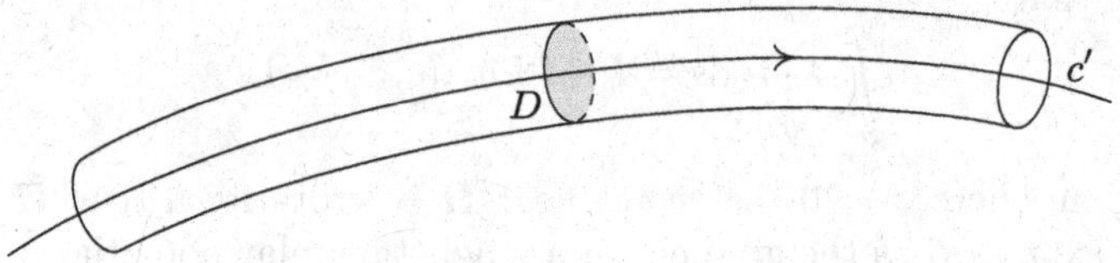

Figure 3.2. Tubular neighborhood of c'.

on c', that $\boldsymbol{r} - \boldsymbol{r}'$ does not vary significantly on D; then (3–10) can be evaluated over D to give the total current I times an integral on c', or

$$\text{(3–11)} \qquad \boldsymbol{H}(\boldsymbol{r}) = \frac{I}{4\pi} \oint_{c'} \frac{d\boldsymbol{r}' \times (\boldsymbol{r} - \boldsymbol{r}')}{|\boldsymbol{r} - \boldsymbol{r}'|^3}$$

which is the Biot–Savart law. Putting (3–11) into (3–6), we get

$$\text{(3–12)} \qquad \psi(p) - \psi(p_0) = -\frac{I}{4\pi} \int_{p_0}^{p} \oint_{c'} \frac{[(\boldsymbol{r} - \boldsymbol{r}') \times d\boldsymbol{r}'] \cdot d\boldsymbol{r}}{|\boldsymbol{r} - \boldsymbol{r}'|^3}.$$

Note also, the related expression for the linking number obtained by putting Equation (3–11) into (3–5) and canceling I on each side of the resulting equation:

$$\text{(3–13)} \qquad \operatorname{Link}(c, c') = \frac{1}{4\pi} \oint_{c} \oint_{c'} \frac{[d\boldsymbol{r}' \times (\boldsymbol{r} - \boldsymbol{r}')] \cdot d\boldsymbol{r}}{|\boldsymbol{r} - \boldsymbol{r}'|^3},$$

which is due to Gauss [Gau77]. Equation (3–12) is an exact formula if we started with a current-carrying knot and (3–13) always yields an integer if c' and c do not intersect.

Gauss approached Equation (3–13) through the notion of a *solid angle*, which we discuss here in order to develop a geometric understanding of the linking number. The solid angle Ω is defined as the area on the sphere S^2 bounded by the intersection of S^2 and a conical surface with vertex at the center of S^2 (see Figure 3.3). If c' bounds a surface s, then the solid angle at an observation point $\boldsymbol{r}$ is easily shown to be [Cou36]

$$\Omega = \int_{s} \frac{\cos\theta}{|\boldsymbol{r} - \boldsymbol{r}'|^2} \, ds$$

where θ is the angle between $\boldsymbol{r} - \boldsymbol{r}'$ and the normal to s. Thus the equation above can be written as

$$\Omega = \int_{s} \frac{(\boldsymbol{r} - \boldsymbol{r}') \cdot d\boldsymbol{s}}{|\boldsymbol{r} - \boldsymbol{r}'|^3}. \tag{3–14}$$

Suppose the observation point $\boldsymbol{r}$ is moved by an amount $d\boldsymbol{r}$. This is the same as moving the circuit by $-d\boldsymbol{r}$, whereby the shifting circuit sweeps out an area $|d\boldsymbol{s}| = |d\boldsymbol{r} \times d\boldsymbol{r}'|$ where $d\boldsymbol{r}'$ is on c'. So the change in solid angle is, from (3–14),

$$d\Omega = \oint_{c'} \frac{(\boldsymbol{r} - \boldsymbol{r}') \cdot (d\boldsymbol{r} \times d\boldsymbol{r}')}{|\boldsymbol{r} - \boldsymbol{r}'|^3} = \oint_{c'} \frac{[(\boldsymbol{r} - \boldsymbol{r}') \times d\boldsymbol{r}'] \cdot d\boldsymbol{r}}{|\boldsymbol{r} - \boldsymbol{r}'|^3}$$

(see [PP62]). If the observation point is moved through a closed path c, the total change in Ω is the expression for the linking number given by Equation (3–13) up to a factor of $1/(4\pi)$. The expression (3–13) is symmetric so that, up to a sign, integration on either c' or c gives the same result. A further development of the linking number can be found in [Spi65].

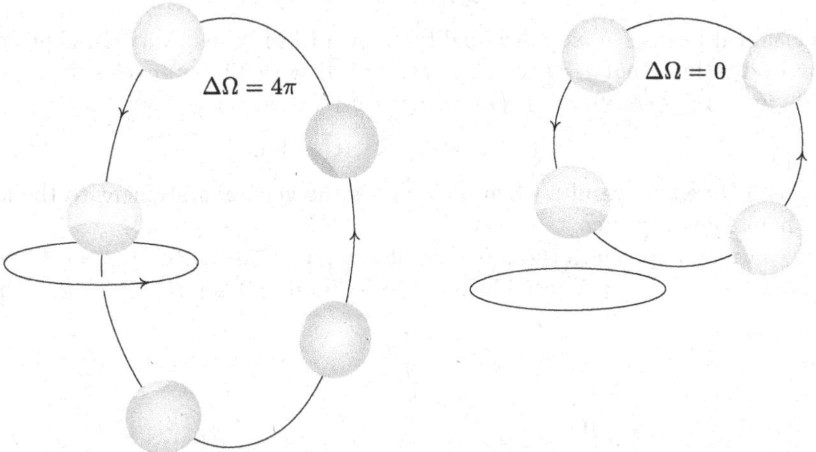

Figure 3.3. Solid angle on the unit sphere S^2, showing $\Delta\Omega$ for pairs of linked and unlinked closed paths.

Figure 3.3 shows that the solid angle provides some geometric insight into the linking number via multivalued scalar potentials, by showing that Ω increases by 4π each time the paths link.

Formal interpretations of the linking number, as discussed in the following section, lead to an understanding of cuts and the intuition behind what is needed to formulate an algorithm for computing cuts. Other uses of the linking number in magnetics can be found in [AK98, Chapter 3] and [KG90].

Linking Numbers and Cuts. We would like to add to the geometric intuition gained from Figure 3.3. As before, consider two closed, nonintersecting, oriented curves c and c' in $\mathbb{R}^3$ such as those in Figure 3.1. One of the curves, say c, can be expressed as $c = \partial S'$, where S' is a two-sided (orientable) surface. Then the linking number can be found by taking the sum of oriented intersections of S' and c',

$$\text{Int}\,(S', c') = \sum_{c' \cap S'} \pm 1 = \text{Link}(c, c')$$

where Int $(\cdot, \cdot)$ denotes the intersection number.

The intersection number Int $(S', c') =$ Int (S, c) where $\partial S = c'$. In the case of Figure 3.1, S can be deformed so that it is simply a disc. Figure 1.8 showed S for a current-carrying trefoil knot. For a knot the surface S exists though it is not always intuitive. In any case, an algorithm for the construction of the surface results from any constructive proof of the fact that such a surface is realizable as a compact, orientable manifold [Kot89a]. It will turn out that S is the cut needed to make the magnetic scalar potential single-valued.

If $\mathbb{R}^3$ is separated into a nonconducting region V and a conducting region $\mathbb{R}^3 - V$, the ranks β_k of the kth homology groups in each region are related thus:

$$\beta_1(V) = \beta_1(\mathbb{R}^3 - V),$$
$$1 + \beta_p(V) = \beta_{2-p}(\mathbb{R}^3 - V) \quad \text{for } p \neq 1.$$

This classical form of Alexander duality, known to Maxwell [Max91], is brought about by the fact that p-cycles in V are linked with $(2-p)$-cycles in $\mathbb{R}^3 - V$ [ST80], and is the corollary of a more general form which states:

$$H^p(V) = H_{3-p}(\mathbb{R}^3, \mathbb{R}^3 - V).$$

The classical version results when one applies the general statement to the long exact homology sequence.

We can now appreciate the meaning of cuts. Consider a set $\{c_i\}$ of *1*-cycles in V, for $1 \leq i \leq \beta_1(V)$, as illustrated in Figure 3.4 for $\beta_1(V) = 2$. These

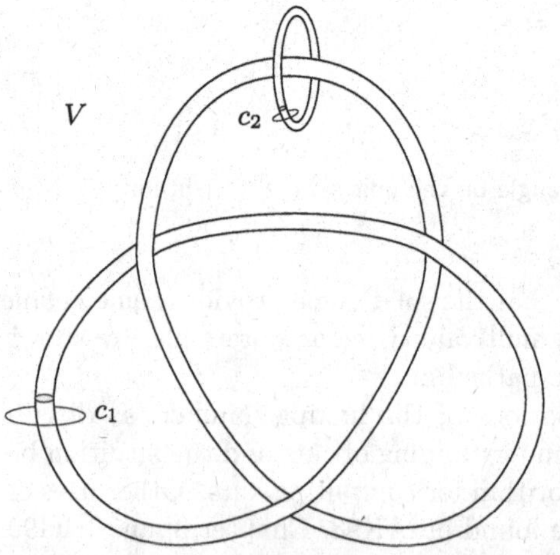

Figure 3.4. The *1*-cycles c_1 and c_2 are generators of $H_1(V; \mathbb{Z})$ while curves c_1' and c_2' on the knot and loop (not shown) are generators of $H_1(\mathbb{R}^3 - V; \mathbb{Z})$. So $\beta_1(V) = 2$. Cut for the knot is same as that in Figure 1.8. Cut for the loop is an annulus.

are generators of $H_1(V;\mathbb{Z})$ and comprise the set of interesting curves used to evaluate the left-hand side of Ampère's law (3–5) since they link current. They are boundaries $c_i = \partial S_i'$, of surfaces S_i' in $\mathbb{R}^3 - V$ used to measure current flux.

Now consider the set $\{c_j'\}$, $1 \leq j \leq \beta_1(\mathbb{R}^3 - V)$, which forms a basis for $H_1(\mathbb{R}^3 - V)$, and $\text{Link}(c_i, c_j')$, the intersection number of S_i' with c_j'. Alexander duality guarantees that the $\beta_1 \times \beta_1$ intersection matrix which has $\text{Link}(c_i, c_j')$ for its ijth entry is nonsingular.

The symmetry of the linking number suggests that $\{c_j'\}$ are also boundaries, $c_j' = \partial S_j$, where S_j is a surface in V. The *1*-cycles which link current are generators of $H_1(V;\mathbb{Z})$ and intersect surfaces S_j which are generators of equivalence classes in $H_2(\mathbb{R}^3, \mathbb{R}^3 - V; \mathbb{Z})$. Alexander duality guarantees that the S_j are dual to $H^1(V;\mathbb{Z})$, insuring that $\boldsymbol{H}$ can be expressed as the gradient of a single-valued scalar potential. If the scalar potential has a discontinuous jump I_j across S_j, then S_j is the cut surface. Furthermore S_j can also be used as a surface for calculating magnetic flux. The set $\{S_j\}$ of cuts in V allow ψ to be single-valued on $V - \bigcup_j S_j$. Closed curves in $V - \bigcup_j S_j$ link zero current as illustrated in Figure 1.8. Note, however, that there is no guarantee that $V - \bigcup_j S_j$ is simply connected! Section MA-I shows that these cuts are compact, orientable, embedded surfaces.

While Alexander duality provides an intuitive way of defining cuts, it is phrased in terms of the current-carrying region, making it useless for a finite elements algorithm which must be expressed entirely in terms of V and ∂V, the nonconducting region and its boundary. To express cuts in terms of a mesh which represents the current-free region, a limiting process [GH81] takes Alexander duality to Lefschetz duality:

$$H^p(V) \cong H_{n-p}(V, \partial V),$$

a duality theorem expressed in terms of the region and its boundary. A constructive proof that generators of $H_2(V, \partial V)$ are realizable as compact, orientable, embedded manifolds then gives rise to an algorithm for finding cuts which will be discussed at length in Chapter 6.

3D. Lack of Torsion for Three-Manifolds with Boundary

Consider Ω a compact 3-dimensional manifold with boundary and hence it is a closed subset of $\mathbb{R}^3$ which is "tightly embedded" and has finitely generated integral homology groups. Hence Alexander duality is applicable [Mas80]. We would like to show that the integral (co)homology groups of Ω have no torsion groups. To do this, let S^3 be the unit sphere in $\mathbb{R}^4$ and map Ω into S^3 by stereographic projection. Call the image of Ω under this map Ω_c. The problem now reduces to proving that Ω_c, as a closed subset of the compact orientable 3-dimensional manifold S^3, has torsion-free (co)homology groups. This result can be demonstrated by using the Alexander duality theorem and the universal coefficient theorem for cohomology while working with $S^3 - \mathring{\Omega}_c$, the complement of the interior of Ω_c ($\mathring{X}$ denotes the interior of X). Since S^3 is orientable, the

Alexander duality theorem applies and states that

$$\widetilde{H}^{3-q-1}(S^3 - \mathring{\Omega}_c; \mathbb{Z}) \simeq \widetilde{H}_q(\mathring{\Omega}_c; \mathbb{Z}) \quad \text{for } q = 0, 1, 2,$$

where the tilde indicates that reduced (co)homology groups are used to condense the statement of the theorem. If we decompose a given (co)homology group $H(\,\cdot\,; \mathbb{Z})$ into a direct sum of free and torsion subgroups and denote them by $F(\,\cdot\,)$ and $T(\,\cdot\,)$ respectively, the Alexander duality theorem then says the following about torsion subgroups:

$$T^{2-q}(S^3 - \mathring{\Omega}_c) \simeq T_q(\mathring{\Omega}_c), \quad \text{for } q = 0, 1, 2. \tag{3–15}$$

A corollary of the universal coefficient theorem for cohomology [BT82] asserts that

$$T^p(\cdot) \simeq T_{p-1}(\cdot), \quad \text{for } p = 0, 1, 2, 3. \tag{3–16}$$

By definition, homology groups are trivial in negative dimensions, so that, by (3–16),

$$T^0(S^3 - \mathring{\Omega}_c) \simeq 0$$

and, by (3–15),

$$T_2(\mathring{\Omega}_c) \simeq 0. \tag{3–17}$$

Also, since zeroth homology groups are always torsion-free, the following chain of reasoning reveals that certain higher homology torsion subgroups vanish:

$$\begin{aligned} T_0(S^3 - \mathring{\Omega}_c) \simeq 0 &\Longrightarrow T^1(S^3 - \mathring{\Omega}_c) \simeq 0 && \text{(by (3–16))} \\ &\Longrightarrow T_1(\mathring{\Omega}_c) \simeq 0 && \text{(by (3–15))}. \end{aligned} \tag{3–18}$$

Thus by (3–17) and (3–18) we have

$$\begin{aligned} T_q(\mathring{\Omega}_c) &\simeq 0, \quad \text{for } q = 0, 1, 2, \\ T^q(\mathring{\Omega}_c) &\simeq 0, \quad \text{for } q = 0, 1, 2, 3 \quad \text{(by (3–16))}. \end{aligned} \tag{3–19}$$

Thus, (3–19) shows that, with the exception of the third homology group, all of the (co)homology groups of $\mathring{\Omega}_c$ are torsion-free. The image of Ω under stereographic projection can be contained in a neighborhood of $\mathring{\Omega}_c$ of which $\mathring{\Omega}_c$ is a deformation retract, so one may substitute Ω for $\mathring{\Omega}_c$ in (3–19), obtaining

$$\begin{aligned} T_q(\Omega) &\simeq 0 \quad \text{for } q = 0, 1, 2, \\ T^q(\Omega) &\simeq 0 \quad \text{for } q = 0, 1, 2, 3. \end{aligned} \tag{3–20}$$

Now suppose that Ω is a compact manifold with boundary embedded in $\mathbb{R}^3$, and let Ω be the union of disjoint connected submanifolds Ω_i. On each Ω_i there is a volume form which is inherited from $\mathbb{R}^3$. For each i this volume form can be multiplied by a scalar function which is positive on the interior of Ω_i and zero on $\partial\Omega_i$ to give a cohomology class in $H^3(\Omega_i, \partial\Omega_i; \mathbb{R})$. Since Ω_i is connected, we have

$$H^3(\Omega_i, \partial\Omega_i; \mathbb{Z}) \simeq \mathbb{Z}.$$

This ensures the existence of an orientation class in integer homology for each Ω_i which means that the Lefschetz duality theorem [Mas80] applies for each connected component of Ω and hence for Ω as a whole. Thus:

$$H^q(\Omega, \mathbb{Z}) \simeq H_{3-q}(\Omega, \partial\Omega; \mathbb{Z}) \quad \text{for } q = 0, 1, 2, 3,$$
$$H_{3-q}(\Omega; \mathbb{Z}) \simeq H^q(\Omega, \partial\Omega; \mathbb{Z}) \quad \text{for } q = 0, 1, 2, 3.$$

Breaking these groups into direct sums of free and torsion subgroups, we see that the torsion subgroups satisfy

$$\begin{aligned} T^q(\Omega) &\simeq T_{3-q}(\Omega, \partial\Omega) \quad \text{for } q = 0, 1, 2, 3, \\ T_q(\Omega) &\simeq T^{3-q}(\Omega, \partial\Omega) \quad \text{for } q = 0, 1, 2, 3; \end{aligned} \tag{3–21}$$

hence, since $T^0(\Omega, \partial\Omega)$ is always zero, (3–21) tells us that

$$T_3(\Omega) \simeq 0 \simeq T^0(\Omega, \partial\Omega). \tag{3–22}$$

Substituting (3–20) into (3–21) gives

$$\begin{aligned} T_q(\Omega) \simeq 0 &\simeq T^{3-q}(\Omega, \partial\Omega) \quad \text{for } q = 0, 1, 2, \\ T^q(\Omega) \simeq 0 &\simeq T^{3-q}(\Omega, \partial\Omega) \quad \text{for } q = 0, 1, 2, 3, \end{aligned} \tag{3–23}$$

and combining (3–22) and (3–23) we have proved the following lemma:

LEMMA. If Ω is a compact three dimensional manifold with boundary, embedded in $\mathbb{R}^3$, then for $q = 0, 1, 2, 3$

$$0 \simeq T^q(\Omega) \simeq T_q(\Omega) \simeq T^q(\Omega, \partial\Omega) \simeq T_q(\Omega, \partial\Omega).$$

Since (co)homology computed with coefficients in $\mathbb{R}$ yields vector spaces whose dimension is equal to the rank of the corresponding (co)homology group computed with integer coefficients, we have the immediate useful corollary:

COROLLARY. Let Ω be a compact three dimensional manifold with boundary embedded in $\mathbb{R}^3$. For $q = 0, 1, 2, 3$, the groups $H^q(\Omega)$, $H_q(\Omega)$, $H^q(\Omega, \partial\Omega)$, $H_q(\Omega, \partial\Omega)$ with coefficients in $\mathbb{Z}$ are free groups whose rank equals the dimension of the corresponding vector space if the coefficient group is $\mathbb{R}$.

This corollary tells us that there is no loss or gain of information in the transition to and from integer coefficients.

4
The Finite Element Method and Data Structures

This chapter serves two purposes. The first is to point to some applied mathematics, in particular the finite element method and corresponding numerical linear algebra, which belong in a book oriented towards computation. The second purpose is to point out the role of topology, namely simplicial homology of triangulated manifolds, in various aspects of the numerical techniques.

The chapter begins simply enough with an introduction to the finite element method for Laplace's equation in three dimensions, going from the continuum problem to the discrete problem, describing the method in its most basic terms with some indication of its practice. This leads naturally to numerical linear algebra for solving sparse positive-definite matrices which arise from the finite element method. The tie to previous chapters is that we would like to compute scalar potentials for electro- and magnetostatics. At a deeper level, there is a connection to a homology theory for the (finite element) discretized domain, so that useful tools such as exact homology sequences survive the discretization. We will see that in addition to everything discussed in the first chapters, the Euler characteristic and the long exact homology sequence are useful tools for analyzing algorithms, counting numbers of nonzero entries in the finite element matrix, and for constructing the most natural data structures.

The purpose here is to draw out some connections between the finite element method and the relevant homological tools. The reader interested in more on the finite element method is referred to [SF73]. In addition [SF90] is a good introduction to the finite element method in electrical engineering while [Bos98] provides a more current and advanced view on finite elements for magnetostatics and eddy current problems.

4A. The Finite Element Method for Laplace's Equation

We commence with the continuum problem. The one which is central in this book is quite straightforward: the field quantity of interest is described by a scalar potential, say $\boldsymbol{E} = \operatorname{grad} u$, and u satisfies Laplace's equation, $\nabla^2 u = 0$, in a bounded three-dimensional domain Ω. The boundary consists of two parts S_1 and S_2,

$$\partial\Omega = S_1 \cup S_2,$$

such that $S_1 \cap S_2$ has no area. There is a prescribed constant potential on each connected component of S_1 (Dirichlet boundary condition), and on S_2 the normal derivative

$$\operatorname{grad} u \cdot \boldsymbol{n} = 0$$

(Neumann condition). The second boundary condition typically arises when symmetry in the geometry of the original problem domain can be exploited in order to reduce to a new domain with reduced geometry. When using cuts to compute a magnetic scalar potential there are some additional conditions across cuts in the interior of the domain, but the essence of the problem is as stated above. The special conditions related to cuts will be fully treated later.

The Ritz method starts from a minimum principle which requires that the potential distribution in Ω be such that the energy of the field associated with the potential be minimized. This minimum principle is the requirement that the functional

$$P(u) = \frac{1}{2}\int_\Omega |\operatorname{grad} u|^2 \, dV \tag{4–1}$$

be minimized over all functions which satisfy the boundary conditions. It turns out that the variational problem of minimizing this functional is equivalent to Laplace's equation.

By assuming that u can be expressed in terms of a set of piecewise polynomials over Ω, it is possible to construct an approximate expression for $P(u)$. Minimization of the approximate expression for $P(u)$ determines the coefficients of the polynomials and will approximate the potential. This is the approach behind the finite element method and will be described in more detail.

First we begin by arguing that the function u which satisfies Laplace's equation also minimizes energy. Suppose u satisfies $\nabla^2 u = 0$ in Ω. Let h be a (once) differentiable function in Ω, $h \in L^2(\Omega)$ which satisfies $h = 0$ on S_1. When $P(h)$ is finite, the function h is called an admissible function. Let $\theta \in \mathbb{R}$ be a parameter. To u we add θh, so that the energy is

$$P(u + \theta h) = P(u) + \tfrac{1}{2}\theta^2 \int_\Omega |\operatorname{grad} h|^2 \, dV + \theta \int_\Omega \operatorname{grad} u \cdot \operatorname{grad} h \, dV.$$

The last term can be rewritten using integration by parts, so that

$$P(u + \theta h) = P(u) + \theta^2 P(h) + \theta \int_{\partial\Omega} h \operatorname{grad} u \cdot \boldsymbol{n} \, dS - \theta \int_\Omega \operatorname{grad}^2 u \, dV.$$

The third term is always zero since h vanishes on S_1 and $\operatorname{grad}\phi \cdot \boldsymbol{n}$ vanishes on S_2. The fourth term is zero since u satisfies Laplace's equation. The second term is positive, so $P(u)$ is the minimum reached when $\theta = 0$ and h is an admissible function.

The error in the energy depends on θ^2. If θ is small, the error in energy can be small even while the error in the potential is relatively large. This is a practical advantage since quantities of interest such as capacitance and inductance which are related to energy can be accurately estimated even while the estimate for the potential has considerable error.

The Ritz Method. The continuum problem we have discussed so far is infinite-dimensional (in the space of admissible functions). The Ritz method builds a finite-dimensional discrete problem for the minimum principle. By choosing a finite number of "trial" functions from a narrow class of admissible functions, it is possible to compute an approximation to the potential. This makes the Ritz method invaluable for practical computational methods.

Particular instances of trial functions will be briefly addressed in 4A, but first we will see how the Ritz method leads to a matrix equation. We begin with a set of trial functions $T_i(x_1, x_2, x_3)$, for $1 \leq i \leq n$. The approximation is a linear combination of trial functions

$$U = \sum_{i=1}^{n} U_i T_i,$$

where the coefficients U_i must be determined. In the finite element method the trial functions will be functions (polynomials) on each element of the mesh. Using (4–1), we can write a new energy functional restricted to n degrees of freedom in the basis functions:

$$P(U) = \frac{1}{2}\int_\Omega \sum_{j=1}^{3}\left(\sum_{i=1}^{n} U_i \frac{\partial T_i}{\partial x_j}\right)^2.$$

Collecting terms in this quadratic form, the coefficients of $U_i U_j$ are

$$\mathcal{K}_{kl} = \frac{1}{2}\int_\Omega \sum_{j=1}^{3} \frac{\partial T_k}{\partial x_j}\frac{\partial T_l}{\partial x_j}. \tag{4–2}$$

This matrix, usually called the element stiffness matrix, is symmetric.

The minimization problem has been reduced to finding the minimum with respect to the parameters U_i of the quadratic form $P(U) = \frac{1}{2}U^T\mathcal{K}U$. The minimum is at $\mathcal{K}U = 0$ when boundary conditions are not prescribed. When boundary conditions are given, some of the variables are prescribed, there are fewer equations resulting from the minimization process, and the remaining variables are nontrivial since the linear system of equations has a nonzero right hand side.

Basis Functions and the Finite Element Method. Now we put the Ritz method to work by choosing trial functions. We will look at what is perhaps the simplest case: polynomials on triangulations of Ω. However the simple case goes quite far and has the clearest ties to simplicial homology theory which we address in the second half of the chapter.

The chief difference between the Ritz method and the finite element method has to do with practical issues in choosing trial functions. A closely related question is how the trial functions are matched to or imposed on the problem domain. As a practical matter, the trial functions should be such that the entries of the stiffness matrix $\mathcal{K}_{ij}$ in (4–2) are relatively easy to compute and that $\mathcal{K}$ should be sparse and well-conditioned.

In summary, the approach is as follows:

(1) Begin with a triangulation of Ω, meaning that the 3-dimensional region is approximated by a set of tetrahedra (3-simplexes) joined together along faces under the same requirements imposed for a simplicial complex. In other words, we begin with the triangulation of a manifold. We will have more to say about simplicial complexes and homology in subsequent sections, but will altogether sidestep computational geometry and the process of triangulation.

(2) Choose a set of trial functions by defining a set of polynomials on each tetrahedron. The case considered here is the linear form $U_k = a + bx_1 + cx_2 + dx_3$ on the kth tetrahedron. The potential is approximated element by element in a piecewise linear manner. The main point to note is that the coefficients of the polynomial are determined entirely by the values of the potential at the vertices of the tetrahedron, independent of the choice of coordinate system. This will allow us to pick any convenient coordinate system which provides for other criteria such as sparsity and efficient computation of $\mathcal{K}$. Note also that continuity of the potential between the tetrahedra occurs naturally.

(3) Construct a global stiffness matrix $\mathcal{K}$ for the entire finite element mesh based on the preceding steps, and solve the resulting matrix equation for U. Construction of the global stiffness matrix involves computing a stiffness matrix for each element and using the incidence matrix which describes how elements in the mesh are connected to assemble a stiffness matrix for the entire mesh.

Simplexes and Barycentric Coordinates. It is convenient to use coordinates independent of mesh orientation, structure, or rotation of coordinate system for the trial functions. For this, we introduce barycentric coordinates.

Let $\{v_0, \ldots, v_p\}$ be an affine independent set of points in $\mathbb{R}^n$ with coordinates $v_i = (x_1^i, \ldots, x_n^i)$. Affine independence means that $v_1 - v_0, \ldots, v_p - v_0$ are linearly independent in sense of linear algebra (thus $0 \leq p \leq n$). Let $\lambda_0, \ldots, \lambda_p$ be real scalars. A *p-simplex* with vertices v_i, denoted by $\langle v_0, \ldots, v_p \rangle$ is defined as the subset of points of $\mathbb{R}^n$ given by

$$\langle v_0, \ldots, v_p \rangle = \left\{ \sum_{i=0}^{p} \lambda_i v_i \;\middle|\; \sum_{i=0}^{p} \lambda_i = 1,\ \lambda_i \geq 0 \right\}.$$

A 0-simplex is a point $\langle v_0 \rangle$, a 1-simplex is a line segment $\langle v_0, v_1 \rangle$, a 2-simplex is a triangle $\langle v_0, v_1, v_2 \rangle$, a 3-simplex is a tetrahedron $\langle v_0, v_1, v_2, v_3 \rangle$, etc. Indeed, the p-simplex is the most elementary convex set, namely the convex set spanned by $v_0, \cdots, v_p$. If $s = \langle v_0, \ldots, v_p \rangle$ is a p-simplex, then every subset $\langle v_{i_0}, \ldots, v_{i_q} \rangle$ of s is a q-simplex called a *q-face* of s. The fact that any q-face can be referred to simply by a subset of vertices of s allows for a very clean way to develop chain and homology groups in the context of complexes where p-simplexes are attached to each other along $(p-1)$-faces. We take simplexes to be the elements of the finite element method. If the elements are not simplexes, their generalization as convex cells still give analogous tools. In any case, the definition of simplexes gives a geometric footing for our present discussion, and will re-emerge later for the algebraic and topological discussion.

Under the constraints imposed on the scalars λ_i, each point in the subset can be written uniquely so that $(\lambda_0, \cdots, \lambda_p)$ are coordinates, called *barycentric coordinates*, determined by the vertices. The following is an explicit construction of barycentric coordinates.

Consider an n-simplex $s = \langle v_0, \ldots, v_n \rangle$ in $\mathbb{R}^n$. In practice, barycentric coordinates are calculated by recognizing that they are related to the n-dimensional volume of s, given by the determinant

$$\gamma(s) = \frac{1}{n!} \begin{vmatrix} 1 & x_1^0 & x_2^0 & \cdots & x_n^0 \\ 1 & x_1^1 & x_2^1 & \cdots & x_n^1 \\ & & \vdots & & \\ 1 & x_1^n & x_2^n & \cdots & x_n^n \end{vmatrix}. \tag{4–3}$$

Consider a point $v = (x_1, \ldots, x_n)$ in s, and regard it as a new vertex so that s is subdivided into $n+1$ new n-simplexes $s_k = \langle v_0, \ldots, v_{k-1}, v, v_{k+1}, \ldots, v_n \rangle$ for $0 \le k \le n$. Then

$$\gamma(s) = \sum_{k=0}^{n} \gamma(s_k).$$

The barycentric coordinates corresponding to v are defined by

$$\lambda_k = \frac{\gamma(s_k)}{\gamma(s)}$$

and from the expression above it follows that

$$\sum_{k=0}^{n} \lambda_k = 1 \qquad \text{and} \qquad \sum_{k=0}^{n} \operatorname{grad} \lambda_k = 0.$$

Barycentric coordinates can be written somewhat more explicitly as

$$\lambda_i(s) = \frac{1}{\gamma(s)} \begin{vmatrix} 1 & x_1^0 & x_2^0 & \cdots & x_n^0 \\ & & \vdots & & \\ 1 & x_1^{i-1} & x_2^{i-1} & \cdots & x_n^{i-1} \\ 1 & x_1 & x_2 & \cdots & x_n \\ 1 & x_1^{i+1} & x_2^{i+1} & \cdots & x_n^{i+1} \\ & & \vdots & & \\ 1 & x_1^n & x_2^n & \cdots & x_n^n \end{vmatrix} = a_0^i + \sum_{j=1}^{n} a_j^i x_j,$$

where the determinant is expanded about the ith row and real coefficients a_k^j are cofactors of the determinant. Note that $a_0^i = n!\gamma(s_i)$ and a_j^i can be interpreted as areas of projections of s_i onto the coordinate planes. For a 2-simplex λ_i can be interpreted, by definition, as the ratio of the height of s_i to the height of

s. Equipotential surfaces of λ_i are parallel to the face which s_i and s have in common, and $\nabla\lambda_i$ is normal to that face.

Approximation functions. The basis functions are polynomials, and these are defined on each tetrahedron of the triangulation. There is a choice between describing each basis function T_i and describing U as a polynomial on each tetrahedron. The choices are equivalent, but the latter involves less work. However, to make part of the connection between trial functions and finite element approximation functions in the simple case of linear interpolation, we begin with trial functions on tetrahedral elements. Even this is already more than the Ritz method since the Ritz method says nothing about the discretization of the problem domain.

The simplest trial functions which can satisfy the boundary conditions are linear functions between vertices and satisfy

$$T_i(v_j) = \delta_{ij}, \quad 1 \le i, j \le 4,$$

where v_j is the jth vertex. This has the advantage that when U is written as the linear combination $U = \sum U_i T_i$, the coefficients U_i are simply the values of U at each vertex. For readers familiar with finite difference methods, the resulting matrix equation will look like it came from a finite difference equation.

The main point is that the linear combination of these trial functions is a linear function $U_k = a_0 + a_1x_1 + a_2x_2 + a_3x_3$ which must satisfy

$$\begin{bmatrix} U_0 \\ U_1 \\ U_2 \\ U_3 \end{bmatrix} = \begin{pmatrix} 1 & x_1^0 & x_2^0 & x_3^0 \\ 1 & x_1^1 & x_2^1 & x_3^1 \\ 1 & x_1^2 & x_2^2 & x_3^2 \\ 1 & x_1^3 & x_2^3 & x_3^3 \end{pmatrix} \begin{bmatrix} a_0 \\ a_1 \\ a_2 \\ a_3 \end{bmatrix}.$$

The role of barycentric coordinates is now beginning to emerge. They permit us to rewrite the linear form as $U = \sum \lambda_i u_i$. In fact, in some sense, we could simply regard barycentric coordinates as the trial functions.

U_k is determined by its values on the vertices of the kth tetrahedron, but we have to ensure that U is continuous across element interfaces. Note that three vertices of the tetrahedron are vertices of the interface shared by two simplexes. U is linear on the interface, so it is determined by the values of U at the three vertices. Therefore, U must be the same linear function when the interface is approached from the neighboring tetrahedron.

There is a great deal to say about interpolation beyond linear functions on tetrahedra. The functions vary according to application and questions of numerical analysis. The following examples merely scratch the surface.

Example 4.1 Quadratic interpolation on 2-simplexes. Consider, for a 2-dimensional triangulation, a triangle with vertices $((0,0),(0,1),(1,0))$ in the xy-plane. For quadratic interpolation, U is cast as

$$U = a + bx + cy + dx^2 + exy + fy^2.$$

The trial function associated with the first vertex is

$$T = (1 - x - y)(1 - 2x - 2y).$$

Note that this trial function has zeros at the midpoints of the 1-faces of the 2-simplex. These zeros are additional "nodes of interpolation" for the function, corresponding to the coefficients in the interpolation function. □

Example 4.2 Lagrange polynomials on p-simplexes. Lagrange polynomials are also a popular source of interpolation functions. In this case we start with a family of associated Lagrange polynomials R_i of degree i with a parameter n in terms of barycentric coordinates:

$$R_0(n,\lambda) = 1, \qquad R_m(n,\lambda) = \frac{1}{m!}\prod_{j=0}^{m-1}(n\lambda - j) \quad \text{for } m > 0.$$

Note that R_m has zeroes at $\lambda = 0, \frac{1}{n}, \ldots, \frac{m-1}{n}$, and that $R_m(n, m/n) = 1$. The interpolation function for a p-simplex is defined by a product of these polynomials:

$$\alpha(\lambda_0, \ldots, \lambda_p) = R_{i_0}(n, \lambda_0) \cdots R_{i_p}(n, \lambda_p)$$

where the order of α is $n = \sum i_j$. As in the previous example, these polynomials provide for interpolation points in addition to the vertices. The locations of the interpolation nodes are given by the zeroes of the polynomial. In practice, a numbering scheme must be implemented for interpolation node accounting. □

These examples point to the possibility of raising accuracy in the solution by increasing polynomial order, though the computational cost of amount of accuracy may not be worthwhile. Interpolation on linear functions may be inadequate in most applications, however for topological questions the extra interpolation nodes required for higher-order polynomials somewhat obscure the picture. In any case, whatever polynomials one uses, once they are defined in terms of barycentric coordinates, an analysis can be made for a "standard" simplex, and the procedure of computing the stiffness matrix can be easily automated.

Assembly of the stiffness matrix. Assembly of the global stiffness matrix is a final unique aspect of the finite element method which merits discussion. As mentioned previously, assembly refers to the process of computing stiffness matrices for each 3-simplex and using the matrix which describes how 3-simplexes are connected to their neighbors (via 0-simplexes) to construct a stiffness matrix for the entire simplicial mesh which represents the computational domain. This will lead to the quadratic matrix form mentioned earlier which is minimized at the solution for U. The present purpose is to elaborate on the connection between simplicial complexes, topology, and the data structures of the finite element mesh. Assembly of the stiffness matrix is discussed at length in Sections 5E (starting on page 153) and 6D.

The procedure can be formulated generally for interpolation in n dimensions, however we will write everything for $n = 3$ and linear interpolation.

4B. Finite Element Data Structures

In many algorithms for finite element applications there are computations which do not depend on the metric of the space. In these cases, once the vertex coordinates are used to ascribe an orientation (± 1) to each element of the mesh by taking the sign of (4–3), it is only the finite element connection matrix which is left to play a purely integer combinatorial role in the computation which handles the topological business of the algorithm. The connection matrix alone contains a wealth of topological information about the discretized region, realizing its simplicial or cellular complex [Mun84].

The complex is an old idea in algebraic topology and electrical circuit theory [Kro59], but has not attracted much attention in 3-dimensional finite elements theory. Moreover, it is intimately connected to homology and cohomology theories which are an algebraic expression of how the topology of a region is tied to fields in the region and is the formalism which links fields in the continuum to lumped circuit parameters [Kot84]. Nevertheless, the profound consequences of this algebraic structure can be seen in the computation of cuts for magnetic scalar potentials [GK01b, Bro84], discretization of the magnetic helicity functional [CK96], and Whitney forms for finite elements [Mül78, Dod76, Bos98]. More generally, the same data structures are related to presentations of the fundamental group of the triangulated region, sparse matrix techniques for 3-d finite elements, and, when the metric of the space is introduced, general 3-d finite element computations.

We will consider finite element data structures which incarnate the simplicial chain and cochain complexes and point out a duality theorem which is a useful tool in algorithm development. In fact, the data structures are not new – some likeness of the structures described here is often found in computer graphics [Bri93] – however we are attempting to give the motivation for their existence and a connection to the relevant physics. Furthermore, while the structures are primarily described in the context of tetrahedral discretizations, they extend to cellular discretizations (e.g. hexahedral meshes) at slightly higher time and storage complexity. The discussion here emphasizes only "first-order" elements since we want to see the underlying connections through the simplest and most elegant data structure. The entire discussion can be repeated for higher-order elements.

The Complex Encoded in the Connection Data. This section begins with some definitions needed for the algebraic framework, leading to the simplicial chain complex. Since there are many good references [Cro78, Mun84, Rot88], we do not elaborate on the technical details. Following the definitions, we show that bases for the chain groups of the complex can be constructed in a simple hierarchy of data extracted from the finite element connection matrix.

Background and definitions. In section 4A, a p-simplex

$$\sigma_p = \langle v_0, \dots, v_p \rangle$$

was defined as the convex set spanned by vertices $\{v_0, \dots, v_p\}$. The representation is unique up to a sign which can be assigned to the permutations of the vertex indices. As illustrated in Figure 4.1, a tetrahedron is a 3-simplex, and its faces, edges, and nodes are 2-, 1-, and 0-simplexes, respectively.

Formally, a simplex $\langle v_0, \dots, v_q \rangle$ spanned by a proper subset of $q+1$ vertices of σ_p is called a *q-face* of σ_p. A formal linear combination of p-simplexes is

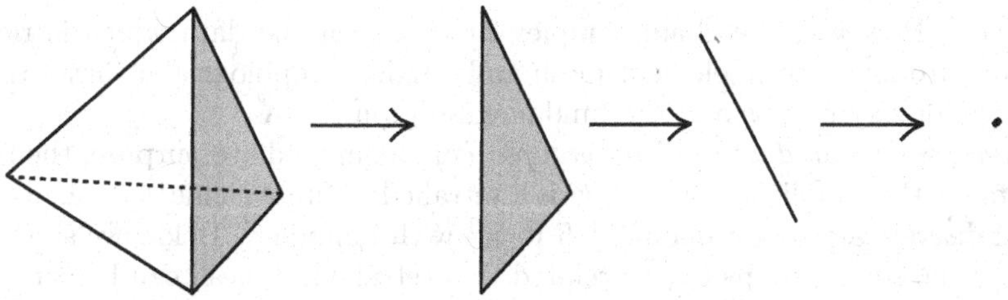

Figure 4.1.

called a *p-chain.* A simplex can be assigned an orientation which is induced by the permutation of vertex order in $\langle v_0, \dots, v_p \rangle$, odd permutation giving negative orientation and even permutation giving positive orientation. The *boundary* of a p-simplex is the $(p-1)$-chain which is the following alternating sum of $(p-1)$-faces:

$$\partial_p \sigma_p = \partial(\langle v_0, \dots, v_p \rangle) = \sum_{i=0}^{p} (-1)^i \langle v_0, \dots, v_{i-1}, v_{i+1}, \dots, v_p \rangle.$$

Note that $\langle v_0, \dots, v_{i-1}, v_{i+1}, \dots, v_p \rangle$ is the p-face opposite vertex v_i. This definition can be used to find the boundary of a p-chain. From the expression above and direct calculation one can verify that in general $\partial_{p-1}\partial_p(\cdot) = 0$, that is the boundary of the boundary of any chain is zero. The operator ∂_p determines a matrix of incidence of $(p-1)$-simplexes with p-simplexes; we will have more to say about this when discussing the coboundary operator in section 4B.

A *simplicial complex* K is a collection of simplexes such that every face of a simplex of K is in K and the intersection of two simplexes in K is a face of each of the simplexes. For each $p \geq 0$, the structure formed by taking p-chains with integer coefficients in a complex K is a finitely generated free abelian group $C_p(K; \mathbb{Z})$ with basis all the p-simplexes in K. This is called the p-chain group of K.

The connection between $C_p(K)$ and $C_{p-1}(K)$ is via the boundary map. Defining the boundary map on a basis of $C_p(K)$, the map extends by linearity to a map

$$\partial_p : C_p(K) \to C_{p-1}(K) \tag{4–4}$$

so that it is a homomorphism between the chain groups. Thus, on a complex of dimension n the collection of abelian groups $C_i(K; \mathbb{Z})$ and boundary homomorphisms give the sequence

$$0 \longrightarrow C_n(K) \xrightarrow{\partial_n} \cdots \xrightarrow{\partial_p} C_{p-1}(K) \xrightarrow{\partial_{p-1}} \cdots \longrightarrow C_1(K) \xrightarrow{\partial_1} C_0(K) \to 0. \tag{4–5}$$

Since $\operatorname{im} \partial_{p+1} \subseteq \ker \partial_p$ (because $\partial_p \partial_{p+1}(\,\cdot\,) = 0$), (4–5) defines the *chain complex* of K, denoted by $(C_*(K), \partial)$ or simply $C_*(K)$.

As described below, the finite element connection matrix contains this basic algebraic structure. It is of interest because while $\operatorname{im} \partial_{p+1} \subseteq \ker \partial_p$, in general $\operatorname{im} \partial_{p+1} \neq \ker \partial_p$ and the part of $\ker \partial_p$ not in the inclusion contains useful information formulated concisely via homology groups and the exact homology sequence. However, the chain complex fits between the data which is readily available from the finite element mesh and "higher" topological structures such as the homology groups and the fundamental group of K.

From connection data to chain groups. For the immediate purpose, the prime example of the simplicial complex K is a tetrahedral finite element mesh, a tetrahedral discretization of a manifold R in $\mathbb{R}^3$ with boundary. Below we show that bases of the chain groups $C_i(K)$ related to a tetrahedral mesh can be computed from the connection matrix by "following the boundary homomorphism down the chain complex".

Consider an n-dimensional simplicial complex K with m_n n-simplexes and m_0 vertices or 0-simplexes. There is a total ordering of the vertices on the index set $\{0,\ldots,m_0-1\}$ called the global vertex ordering. There is a partial ordering of vertices such that vertices of a p-simplex are ordered on the set $\{0,\ldots,n\}$. The connection matrix is the following $m_n \times m_0$ matrix defined in terms of the global and local orderings:

(4–6)

$$C^i_{n,j,k} = \begin{cases} 1 & \text{if global vertex } k \text{ is the } j\text{th local vertex of the } i\text{th } n\text{-simplex,} \\ 0 & \text{otherwise.} \end{cases}$$

For a 3-d finite element mesh, $C^i_{3,j,k}$ is simply the connection matrix which is typically the output of a mesh generator. $C^i_{3,j,k}$ is an $m_3 \times m_0$ matrix, and typically $m_3 = km_0$ where k is approximately 5 to 6. Also note that since a tetrahedron has four vertices, the connection matrix $C^i_{3,j,k}$ has $4m_3$ nonzero entries so that the matrix is sparse.

In general, (4–6) also describes the connection for lower-dimensional subcomplexes or p-skeletons of a mesh where $0 \le p \le n$. Because this matrix is sparse, only the nonzero entries of matrix $C^i_{p,j,k}$ are stored in an $m_p \times (p+1)$ array such that the ith row of the array gives the (global) indices of the vertices $\sigma_{0,k}$ which define the ith p-simplex:

$$C^i_{p,k} = \{\sigma_{0,k_0},\ldots,\sigma_{0,k_p}\},$$

where $0 \le i \le m_p - 1$ and m_p is the number of p-simplexes in the mesh. This is an efficient way of storing (4–6), and in this form $C^i_{p,j,k}$ also resembles a basis for the chain group $C_p(K)$ (the notation intentionally takes them to be the same). It must be emphasized that referring to the nonzero entries of the matrix in a table of pointers to the global vertex ordering gives computational efficiency and a direct link to the maps and definitions of Section 4B.

Consider the following map, which extracts the jth p-face of the kth $(p+1)$-simplex:

$$f_j(\sigma_{p+1,k}) = \langle \sigma_{0,0},\ldots,\widehat{\sigma_{0,j}},\ldots,\sigma_{0,p+1}\rangle, \tag{4–7}$$

where $1 \le i \le m_p$, $1 \le k \le m_{p+1}$, $0 \le j \le p+1$, and $\widehat{\sigma_{0,j}}$ denotes that vertex $\sigma_{0,j}$ is omitted. Note that j and k do not uniquely specify the p-simplex since $\langle \sigma_{0,0},\ldots,\widehat{\sigma_{0,j}},\ldots,\sigma_{0,p+1}\rangle$ may be a p-face in more than one $(p+1)$-simplex. The representation of the p-simplex by vertex ordering is unique up to orientation, but the orientation induced from the $(p+1)$-simplex can always be adopted in order to maintain consistency. In any case, the map gives the p-faces of the (p+1)-simplex when used for $0 \leq j \leq p+1$.

To build C_p, Equation (4–7) can be used $p+2$ times on each (p+1)-simplex in C_{p+1} (effectively taking the boundary of each (p+1)-simplex in C_{p+1}). In each instance this requires that an algorithm which extracts the p-simplexes determine from the existing data for C_p whether the result of applying (4–7) is a new p-simplex or one that has already been extracted. Thus, starting with C_n, it is possible to go down the complex (4–5) and extract all the tables C_p.

ALGORITHM 4.1. Extraction of C_p from C_{p+1}

Set C_p to be empty.
for each $\sigma_{p+1} \in C_{p+1}$
    for each p-face of σ_{p+1}
        if p-face is not in C_p **then** add p-face to C_p.

The decision at the inner loop requires a search through C_p but if implemented in an efficient data structure such as a linked list, the search is bounded by the number of times any vertex of the p-face is a vertex in a p-simplex.

Considerations for cellular meshes. While we focus primarily on simplicial complexes, all of the algebraic structure described is consistent for cellular complexes (e.g., hexahedral meshes). In practice, the data structures are somewhat more complicated because the vertices, while ordered on a cell, are not generally permutable. This affects the definition of (4–7) so that some additional information about the ordering of vertices may have to be preserved at every step of the algorithm. This also influences the way in which the algorithms are implemented – in particular, depending on application, it is most efficient to extract the 1- and 2-subcomplexes simultaneously, 3-cell by 3-cell, in order to avoid storing extra information about vertex ordering.

The Cochain Complex. In de Rham theory, integration on manifolds in $\mathbb{R}^n$ is formulated as an algebraic structure which pairs p-chains with differential p-forms. The algebras of differential p-forms are related in a (de Rham) complex, and the related (co)homology groups are the link between lumped field parameters and topological invariants of the manifold in question [Kot84]. In the discrete setting (e.g., triangulated manifolds), cochains play a role analogous to differential forms and since simplicial (co)homology satisfies the same axioms as the de Rham cohomology, the theories are equivalent [ES52]. In this section we define cochains and their algebraic structure. The algebra is dual to the chain complex. Then we see how this structure also comes out of the connection matrix.

Simplicial cochain groups and the coboundary operator. Formally, the *simplicial p-cochain group* $C^p(X;\mathbb{Z})$ is the group of homomorphisms from p-chains to (for the present purpose) the integers:

$$C^p(X;\mathbb{Z}) = \hom(C_p, \mathbb{Z}).$$

$C^p(X;\mathbb{Z})$ is not a vector space, but one can regard the homomorphisms as functionals on chains and denote the operation of a cochain $c^p \in C^p(X;\mathbb{Z})$ on a chain

$c_p \in C_p(X;\mathbb{Z})$, by functional notation:

$$c^p(c_p) = \langle c^p, c_p \rangle. \tag{4–8}$$

The *p-coboundary operator* d^p is the adjoint of theboundary operator. It is defined by

$$\langle d^p c^p, c_{p+1} \rangle = \langle c^p, \partial_{p+1} c_{p+1} \rangle, \tag{4–9}$$

so that

$$d^p = (\partial_{p+1})^T : C^p(X;\mathbb{Z}) \longrightarrow C^{p+1}(X;\mathbb{Z}).$$

From this point the p-coboundary operator is always written explicitly as the adjoint operator ∂^T_{p+1}. Equation (4–9) is simply a discrete rendition of Stokes theorem on manifolds:

$$\int_\Omega d\omega = \int_{\partial\Omega} \omega,$$

where ω is a differential p-form, $d\omega$ is a $(p+1)$-form, Ω is a $(p+1)$-chain, and $\partial\Omega$ is its boundary. This "generalized Stokes theorem" can be called the fundamental theorem of multivariable calculus. Since $\partial^2 = 0$, $\partial^T_{p+1}\partial^T_p(\cdot) = 0$, and there is a cochain complex:

$$0 \longleftarrow C^n(K) \xleftarrow{\partial^T_n} \cdots \xleftarrow{\partial^T_p} C^{p-1}(K) \xleftarrow{\partial^T_{p-1}} \cdots \longleftarrow C^1(K) \xleftarrow{\partial^T_1} C^0(K) \longleftarrow 0.$$

Coboundary data structures. Since the coboundary operator is the adjoint of the boundary operator, it can be formulated in terms of pairs of simplexes (σ_p, σ_{p-1}) and their "incidence numbers". Consider a p-simplex $\sigma_p = \langle v_o, \dots, v_p \rangle$ and a $(p-1)$-face of σ_p, $\sigma_{p-1} = \langle v_0, \dots, \widehat{v}_j, \dots, v_p \rangle$. Let π be a permutation function on $\{0, \dots, p\}$, then

$$(\operatorname{sign}\pi)\sigma_p = \langle v_{\pi 0}, \dots, v_{\pi p} \rangle = \langle v_j, v_0, \dots, v_{j-1}, v_{j+1}, \dots, v_p \rangle,$$

where $\operatorname{sign}\pi = \pm 1$ depending on the parity of π. When $\sigma_{p-1,j}$ is a face of $\sigma_{p,i}$, $\operatorname{sign}\pi_{ij}$ is a nonzero entry in a p-simplex–$(p-1)$-simplex incidence matrix.

The coboundary operator ∂^T_p can be represented by storing only the nonzero entries of ∂^T_p, $\operatorname{nz}(\partial^T_p)$, and referencing each $(p-1)$-simplex to the p-simplexes in which the $(p-1)$-simplex is a face in sets of pairs

$$\operatorname{nz}(\partial^T_p(\sigma_{p-1,j})) = \{(\sigma_{p,i}, \operatorname{sign}\pi_{ij}) | \sigma_p \in C_p(K)\}, \tag{4–10}$$

where $\sigma_{p,j}$ need only be referenced by its global number j. Since a p-simplex has $p+1$ $(p-1)$-faces, every p-simplex is found in $p+1$ of the sets described in (4–10). This is equivalent to saying that there are $p+1$ nonzero entries per column in ∂^T_p.

In general (4–10) can be implemented efficiently in a linked list so that ∂^T_p becomes a list of linked lists. In the codimension 1 case (∂^T_n), an $(n-1)$-simplex is shared by at most two n-simplexes and there is no need to store $\operatorname{sign}\pi_{ij}$ explicitly since the data can be indexed by the incidence number as follows:

$$\operatorname{nz}(\partial^T_n(\sigma_{n-1,j})) = \{(\sigma_{n,i} | \operatorname{sign}\pi_{ij} = 1), (\sigma_{n,l} | \operatorname{sign}\pi_{lj} = -1)\}. \tag{4–11}$$

We will simply denote the data structure which contains (4–10) for all $(p-1)$-simplexes as ∂^T_p and generate it by the following algorithm:

ALGORITHM 4.2. Construction of ∂_p^T

Set ∂_p^T to be empty.
for each $\sigma_{p-1} \in C_{p-1}$ (m_{p-1} $(p-1)$-simplexes)
  for each σ_p such that $\operatorname{sign} \pi_{ij} \neq 0$
    augment list for σ_{p-1} with $(\sigma_p, \operatorname{sign} \pi)$.

At first sight, the inner loop of algorithm 4.2 seems to require a search through all of the C_p data structure for each case where σ_{p-1} is a $(p-1)$-face. In practice the searching can be avoided by performing the augmentation procedure each time the $(p-1)$-simplex is encountered in the inner loop of algorithm 4.1. With an efficient data structure implementation, reaching the point where the ∂_p^T list is augmented is bounded by the number of times σ_p is a p-face.

Applications: Simplicial (Co)Homology and Cuts for Scalar Potentials. So far we have spelled the simplicial consequences of the finite element connection matrix. In this section we look at two applications of simplicial (co)chain complexes and Poincaré duality to see how they beneficial in 3-d finite element computation.

Simplicial homology and cohomology. The chain complex is readily available from the connection matrix, but for many purposes it is merely the starting point. In this section we consider how homology groups follow from the chain complex and see how they algebraically expose the lumped parameters of electrical engineering (e.g., current, voltage, flux) which come about from integration on p-chains.

For the boundary homomorphism ∂_p in Equation (4–4), we call $\ker \partial_p = Z_p(K)$ the *p-cycles* in K and $\operatorname{im} \partial_{p+1} = B_p(K)$ the *p-boundaries* in K. Both $Z_p(K)$ and $B_p(K)$ are subgroups of $C_p(K)$, and furthermore, $B_p(K) \subseteq Z_p(K) \subset C_p(K)$. This is true since if $\beta \in B_p(K)$ then $\beta = \partial_{p+1}\alpha$, for some $\alpha \in C_{p+1}(K)$, but $\partial_p \partial_{p+1} \alpha = 0$ says that $\beta \in \ker \partial_p$, i.e., $\beta \in B_p(K)$.

In general it is interesting to ask when is a p-cycle not a p-boundary. This information is summarized in the pth simplicial homology group of K, $p \geq 0$, defined as the quotient group

$$H_p(K) = \frac{Z_p(K)}{B_p(K)} = \frac{\ker \partial_p}{\operatorname{im} \partial_{p+1}}.$$

This quotient group consists of equivalence classes of cycles c such that $\partial c = 0$ but c is not a boundary. Two p-cycles a and b are in the same equivalence class if they satisfy the equivalence relation

$$[a] \sim [b] \iff a - b = \partial c_{p+1},$$

where c is $(p+1)$-chain and $[a]$ denotes the homology class of a. The rank of $H_p(K)$ is the number of independent equivalence classes in the group and is known as the pth Betti number of K, denoted by $\beta_p(K)$; intuitively, $\beta_0(K)$ counts the number of connected components of K, and $\beta_1(K)$ counts the "number of holes in K" [Max91, GK95, GK01b].

	Electrostatics	Magnetostatics
Parameter Rel. cohomology group	voltage, V $H^1(\Omega, \partial\Omega)$	flux, ϕ $H^2(\Omega, \partial\Omega)$
Parameter Abs. cohomology group	charge, Q $H^2(\Omega)$	current, I $H^1(\Omega)$
Entries of an energy quadratic form normalized to charges or currents	$C_{ij}^{-1} = \frac{V}{Q} = \frac{\int_{c_j} \boldsymbol{E}_i \cdot dl}{\oint_{S_i} \boldsymbol{D}_i \cdot ds}$ $[S_i] \in H_2(\Omega)$ $[c_j] \in H_1(\Omega, \partial\Omega)$	$L_{ij} = \frac{\phi}{I} = \frac{\int_{S_i} \boldsymbol{B}_j \cdot ds}{\oint_{c_j} \boldsymbol{H}_j \cdot dl}$ $[S_i] \in H_2(\Omega, \partial\Omega)$ $[c_j] \in H_1(\Omega)$

Table 4.1. Summary of relation between lumped parameters and cohomology.

Calling $Z^p(X;\mathbb{Z}) = \ker \partial_p^T$ the group of p-cocycles, and $B^p(X;\mathbb{Z}) = \operatorname{im} \partial_{p-1}^T$ the group of p-coboundaries, the pth cohomology group is:

$$H^p(X;\mathbb{Z}) = \frac{Z^p(X;\mathbb{Z})}{B^p(X;\mathbb{Z})} = \frac{\ker \partial_{p+1}^T}{\operatorname{im} \partial_p^T}.$$

To make the connection with lumped parameters, we also need to introduce relative homology groups. Let L be a subcomplex of K, that is a simplicial complex contained in K. Then the pth relative simplicial homology group of K "modulo" L is

$$H_p(K, L; \mathbb{Z}) = H_p(C_*(K)/C_*(L); \mathbb{Z}),$$

that is, the homology of the quotient of the two complexes. In particular, if $L = \partial K$, two p-cycles from an equivalence class in $H_p(K, \partial K)$ form a p-boundary in K when taken in combination with a p-chain in ∂K. Relative cochains (with integer coefficients) are defined by

$$C^p(K, L; \mathbb{Z}) = \hom(C_p(K, L), \mathbb{Z}),$$

so that the pth relative cohomology group is

$$H^p(K, L; \mathbb{Z}) = \frac{\ker \partial_p^T}{\operatorname{im} \partial_{p+1}^T},$$

where $\partial_p^T : C^p(K, L; \mathbb{Z}) \to C^{p+1}(K, L; \mathbb{Z})$.

Table 4.1 outlines the relation of these (co)homology groups to "lumped parameters" in electro- and magnetostatics. For electrostatics, Ω is the charge-free region and for magnetostatics, Ω is the region free of conduction currents.

Finally, we note that Poincaré–Lefschetz duality on chains "descends" to the (co)homology groups. This happens through the dual complex to K which will be introduced shortly.

Cuts for magnetic scalar potentials. In "magnetoquasistatics", displacement current is ignored in Ampère's law, and the magnetic field is described by

$$\operatorname{curl} \boldsymbol{H} = \boldsymbol{J}.$$

In nonconducting regions $\boldsymbol{J} = 0$ and one may ask if $\boldsymbol{H} = -\operatorname{grad}\psi$ where ψ is a single-valued scalar potential defined in the nonconducting region. In general ψ is multivalued since Ampère's law shows that

$$I = \oint_{c_i} \operatorname{grad}\psi \cdot dl \neq 0$$

if $I \neq 0$ and c_i is a closed path linking the current I. For reasons of computational cost and numerical analysis, it is still worthwhile to pursue the scalar potential in 3-d if one can introduce cut surfaces and impose a discontinuity across the cuts in order to make the potential single-valued. Informally, cuts are orientable surfaces embedded in the current-free region such that when integrating $\boldsymbol{H} \cdot dl$ around a closed path which links current, the path must pass through the cuts. Cuts coincide with the flux measurement surfaces S_i in the right column of Table 4.1 [GK01b].

The existence of cuts as compact, embedded, orientable manifolds in R can be formulated via a constructive proof which gives an algorithm for computing them on finite element meshes which are triangulations of the nonconducting region [Kot89a]. There are many facets to the algorithm which are discussed in Chapter 6, but here we touch on only one aspect. The first step to computing a set of cuts for a mesh is to compute a set of topological constraints which represent a set of generators for classes in $H_2(K, \partial K)$ [GK01b]. Poincaré–Lefschetz duality provides for rephrasing this problem in terms of finding a basis for cohomology classes of a dual complex of K and reduces to finding a basis of the nullspace of ∂_2^T, or a set of vectors $\{\zeta_1, \ldots \zeta_{\beta_1}\}$ satisfying

$$\partial_2^T \zeta_i = 0,$$

subject to $\operatorname{im} \breve{\partial}_0^T = 0$ [GK01b]. The problem is motivated strictly by topological considerations based on the dual complex and Poincaré duality relation which will be considered shortly. In addition the computational formulation requires the data structures C_3, C_2, ∂_3^T, ∂_2^T, ∂_1^T.

Whitney forms. In recent years, there has been growing interest in so-called Whitney 1-forms or edge elements for a variety of finite element computations. The general idea comes from Whitney [Whi57, Whi50] and was developed in [Dod76, Mül78] to which we refer for proofs. It starts with a linear Whitney map that makes piecewise linear differential q-forms from simplicial q-cochains:

$$W : C^q(K) \to L^2\Lambda^q(X), \tag{4–12}$$

X being a compact oriented C^∞ Riemannian manifold of dimension n, $L^2\Lambda^q(X)$ the space of square-integrable de Rham C^∞ differential q-forms on X, and K a simplicial triangulation of X. Let λ_i be barycentric coordinates corresponding to vertices v_i in K. The basic form $W\sigma \in L^2\Lambda^q(X)$ on a q-simplex σ, is defined by

$$W\sigma = q! \sum_{k=0}^{q} (-1)^k \lambda_{i_k}\, d\lambda_{i_0} \wedge \cdots \wedge \widehat{d\lambda_{i_k}} \wedge \cdots \wedge d\lambda_{i_q} \quad \text{if } q > 0,$$

$$W(v_i) = \lambda_i \quad \text{if } q = 0,$$

where $\wedge$ denotes the wedge product for differential forms and, as in Section 4B, $\widehat{\cdot}$ denotes that the differential is excluded. Note that the construction of this q-form corresponds neatly to the process of extracting $(q-1)$-simplexes from q-simplexes in algorithm 4.1. We mention two properties of the Whitney map:

(1) $W\partial^T c = dWc$ for $c \in C^q(K)$ where $\partial^T c \in C^{q+1}(K)$ is the simplicial coboundary of c. The exterior derivative $d : \Lambda^q \to \Lambda^{q+1}$ applied to Wc is well-defined in this case.

(2) Let $\langle \cdot , \cdot \rangle$ denote the pairing of $C_q(K)$ and $C^q(K)$ as in (4–8). Then

$$\int_{c_q} Wc^q = \langle c^q, c_q \rangle$$

for every cochain $c^q \in C^q$ and chain $c_q \in C_q(K)$.

The first property is significant because it implies that the simplicial cohomology groups of K and the de Rham cohomology group of X are isomorphic.

In addition to the Whitney map (4–12), there is a de Rham map

$$R : L^2\Lambda^q(X) \to C^q(K),$$

which is defined on a basis of chains by

$$\int_{c_q} \omega = \langle c^q, c_q \rangle,$$

and the second property of the Whitney map ensures that

$$RW = I,$$

where I is the identity map. The convergence $WR \to I$ as a mesh is refined is a special case of both finite element theory and Whitney's program, but this obvious connection does not seem to exist outside of computational electromagnetics and the work of Dodziuk [Dod74, Dod76] and Müller [Mül78].

For $c, c' \in C^q(K)$, an inner product can be defined:

$$(c, c') = \int_X Wc \wedge *Wc' = (Wc, Wc').$$

This is nondegenerate by the property of the de Rham map. Although there is no obvious metric inherent to the simplicial complex K, this inner product inherits a metric from X through the Whitney map. As a mesh is refined, the inherited metric in the inner product converges to the (Riemannian) metric on X.

Example 4.3 Whitney form interpolation of the helicity functional. For finite element computations of magnetic fields, a Whitney 1-form can be used to discretize the magnetic field intensity. Namely,

$$\omega = \boldsymbol{H} \cdot dr.$$

Then, on a 1-simplex,

$$h_{ij} = \int_{v_i}^{v_j} \omega$$

define the variables of a 1-cochain for ω. In this formulation, it is interesting to note that the contribution of the so-called helicity density $\omega \wedge d\omega$ to the finite element stiffness matrix is independent of metric and constitutive laws [Kot89b]. In particular, the contribution of the helicity on a tetrahedron to the stiffness matrix is

$$\frac{1}{2} \int_{\sigma_3} \omega \wedge d\omega = \tfrac{1}{6}(h_{01}h_{23} - h_{02}h_{13} + h_{03}h_{12}).$$

The right-hand side is a quadratic form which remains invariant by the action of the Lie group $SL(4, \mathbb{R})$ associated with piecewise linear volume-preserving diffeomorphisms. □

The Dual Complex and Discrete Poincaré Duality For Chains and Cochains. One reason for developing the coboundary data structures is to make use of a duality relation which relates cochains of the simplicial complex K to chains on the dual complex of K. A thorough development of the dual chain complex of K (as in [Mun84]) usually starts with the first barycentric subdivision of K and construction of "blocks" in the subdivision which are dual to the p-simplexes of K, where the dual blocks are unions of certain sets of open simplexes in the subdivision.

Although the definition of the dual complex [Mun84] relies on some geometry, the incidence data for the dual complex can be recovered from the coboundary data structures of the simplicial complex [Whi37], so we will formally define the dual chain complex DK with the following construction. The *dual complex* of a simplicial complex K, is a cell complex DK obtained by identifying p-simplexes on K with $(n-p)$-cells. In general DK is not a simplicial complex, so it is necessary to use the terminology of cells. However it is possible to formulate a complex $C_*(DK)$ as previously done for K in (4–5).

In explicit terms, we identify 3-simplexes with 0-cells (vertices of DK), 2-simplexes with 1-cells (edges), 1-simplexes with 2-cells (faces) and 0-simplexes with 3-cells. Since the coboundary data structures already contain the incidence of p-simplexes in $(p+1)$-simplexes, these can be reinterpreted on the dual complex as the boundaries of $(n-p)$-cells. For example, the entries of (4–11) can be regarded as the 0-cells incident to a 1-cell in DK which passes "through" the barycenter of $\sigma_{2,j}$. Equation (4–10) is interpreted as the boundaries of 2- and 3-cells (∂_1^T and ∂_2^T respectively) in DK associated with $(3-2)$- and $(3-3)$-simplexes, respectively, in K.

A useful form of Poincaré duality formalizes the connection between ∂_p^T and boundaries of $(n-p)$-cells in DK seen in the data structures. It establishes a duality between cochains in K and chains in DK. While the complexes $C_p(K)$ and $C^p(K)$ are duals by definition, there exists a nondegenerate bilinear intersection pairing. Comparison of the chain complexes $C^*(K)$ and $C_*(DK)$ in light of this duality says that boundary and coboundary operators can be identified:

$$\partial_{p+1}^T = \breve{\partial}_{n-p},$$

where $\breve{\partial}$ denotes the boundary operator on DK. The same identification is seen in the data structures. Hence, for the price of extracting the cochain complex from the connection matrix, we have learned everything about the dual chain complex.

Once a tetrahedral finite element mesh is identified as the triangulation of a 3-manifold with boundary, simplicial complexes give a systematic and general way for creating and organizing finite element data structures. The simplicial chain and cochain complexes are the bridge between the topology of the manifold, vector fields in the region, and structures from algebraic topology which are useful for finite element computation. The data structures are the most natural for using Whitney elements, in particular Whitney edge elements. Helicity functionals and cuts for magnetic scalar potentials are good examples of applications where the metric of the space and the topology can be separated. In these contexts, the data structures provide a high degree of computational efficiency.

In Chapter 7 we formulate a paradigm variational problem in terms of differential forms. It has the virtue that it includes all of the variational principles considered so far. Furthermore, its finite element discretization by means of Whitney forms preserves all of the topological, homological, and circuit theoretic features emphasized so far in this book. See Bossavit [Bos98] for an engineer's take on the history of the subject, and Hiptmair [Hip02] for some recent developments.

4C. The Euler Characteristic and the Long Exact Homology Sequence

The Euler characteristic is best known in the context of the topology of polyhedra [Arm83] and in electrical circuit theory. In a more general form it is useful for analyzing numerical algorithms for finite element matrices [Kot91]. Here we discuss how the Euler characteristic can be used to count the number of nonzero entries in the global finite element stiffness matrix. Since the amount of computation required for one iteration of the conjugate gradient (CG) algorithm can be expressed in terms of the number of nonzero entries of the matrix, the topological counting discussed below leads to good *a priori* estimates of computational complexity per iteration. A proof and further discussion of the Euler characteristic theorem can be found in [Mun84].

Let m_i denote the number of i-simplexes in a tetrahedral finite element mesh K. The Euler characteristic $\chi(K)$ is defined as

$$\chi(K) = \sum_{i=0}^{3}(-1)^i m_i.$$

The Euler characteristic is related to the ranks $\beta_i(K)$ of the homology groups of the simplicial complex of K as follows.

THEOREM.

$$\chi(K) = \sum_{i=0}^{3}(-1)^i \beta_i(K).$$

The definition and theorem are the same for a cellular complex where m_i denotes the number of i-cells in the complex, and while the statements above are given for a 3-complex, the definition and theorem are the same for an n-complex.

We will see how the Euler characteristic can be used in the analysis of algorithms. Consider the long exact sequence of the pair $(M, \partial M)$ developed in Section 1H. It can be thought of as a complex with trivial homology and hence, by the theorem above, trivial Euler characteristic. Examining the columns of the long exact sequence, we see that the terms in the Euler characteristic of the long exact sequence are just the alternating sum of the Euler characteristics of ∂M, M, and the pair $(M, \partial M)$. Hence

$$\chi(\partial M) = \chi(M) - \chi(M, \partial M). \tag{4–13}$$

Furthermore, if M is a compact orientable manifold with boundary, the Lefschetz duality theorem implies that

$$\chi(M) = (-1)^n \chi(M, \partial M). \tag{4–14}$$

Combining Equations (4–13) and (4–14) we get

$$\chi(\partial M) = (1 - (-1)^n)\,\chi(M) = \begin{cases} 2\chi(M) & \text{if } n \text{ is odd,} \\ 0 & \text{if } n \text{ is even.} \end{cases}$$

So, if M is three-dimensional,

$$\chi(\partial M) = 2\chi(M). \tag{4–15}$$

Let us now consider how Equation (4–15) is used in the analysis of algorithms by considering the conjugate gradient iteration used for solving linear systems of equations.

A CG iteration involves one matrix-vector multiplication, two inner products and three vector updates. For a given interpolation scheme on a finite element mesh, the number of floating point operations (FLOPS) per CG iteration is [Kot91]

$$F = 5D + X,$$

where D is the number of degrees of freedom and X is the number of nonzero entries in the stiffness matrix. For the scalar Laplace equation, one can write:

$$F_0^s = 5D_0^s + X_0^s, \tag{4–16}$$

where

$$D_0^s = m_0, \qquad X_0^s = m_0 + 2m_1, \tag{4–17}$$

and m_i, for $0 \le i \le 3$, is the number of i-simplexes in the mesh. Similarly, let n_i be the number of i-simplexes in the boundary of the mesh. For a three dimensional simplicial mesh which is the triangulation of a manifold M, we have the following two linear equations

$$4m_3 = 2m_2 - n_2,$$
$$3n_2 = 2n_1.$$

The Euler characteristic applied to M and ∂M provides two additional linear relations. Equation (4–15) above enables us to express $\chi(\partial M)$ in terms of $\chi(M)$. So, the four linear relations can be used to express the four variables m_1, m_2, n_1, and n_2 in terms of the remaining four: m_0, m_3, n_0, and $\chi(M)$. Hence

$$\begin{aligned} m_1 &= m_0 + m_3 + n_0 - 3\chi(R), \\ m_2 &= 2m_3 + n_0 - 2\chi(R), \end{aligned} \tag{4–18}$$

and Equation (4–16) can be rewritten as

$$F_0^s = 8m_0 + 2m_3 + 2n_0 - 6\chi(R). \tag{4–19}$$

For nodal interpolation of three-component vectors,

$$F_0 = 5D_0 + X_0, \tag{4–20}$$

where

$$D_0 = 3D_0^s, \qquad X_0 = 3(3X_0^s). \tag{4–21}$$

Using Equations (4–18) and (4–17), Equation (4–21) can be substituted into (4–20) to yield

$$F_0 = 42m_0 + 18m_3 + 18n_0 - 54\chi(R). \tag{4–22}$$

Similarly [Kot91], for edge interpolation of vectors, we can write

$$F_1 = 5D_1 + X_1,$$

where, by [Kot91] $D_1 = m_1$ and

$$X_1 = m_1 + 2(3m_2) + 6m_3. \tag{4–23}$$

Then, via (4–18), we have

$$F_1 = 6m_0 + 24m_3 + 12n_0 - 30\chi(R). \tag{4–24}$$

Assuming that hexahedral meshes have, on average, as many nodes as elements, a useful heuristic is $m_3 = km_0$ where $k = 5$ or 6 (depending on whether hexahedra are divided into 5 or 6 tetrahedra). So, comparing (4–22) and (4–24) to (4–19), one finds that in the limit, as the mesh is refined ($m_0 \to \infty$),

$$\frac{F_0}{F_0^s} = \begin{cases} 7.5 & \text{if } k = 6, \\ 7.6 & \text{if } k = 5, \end{cases}$$

and

$$\frac{F_1}{F_0^s} = \begin{cases} 10.7 & \text{if } k = 6, \\ 9.7 & \text{if } k = 5. \end{cases}$$

This analysis is used for the comparison presented in Table 6.1. See also [Sai94] for related numerical results.

It could be said if all the text that concerned the application of boundary conditions in electromagnetic problems, and all the topological arguments, were removed from this book, there would be little left. To some extent both topics could be said to be more of an art than a science.

E. R. Laithwaite, *Induction Machines for Special Purposes*, 1966

5

Computing Eddy Currents on Thin Conductors with Scalar Potentials

5A. Introduction

In this chapter we consider a formulation for computing eddy currents on thin conducting sheets. The problem is unique in that it can be formulated entirely by scalar functions — a magnetic scalar potential in the nonconducting region and a stream function which describes the eddy currents in the conducting sheets — once cuts for the magnetic scalar potential have been made in the nonconducting region. The goal of the present formulation is an approach via the finite elements to discretization of the equations which come about from the construction of the scalar potentials. Although a clear understanding of cuts for stream functions on orientable surfaces has been with us for over a century [Kle63] there are several open questions which are of interest to numerical analysts:

(1) Can one make cuts for stream functions on nonorientable surfaces?
(2) Can one systematically relate the discontinuities in the magnetic scalar potential to discontinuities in stream functions by a suitable choice of cuts?
(3) Given a set of cuts for the stream function, can one find a set of cuts for the magnetic scalar potential whose boundaries are the given cuts?

In preceding chapters we have alluded to the existence of cuts, though we have not yet dealt with the details of an algorithm for computing cuts. The algorithm for cuts will wait for Chapter 6, but it is possible to answer the questions above. Section 5B gives affirmative answers to the first two questions by using the existence of cuts for the magnetic scalar potential to show that cuts for stream functions can be chosen to be the boundaries of the cuts for magnetic scalar potentials. Although this need not be the case, this approach enables the computation of cuts for stream functions to be a by-product of the computation of cuts for magnetic scalar potentials. This is an important point since investigations of cuts in three dimensions indicate that the third question cannot be resolved with a polynomial time algorithm.

Problems of eddy current analysis and nondestructive testing do not generally lead to boundary value problems in the usual sense because the excitations are not prescribed current distributions or boundary values but rather lumped parameter "total currents". As noted throughout (as in Example 1.9), jumps in the magnetic scalar potential across cuts relate to these "total currents" in the current-carrying region. Dually, the same cuts are orientable surfaces used for calculating time rates of change of flux linkage which correspond to electromotive forces around closed nonbounding curves in the current-carrying region. With regard to the duality of lumped parameters it is natural to ask whether the topological formalism enables one to relate lumped-parameter currents to statements about voltages and fluxes. This question is considered in section 5B where an intersection matrix (a nondegenerate bilinear pairing between homology groups) is introduced to articulate current constraints and in section 5D where the intersection matrix relates free currents to dual conditions involving voltages. Section 5E considers the discretized finite element problem for computing eddy currents with cuts. The result of this chapter is an outline of how a computer can solve an eddy current problem given only information about circuit parameters (current and voltage) and cuts for the magnetic scalar potential.

5B. Potentials as a Consequence of Ampère's Law

We begin by stating the eddy current problem for thin conductors. Consider $\tilde{R}$, a three-dimensional manifold with boundary which is a subset of $\mathbb{R}^3$. S is a conducting surface embedded in $\tilde{R}$ which may have more than one connected component. Let R be the complement of S relative to $\tilde{R}$, $R = \tilde{R} \cap S^c$ and $\beta_1(R)$ its first Betti number. Then R is the current-free or nonconducting region and in Maxwell's terminology [Max91], $\beta_1(R)$ is the cyclomatic number.

The current sources are taken to be lumped-parameter currents

$$I_i, \quad 1 \le i \le n_c \le \beta_1(R)$$

confined to the region $\tilde{R}^c$, the complement of $\tilde{R}$ relative to $\mathbb{R}^3$, where n_c is the number of prescribed currents. There may also be inhomogeneous boundary conditions. The driving sources lie in the exterior of $\tilde{R}$. We assume that the excitations are such that the resulting magnetic field in R is quasistatic so that displacement currents can be neglected.

If the boundary conditions are homogeneous, there are two types of global topological conditions which can be prescribed. The first type of condition is

$$\oint_{c_i} \boldsymbol{H} \cdot dl = I_i, \quad 1 \le i \le n_c, \tag{5–1}$$

where c_i are generators of classes in the homology groups $H_1(R;\mathbb{Z})$. The second type of condition is

$$\int\!\!\int_{S_j} \boldsymbol{B} \cdot dS = -\frac{dV_j}{dt}, \quad n_c + 1 < j \le n_c + n_v,$$

where S_j are generators of classes in the homology groups $H_2(R, \partial R; \mathbb{Z})$ and n_v is the number of voltages prescribed for the problem. As brought forth in the first chapter these conditions can contribute nontrivial solutions of the boundary value problem.

As outlined in Chapter 2, the integral form of Ampère's law implies that

$$\text{curl}\, \boldsymbol{H} = 0 \quad \text{in } R, \tag{5–2}$$

$$\boldsymbol{n} \times (\boldsymbol{H}_a - \boldsymbol{H}_b) = \boldsymbol{K} \quad \text{on } S, \tag{5–3}$$

$$\text{div}_S\, \boldsymbol{K} = 0 \quad \text{on } S, \tag{5–4}$$

where $\boldsymbol{n}$ is the normal to surface S and $\boldsymbol{K}$ describes current flowing on S. The surface current density vector $\boldsymbol{K}$ is defined as

$$\boldsymbol{K} = -D\, \boldsymbol{n} \times (\boldsymbol{n} \times \boldsymbol{J}_{\text{av}}) \text{ on } S,$$

where D is the local sheet thickness or skin depth, $\boldsymbol{J}_{\text{av}}$ is an average current density in the effective depth of the current sheet, and div_S refers to the divergence operator in surface S. Since S may be nonorientable, $\boldsymbol{K}$ is defined locally on S, and subscripts a and b can only be defined locally on S.

Equations (5–2)–(5–4) give the local consequences of Ampère's law. Equation (5–2) suggests using a magnetic scalar potential to describe $\boldsymbol{H}$ in R and Equation (5–4) suggests using a stream function to describe $\boldsymbol{K}$ on S. The introduction of these potentials has topological consequences which cannot be deduced from Equations (5–2)–(5–4) but rather from the integral form of Ampère's law and the systematic use of (co)homology groups. In order to globally define a single-valued magnetic scalar potential we need the framework established in earlier chapters and summarized below:

(1) the cohomology group $H^1(R; \mathbb{R})$: equivalence classes of irrotational vector fields (with equivalence defined in the sense that the fields in a class differ by the gradient of a single-valued function);
(2) the homology group $H_1(R; \mathbb{Z})$, associated with equivalence classes of closed curves in R which "link" current paths;
(3) the relative homology group $H_2(\tilde{R}, S \cup \partial\tilde{R}; \mathbb{Z})$, associated with equivalence classes of cuts which enable one to use a single-valued scalar potential.

The relation between each of the two homology groups to the cohomology group is clear, however for geometric modeling and for the visualization of cuts, it is useful to introduce an intersection matrix as a way of seeing the relationship between the two homology groups. Assuming for the moment that cuts can be represented by embedded, orientable submanifolds, let $\{S_i\}$ be a set of cuts whose homology classes are a basis of $H_2(\tilde{R}, S \cup \partial\tilde{R}; \mathbb{Z})$ and $\{c_j\}$ be a set of closed curves whose homology classes are a basis for $H_1(R; \mathbb{Z})$ where $1 \leq (i, j) \leq \beta_1$. S_i and c_j intersect at a finite number of points, and the kth intersection point is denoted by p_{ij}^k. Furthermore, suppose that at p_{ij}^k, S_i has normal vector $\boldsymbol{n}_{ij}^k$, c_j has tangent vector $\boldsymbol{t}_{ij}^k$, and that S_i or c_j have been perturbed locally so that $\boldsymbol{n}_{ij}^k \cdot \boldsymbol{t}_{ij}^k \neq 0$. The intersection matrix $\mathcal{I}$ is defined by letting its ijth entry be the oriented intersection number of S_i and c_j. That is

$$(\mathcal{I})_{ij} = \operatorname{Int}(S_i, c_j) \triangleq \sum_k \operatorname{sign}(\boldsymbol{n}_{ij}^k \cdot \boldsymbol{t}_{ij}^k). \tag{5–5}$$

It can be shown that this integer-valued matrix $\mathcal{I}$ is defined independent of how the cuts $\{S_i\}$ and curves $\{c_j\}$ are chosen to represent their homology classes. Furthermore, since the homology groups involved are torsion-free one can choose the cuts and curves so that $\mathcal{I}$ is unimodular ($\det \mathcal{I} = 1$) and this in turn indicates that cuts and curves can be chosen so that $\mathcal{I}$ is the identity matrix. However, it is not clear that there is an algorithm which can make this choice.

Having defined the curves $\{c_j\}$, the cuts $\{S_i\}$ and the intersection matrix $\mathcal{I}$ we can proceed to the magnetic scalar potential. First note that

$$\oint_{c_i} \boldsymbol{H} \cdot d\boldsymbol{l} = I_i, \tag{5–6}$$

where the $\{I_i\}$ are currents flowing outside R. In the nonconducting region exclusive of the cuts, the magnetic field can be expressed in terms of a scalar potential:

$$\boldsymbol{H} = -\operatorname{grad}\psi \quad \text{in } R \cap \textstyle\bigcap_j S_j^c, \tag{5–7}$$

where ψ is single-valued. This scalar potential has jumps $[\psi]_{S_j}$ across S_j which can be determined from the linear system

$$I_i = \sum_{j=1}^{\beta_1} [\psi]_{S_j} (\mathcal{I})_{ji}, \quad 1 \le i \le \beta_1(R). \tag{5–8}$$

This matrix equation involving the intersection matrix arises from the application of Ampère's law:

$$\begin{aligned}
I_i = \oint_{c_i} \boldsymbol{H} \cdot d\boldsymbol{l} &= \int_{c_i \cap \bigcap_j S_j^c} \boldsymbol{H} \cdot d\boldsymbol{l} && \text{by (5–6)} \\
&= -\int_{c_i \cap \bigcap_j S_j^c} \operatorname{grad}\psi \cdot d\boldsymbol{l} && \text{by (5–7)} \\
&= -\psi|_{\partial(c_i \cap \bigcap_j S_j^c)} && \text{by Stokes' theorem} \\
&= \sum_{j=1}^{\beta_1} [\psi]_{S_j} \sum_{p_{ij}^k \in S_j \cap c_i} \operatorname{sign}(\boldsymbol{n}_{ij}^k \cdot \boldsymbol{t}_{ij}^k) && \text{by the definition of } p_{ij}^k \text{ and } [\psi]_{S_j} \\
&= \sum_{j=1}^{\beta_1} [\psi]_{S_j} (\mathcal{I})_{ji} && \text{by (5–5).}
\end{aligned}$$

Having described a potential for the magnetic field $\boldsymbol{H}$ we turn our attention to the stream function describing the surface current density $\boldsymbol{K}$. For the purposes of formulating a boundary value problem in R, it is sufficient to assume that the currents are confined to S and the boundary of $\tilde{R}$. That is, the support of $\boldsymbol{K}$ is

$\partial\tilde{R} \cup S$. Equation (5–4) ensures the existence (locally) of a stream function χ which, for flat surfaces, is related to $\boldsymbol{K}$ by

$$\boldsymbol{K} = -\boldsymbol{n} \times \operatorname{grad}\chi. \tag{5–9}$$

It is easy to verify that Equation (5–9) is consistent with Equations (5–2)–(5–4) and (5–7). For curved surfaces there are two difficulties with this approach. First, a considerable investment in differential geometry is required in order to have an expression which is coordinate-free. In the notation of Nedelec [Ned78] this is handled by writing

$$\boldsymbol{K} = -\overline{\operatorname{curl}}\,\chi \tag{5–10}$$

where $\overline{\operatorname{curl}}$ is the formal adjoint to the curl operator in the surface. That is,

$$\int_S \chi \cdot \operatorname{curl}\boldsymbol{F} \cdot \boldsymbol{n} dS = \oint_{\partial S} \chi\boldsymbol{F} \cdot d\boldsymbol{l} - \int_S \boldsymbol{F} \cdot \overline{\operatorname{curl}}\,\chi\, dS \tag{5–11}$$

for any admissible χ and $\boldsymbol{F}$. The second problem is that for nonorientable surfaces the normal vector in (5–9) and the integration by parts formula (5–11) leading to (5–10) are not well-defined. Hence the problem of defining χ on a nonorientable surface and the problem of making cuts for χ on orientable and nonorientable surfaces must be resolved.

A systematic way out of these difficulties begins with observing that everything that everything developed so far is a consequence of Ampère's law and not of Maxwell's remaining equations nor of metric notions such as constitutive relations. If the metric-free formalism of differential forms is used, it becomes clear that generalizing Equation (5–9) to the case at hand is tedious and it is more fruitful to work with Ampère's law and first principles. The argument is summarized in ordinary vector notation as follows. Let S' be that part of the boundary of $\tilde{R}$ and the current sheets which does not touch the boundaries of the cuts for the magnetic scalar potential. That is,

$$S' = (S \cup \partial\tilde{R}) \cap \left(\bigcup_{j=1}^{\beta_1} S_j\right)^c.$$

PROPOSITION. *There is a single-valued stream function defined on each connected component of S'. In particular, each connected component of S' is orientable.*

To see why this is so, consider a connected component S'_c of S' and a point $p_o \subset S'_c$. In a sufficiently small neighborhood N of p_o one can define an orientation (a consistent choice of normal vector). Next, recall from (5–7) that the magnetic scalar potential is single-valued in $R \cap \left(\bigcup_{j=1}^{\beta_1} S_j\right)^c$. Hence on the set N one can define a single-valued stream function

$$\chi = \psi^a - \psi^b = [\psi]_{S'_c}, \tag{5–12}$$

where ψ^a is taken to be zero on $\partial\tilde{R}$. ($\partial\tilde{R}$ is always orientable; see [GH81].)

This definition is consistent with Ampère's law, equations (5–2)–(5–4), and (5–8), (5–9), and (5–10). The physical interpretation of differences in a single-valued stream function is that of net current flowing across a curve between two points. Next, note that $\psi_a - \psi_b$ is well-defined (single-valued) on all of S'_c. Furthermore since this result is valid for any solenoidal distribution of current on $S \cup \partial\tilde{R}$, by Equation (5–12) there is a well-defined χ on all of S'_c. This proves the proposition.

So far we have succeeded in breaking $S \cup \partial\tilde{R}$ into orientable patches whose union is the set-theoretic complement of a set of curves which are the boundaries of cuts for the magnetic scalar potential. On each of these patches we have an orientation and a stream function related to the magnetic scalar potential by Equation (5–12). From this we conclude that the cuts for the stream function are the curves ∂S_i whose orientation is induced from S_i. This is clear because, as sets,

$$\mathcal{S} \cup \bigcup_{i=1}^{\beta_1} \partial S_i = (S \cup \partial\tilde{R}) \cap \bigcup_{j=1}^{\beta_1} S_j. \tag{5–13}$$

Here $\mathcal{S}$ is where $\bigcup_{j=1}^{\beta_1} S_j$ "passes through" S rather than just meeting it, and the term on the right-hand side is where χ is discontinuous. Note that we can always cut the S_j along $\mathcal{S}$ and perturb them so that there is no loss of generality in assuming that $\mathcal{S}$ consists of isolated points. To find the constraints which must be imposed on χ at ∂S_j, consider S_j meeting S or $\partial\tilde{R}$. Recall that S_j is orientable and call the normal vector $\boldsymbol{n}_j$. This induces an orientation (tangent vector $\boldsymbol{t}_j$) on ∂S_j. Let $\boldsymbol{n}_{ck}$ be the normal vector to S'_{ck} where S'_{ck} is the kth connected component of S'. Locally, on either side of the cut the following quantity is well-defined:

$$\varepsilon_{jk}|_{\partial S'_{ck} \cap \partial S_j} = \text{sign}\ \det(\boldsymbol{t}_j, \boldsymbol{n}_j, \boldsymbol{n}_{ck}). \tag{5–14}$$

Note that $\varepsilon_{jn} + \varepsilon_{jm} = 0$ if the orientation of S changed when the cut is crossed. Equation (5–14) is defined locally and can do unexpected things as one moves along ∂S_j (consider a Möbius band where $S'_{cn} = S'_c = S'_{cm}$, $\varepsilon_{jn} + \varepsilon_{jm} = 0$, and where ∂S_j necessarily touches the edge of the band). Given the preceding notation, by Equations (5–13) and (5–14) we have:

$$0 = \lim_{\text{length } c \to 0} \oint_c \boldsymbol{H} \cdot dl = \varepsilon_{jn}[\psi]_{S'_{cn}} - [\psi]_{S_j} - \varepsilon_{jm}[\psi]_{S'_{cm}},$$

so that by (5–12) we have

$$[\psi]_{S_j} = \varepsilon_{jn}\chi|_{S'_{cn}} - \varepsilon_{jm}\chi|_{S'_{cm}} \quad \text{along each } \partial S_j, \text{ for } 1 \le j \le \beta_1. \tag{5–15}$$

Expression (5–15) is the constraint on the stream function at a cut. In conclusion, (5–7) and (5–12) define the potentials while (5–15) gives a topological consistency condition. External "forcing currents" can be imposed in terms of the magnetic scalar potential through use of (5–8) or in terms of the stream function by using (5–15) to eliminate $[\psi]_{S_j}$ in (5–8).

An elegant development of the above can be obtained by working "dually" in terms of the cohomology and "twisted differential forms". Since no Poincaré-like duality theorem on S is ever used, this approach is equivalent to the one taken above because the (co)homology groups of R are torsion-free as discussed in 3D.

5C. Governing Equations as a Consequence of Faraday's Law

In this section we derive the equations satisfied by the potentials. These are the equations which we expect would be the Euler–Lagrange equations if a variational principle exists. We introduce constitutive relations

$$\boldsymbol{B} = \mu\left(|\boldsymbol{H}|^2\right)\boldsymbol{H} \quad \text{in } R \tag{5–16}$$

and

$$\boldsymbol{E}_S = \frac{1}{\sigma D}\boldsymbol{K} \quad \text{on } S. \tag{5–17}$$

Here $\boldsymbol{E}_S$ is the projection of the electric field onto the surface S. For simplicity, the magnetic constitutive relation is assumed to be nonhysteretic, isotropic, and monotonic. This explicit dependence of μ on $\boldsymbol{H}$ will be assumed in the following. These constitutive relations and the differential version of Faraday's lawfor stationary media can be used to express the vectors $\boldsymbol{B}$ and $\boldsymbol{E}_S$ in terms of ψ and χ as follows. Defining a dummy index $1 \le j \le \beta_1(R)$, the equations describing the magnetoquasistatic field are

$$\begin{cases} \operatorname{div}\dfrac{\partial}{\partial t}\left(\mu \operatorname{grad}\psi\right) = 0 & \text{in } R \cap \left(\bigcup_{j=1}^{\beta_1} S_j\right)^c, \\ \dfrac{d}{dt}\left(\mu^a\dfrac{\partial\psi^a}{\partial \boldsymbol{n}} - \mu^b\dfrac{\partial\psi^b}{\partial \boldsymbol{n}}\right) = 0 & \text{on each } S'_{cn} \text{ and } S_j, \\ \operatorname{curl}\left(\dfrac{1}{\sigma D}\,\overline{\operatorname{curl}}\,\chi\right) = -\mu\dfrac{\partial}{\partial t}\dfrac{\partial\psi}{\partial \boldsymbol{n}} & \text{on each } S'_{cn}. \end{cases} \tag{5–18}$$

These equations, being in some sense dual to Equations (5–2)–(5–4), are incomplete since the differential laws lack the information encoded in integral laws by cohomology considerations. The missing information pertaining to voltage drops around "closed circuits" is dual to Equations (5–8) and (5–15) and is a consequence of the integral form of Faraday's law:

$$\int_{\partial S_i}\frac{1}{\sigma D}\,\overline{\operatorname{curl}}\,\chi\cdot d\boldsymbol{l} = -\frac{d}{dt}\int_{S_i}\mu\frac{\partial\psi}{\partial \boldsymbol{n}}dS, \quad 1 \le i \le \beta_1. \tag{5–19}$$

The dual role played by the homology group $H_2(\tilde{R}, S \cup \partial\tilde{R}; \mathbb{Z})$ is made clear in the preceding expression. Expressions (5–18), (5–19), (5–12), and (5–15) form a complete set of equations which describe the problem. Note that this "boundary value problem" has no volume sources or prescribed boundary conditions. The "excitations" which give nontrivial solutions are given by prescribing currents via (5–8).

5D. Solution of Governing Equations by Projective Methods

For a numerical solution it is best to develop an energy or power dissipation functional which can be used in conjunction with Galerkin's method to solve Equations (5–18) and (5–19). Although there is a variational principle involving complex-valued functionals for linear time-harmonic problems, there is no energy-based variational principle for the problem at hand since the equations have first order derivatives in time. We will consider a projective method based on the magnetoquasistatic version of Poynting's theorem. Although this approach is not new, we examine it here because global cohomological (lumped parameter) aspects have never been systematically formulated. The magnetoquasistatic form of the Poynting energy theorem is

$$0 = \int_{S \cup \partial \tilde{R}} \boldsymbol{E}_S \cdot \boldsymbol{K} dS + \int_R \boldsymbol{H} \cdot \frac{\partial \boldsymbol{B}}{\partial t}\, dV.$$

The starting point for a projective solution is the equation

$$(5\text{–}20) \quad 0 = \int_{S \cup \partial \tilde{R}} \boldsymbol{E}_s(\boldsymbol{K}) \cdot \delta \boldsymbol{K}\, dS + \int_R \delta \boldsymbol{H} \cdot \frac{\partial}{\partial t} \boldsymbol{B}(\boldsymbol{H})\, dV \quad \text{for all } \delta \boldsymbol{H}, \delta \boldsymbol{K},$$

where $\boldsymbol{H}$ and $\boldsymbol{K}$ are subject to Equations (5–7), (5–8), (5–10), (5–12), and (5–15), while $\delta\boldsymbol{H}$ and $\delta\boldsymbol{K}$ are test functions which satisfy analogous equations:

$$(5\text{–}21) \quad \begin{aligned} \delta \boldsymbol{H} &= -\operatorname{grad} \delta\psi \quad \text{in } R \cap (\textstyle\bigcup_j S_j)^c, \\ \delta \boldsymbol{K} &= -\overline{\operatorname{curl}}\, \delta\chi \quad \text{on } S', \end{aligned}$$

$$(5\text{–}22) \quad \begin{aligned} \delta\chi &= \delta\psi^a - \delta\psi^b && \text{on } S', \\ \delta\psi^a &= 0 && \text{on } \partial\tilde{R}, \\ [\delta\psi]_{S_j} &= \varepsilon_{jn}\delta\chi|_{S'_{cn}} - \varepsilon_{jm}\delta\chi|_{S'_{cm}} && \text{along each } \partial S_j, \end{aligned}$$

and

$$(5\text{–}23) \quad \delta I_i = \sum_{j=1}^{\beta_1} [\delta\psi]_{S_j} \mathcal{I}_{ji},$$

where $\delta I_i = 0$ if I_i is prescribed. Using the constitutive laws (5–16) and (5–17), Equation (5–20) can be rewritten in terms of potentials:

$$(5\text{–}24) \quad 0 = \int_{S'} \frac{1}{\sigma D} \overline{\operatorname{curl}}\, \chi \cdot \overline{\operatorname{curl}}\, \delta\chi\, dS + \int_{R \cap (\bigcup_j S_j)^c} \operatorname{grad} \delta\psi \cdot \frac{d}{dt}(\mu \operatorname{grad} \psi)\, dV$$

for all $\delta\psi$ and $\delta\chi$ satisfying Equations (5–21) and (5–22). Here ψ and χ satisfy Equations (5–12) and (5–15). Excitations are prescribed through (5–8). For numerical computations one minimizes the number of "degrees" of freedom required to describe $\boldsymbol{H}$ and $\boldsymbol{K}$. This is accomplished by Equations (5–12), (5–15), and (5–22) to eliminate χ and $\delta\chi$ from the above equation to obtain

$$(5\text{–}25) \quad \begin{aligned} 0 = &\int_{S'} \frac{1}{\sigma D} \overline{\operatorname{curl}}(\psi^a - \psi^b) \cdot \overline{\operatorname{curl}}(\delta\psi^a - \delta\psi^b)\, dS \\ &+ \int_{R \cap (\bigcup_j S_j)^c} \operatorname{grad} \delta\psi \cdot \frac{d}{dt}(\mu \operatorname{grad} \psi)\, dV \end{aligned}$$

for all eligible $\delta\psi$ where ψ and $\delta\psi$ are continuous except on $S \cup \bigcup_j S_j$. For each j, $[\psi]_{S_j}$ is a constant which satisfies (5–8) when current excitations are prescribed and $[\delta\psi]_{S_j}$ is a constant which satisfies (5–23).

It remains to show that the proposed formulation does indeed yield Faraday's law. That is, we must show that Equations (5–18) and (5–19) follow from (5–25).

This is most easily accomplished by using (5–11) on S' and the divergence theorem in $R \cap (\bigcup_j S_j)^c$ to integrate both terms in (5–24) by parts, so that

$$(5\text{–}26)\quad 0 = -\int_{S'} \delta\chi \operatorname{curl}\Big(\frac{1}{\sigma D}\,\overline{\operatorname{curl}}\,\chi\Big)\,dS + \oint_{\partial S'} \frac{\delta\chi}{\sigma D}\,\overline{\operatorname{curl}}\,\chi\cdot d\boldsymbol{l}$$
$$-\int_{S'\cup(\bigcup_j S_j)} \Big(\delta\psi^b \frac{d}{dt}(\mu^b \operatorname{grad}\psi^b) - \delta\psi^a \frac{d}{dt}(\mu^a \operatorname{grad}\psi^a)\Big)\cdot \boldsymbol{n}\,dS$$
$$-\int_{R\cap(\bigcup_j S_j^c)} \delta\psi \operatorname{div}\Big(\frac{d}{dt}\mu \operatorname{grad}\psi\Big)\,dV.$$

Equation (5–22) can be used to eliminate $\delta\psi^b$ on S' and $\delta\chi$ on $\partial S'$ (that is, ∂S_j), so that (5–26) is rewritten

$$(5\text{–}27)\quad 0 = \int_{R\cap(\bigcup_j S_j)^c} \delta\psi \Big(\operatorname{div}\frac{d}{dt}(\mu \operatorname{grad}\psi)\Big)\,dV - \int_{S'\cup\bigcup_j S_j} \delta\psi^a \frac{d}{dt}\Big(\mu^a\frac{\partial\psi^a}{\partial \boldsymbol{n}} - \mu^b\frac{\partial\psi^b}{\partial \boldsymbol{n}}\Big)\,dS$$
$$-\int_{S'} \delta\chi\Big(\operatorname{curl}\Big(\frac{1}{\sigma D}\,\overline{\operatorname{curl}}\,\chi\Big) + \frac{d}{dt}\Big(\mu^b\frac{\partial\psi^b}{\partial \boldsymbol{n}}\Big)\Big)\,dS$$
$$+\sum_{j=1}^{\beta_1}[\delta\psi]_{S_j}\Big(\int_{\partial S_j}\frac{1}{\sigma D}\,\overline{\operatorname{curl}}\,\chi\cdot d\boldsymbol{l} + \frac{d}{dt}\int_{S_j}\mu^b\frac{\partial\psi^b}{\partial \boldsymbol{n}}dS\Big).$$

Ignoring issues of functional analysis, it is clear that Equations (5–18) are recovered from (5–27) by the projection method. What is less clear is the sense in which Equation (5–19) is recovered. For nontrivial solutions to exist some currents must be prescribed via (5–8). Equation (5–23) then shows that the independent variations are not the $[\delta\psi]_{S_j}$ but rather the nonzero δI_i. Expressing $[\delta\psi]_{S_j}$ in terms of the δI_i we have

$$[\delta\psi]_{S_j} = \sum_{l=1}^{\beta_1}\delta I_l(\mathcal{I}^{-1})_{lj}.$$

Hence, if I_l is not prescribed then δI_l is nonzero and from (5–27),

$$(5\text{–}28)\qquad \sum_{j=1}^{\beta_1}(\mathcal{I}^{-1})_{lj}\Big(\int_{\partial S_j}\frac{1}{\sigma D}\,\overline{\operatorname{curl}}\,\chi\cdot d\boldsymbol{l} + \frac{d}{dt}\int_{S_j}\mu^b\frac{\partial\psi^b}{\partial \boldsymbol{n}}\,dS\Big) = 0.$$

Equation (5–28) is reinforced by the intuition, from electrical network theory, that Kirchhoff's voltage law, applied to a loop containing a branch with a current source does not yield an independent network equation. A more formal analogy comes from considering the interplay between principal and natural conditions in the calculus of variations. It is clear that prescribing r total currents via (5–8), $0 \le r \le \beta_1$, is analogous to prescribing "principal conditions". Equation (5–28) then gives $\beta_1 - r$ "natural conditions". This makes sense since there is a total of β_1 topological degrees of freedom. Thus we see that a systematic use of algebraic topology shows that Equation (5–19) is a consequence of the projection method only in trivial cases and that the correct expression (5–28) necessarily involves the intersection matrix. The results of this formal approach are quite intuitive only in the case of orientable current carrying surfaces where one can resort to the Poincaré duality theorem or methods of Riemann surface theory [Kle63, Spr57].

Finally, in order to have a unique solution for the magnetic field, one must specify a periodicity constraint or the magnetic field at an initial time. In either case, the topological view remains unchanged.

5E. Weak Form and Discretization

Before proceeding to development of the discretized equations, we review the formulation to this point. For the moment we are assuming that cuts which make the magnetic scalar potential single-valued in multiply connected regions exist and there is an algorithm for computing them. Namely, there is an algorithm to compute a set of compact, orientable surfaces S_j, $1 \leq j \leq \beta_1(R)$ embedded in R which are a basis for homology classes which generate $H_2(R, \partial R; \mathbb{Z})$, the second relative homology group of R modulo ∂R. These surfaces comprise a sufficient set of cuts for the region R so that $\boldsymbol{H}$ can be expressed in terms of a single-valued scalar potential ψ,

$$\boldsymbol{H} = -\operatorname{grad}\psi \quad \text{in } R - \bigcup_{i=1}^{\beta_1(R)} S_i. \tag{5–29}$$

The scalar potential has discontinuities equal to the lumped parameter currents in $\tilde{R}^c$,

$$[\psi]_{S_i} = I_i, \tag{5–30}$$

imposed across each cut S_i corresponding to a lumped-parameter current I_i. Thus, ψ is single-valued on that part of R which is complementary to the cuts.

Furthermore, the boundaries of the cuts, ∂S_j, are generators of homology classes in the group $H_1(\partial\tilde{R} \cup S)$ and form a set of (one-dimensional) cuts for a single-valued stream function

$$\chi = \psi^+ - \psi^- \quad \text{in } S' = (\partial\tilde{R} \cup S) \cap \left(\bigcup_{i=1}^{\beta_1} S_i\right)^c. \tag{5–31}$$

Here ψ^+ and ψ^- are the boundary values of ψ on the + and − sides of the orientable surface S with respect to a normal defined on S. The stream function is single-valued on S' which is that part of the boundary of $\tilde{R}$ and the current sheet(s) which does not include the boundaries of the cuts for the magnetic scalar potential. Thus, cuts for this stream function are a byproduct of the choice of cuts for the magnetic scalar potential and discontinuities in the two functions can be related in a systematic fashion.

If σ is the conductivity of the surface S and D is the local thickness or skin depth of S, then the surface current in S is related to the electric field $\boldsymbol{E}_S$ in the conducting surface by the constitutive law

$$D\sigma\boldsymbol{E}_S = \boldsymbol{K} = -\overline{\operatorname{curl}}\,\chi \quad \text{on } S', \tag{5–32}$$

where, for admissible χ and $\mathbf{F}$, $\overline{\text{curl}}$ is the formal adjoint to the curl operator in the surface S and is defined by

$$\int_S \chi \operatorname{curl} \mathbf{F} \cdot \boldsymbol{n}\, dS + \int_S \mathbf{F} \cdot \overline{\operatorname{curl}}\, \chi\, dS = \oint_{\partial S} \chi \mathbf{F} \cdot dl.$$

For the magnetic field, the constitutive law is taken to be

$$\boldsymbol{B} = \mu(|\boldsymbol{H}|^2)\boldsymbol{H} \quad \text{in } R, \tag{5–33}$$

and this relation is assumed to be nonhysteretic, isotropic, and monotonic. For simplicity of exposition we assume that it suffices to deal with the linear case and use Newton's method otherwise. The formulation is general in the sense that cuts for stream functions can be made on nonorientable surfaces.

Substitution of the above representation of the magnetic fields and currents into the magnetoquasistatic form of Poynting's energy theorem yields a weak Galerkin form which is amenable to direct variational methods. Next we develop a finite element discretization of interelement continuity constraints which incorporates (5–1) and (5–30) and yields a finite element matrix equation which naturally partitions into 36 blocks. We discuss the matrix equation and define each piece of the equation in terms of the finite element assembly procedure.

Formulation. The equations which describe the problem are given by Faraday's law subject to (5–29) and constitutive relations (5–32) and (5–33):

$$\begin{aligned} \operatorname{div} \frac{\partial}{\partial t}(\mu \operatorname{grad} \psi) &= 0 && \text{in } R - \bigcup_{i=1}^{\beta_1(R)} S_i, \\ \frac{d}{dt}\Big(\mu^+ \frac{\partial \psi^+}{\partial \boldsymbol{n}} - \mu^- \frac{\partial \psi^-}{\partial \boldsymbol{n}}\Big) &= 0 && \text{on } S_j \text{ and } S, \\ \operatorname{curl}\Big(\frac{1}{\sigma D}\, \overline{\operatorname{curl}}\, \chi\Big) &= -\mu \frac{\partial}{\partial t}\frac{\partial \psi}{\partial \boldsymbol{n}} && \text{on } S. \end{aligned}$$

This is the starting point for construction of the weak Galerkin form and discretization.

For the region R, the weak form based on Poynting's theorem for magnetoquasistatics describes the magnetic energy in R and power dissipation in the conducting surface S by

$$0 = \int_{\partial \tilde{R} \cup S} \boldsymbol{E}_S \cdot \delta \boldsymbol{K}\, dS + \int_R \delta \boldsymbol{H} \cdot \frac{\partial \boldsymbol{B}}{\partial t}\, dV. \tag{5–34}$$

The approach in [Kot89c] is to take variations $\delta \boldsymbol{H}$ and $\delta \boldsymbol{K}$ subject to Equations (5–29), (5–31), and (5–30), integrate by parts, and show that the resulting equation satisfies Faraday's lawwith the given constitutive relations. Discretization of the problem begins with

$$0 = \frac{\delta P_J}{\delta u} + \frac{d}{dt}\frac{\delta E_M}{\delta u}, \tag{5–35}$$

where P_J is current power dissipation, E_M is magnetic energy, and variations are taken with respect to test functions u for potentials ψ, χ, and currents I.

Discretization and Data Structures. Now we develop the linear equations describing the system above on a finite element mesh $\tilde{K}$ with boundary $\partial\tilde{K}$ which is a discretization of $\tilde{R}$. In particular, the finite element connection matrix is defined as follows. Consider an n-dimensional simplicial finite element mesh with m_n n-simplices and m_0 vertices. There is a total ordering of the vertices on $\{1, \ldots, m_0\}$ called the global node ordering. The vertices of an n-simplex are locally ordered on $\{1, \ldots, n+1\}$. The connection or assembly matrix is the following $m_n \times m_0$ matrix, originally defined in Equation (4–6):

$$C^i_{n,jk} = \begin{cases} 1 & \text{if global node } k \text{ is the } j\text{th local node of the } i\text{th } n\text{-simplex,} \\ 0 & \text{otherwise.} \end{cases}$$

The definition is also valid for lower dimensional submeshes or p-skeleta $(0 \le p < n)$ of the n-dimensional mesh and subsets of these. This makes it possible to use the same terminology to refer to the discretizations of $\tilde{K}$, S, S_i, and ∂S_i where the "meshes" of the surfaces are embedded in $\tilde{K}$. In particular, 3-, 2-, and 1-simplices or "tetrahedra", "triangles", and "edges", are used throughout. In practice, the connection matrices of p-dimensional submeshes, $p < n$, are stored as $(p+1)$–p-incidence matrices so that all the data is referenced to C_3 and information of the embedding of the submesh in C_3 is not lost.

Since there are discontinuities in ψ and χ across their respective cuts, the standard finite element connection process must be modified to handle the discontinuities. This is achieved via *global* and *local* variables on vertices or nodes of the mesh as follows:

DEFINITION 1. A *nodal variable* u_c is taken to be a global variable when it is indexed on the global indices $\{1, \ldots, m_0\}$. A *local variable* u_d is a nodal variable on an n-simplex, indexed on the local indices $1 \le j \le n+1$ such that

$$u^i_{d,j} = \sum_{k=1}^{m_0} C^i_{p,jk} u^i_{c,k}.$$

One may say that a local variable is assembled from the connection matrix and corresponding global variable(s) through a relation taking the form of the expression above.When $p = 3$, this expression corresponds to the typical finite element assembly process for a 3-dimensional mesh, but now the process will be altered to accommodate cuts and coupling to the stream function.

Discontinuities in ψ and χ due to cuts are handled by writing each local potential as a sum of the corresponding global potential, contributions from cuts, and coupling to another potential. For ψ_d this means discontinuities in ψ across cuts from (5–30) and a contribution due to eddy currents on S which comes through χ in (5–31):

$$\psi_d = C^E_\psi \psi_c + C^E_\chi \chi_c + C^E_I I, \tag{5–36}$$

where I is a vector of lumped parameter currents and each C^E denotes, for its corresponding variable, the connection matrix used for the magnetic energy term of Poynting's theorem in R. In particular:

C^E_ψ is the connection matrix for $\tilde{K}$;

C_χ^E is the connection matrix for conducting surface S, taken as a submesh of $\tilde{K}$;
C_I^E is the connection matrix for the cuts, taken as a submesh of $\tilde{K}$.

The numbers of nonzero entries in C_χ^E and C_I^E are on the order of the number of nodes on the surfaces S_j and S embedded in K, the number of nonzeros in C_ψ^E is $4m_3$.

The connection matrices are sparse, so they are never stored as defined. In practice, only the nonzero entries are stored in arrays of pointers. For C_ψ^E, this can mean using a $4 \times m_3$ array where the ith row contains the global node numbers of 4 vertices of the ith tetrahedron. C_χ^E and C_I^E can be stored as 4-column arrays with rows referenced to rows in C_χ^E and the jth column entry indicating if the jth face of the ith tetrahedron comprises a face of the surface in question. By referencing the data to $\tilde{K}$, the embedding of the surface in $\tilde{K}$ is implicit.

The assembly of χ_d takes into account discontinuities in χ across cuts S:

$$\chi_d = C_\chi^P \chi_c + C_I^P I, \tag{5–37}$$

where

C_χ^P is the connection matrix for surfaces $\partial\tilde{K} \cup S$ (same as C_χ^E);
C_I^P is the connection matrix for the (1-d) cuts for χ on $\partial\tilde{K} \cup S$.

The previous description regarding data structures is relevant here as well.

The Finite Element Equations. We will build up the finite element equations for the eddy current problem in two stages. In order to tie this section more closely to material in Chapter 4 we first look at a formulation which incorporates cuts into the computation for a magnetic scalar potential. This basic technique is subsequently used in the approach which uses cuts for the full eddy currents computation.

Given cuts, the computation of magnetic scalar potentials in multiply connected regions can be viewed as follows. For a real-valued potential ψ there is a functional

$$F(\psi) = \frac{1}{2}\int_R \mu \operatorname{grad}\psi \cdot \operatorname{grad}\psi \, dV. \tag{5–38}$$

Given that cuts are orientable surfaces [Kot87] and thus have a well-defined surface normal, let $\psi^+(\psi^-)$ be the value of the potential on the positive side of a cut, denoted by S_i^+ and the negative side, denoted by S_i^-, both with respect to a normal defined on the cut surface. Let ψ_B denote the potential on the boundary of the region. Then (5–38) is subject to

$$\psi^+ - \psi^- = I_i \quad \text{on } S_i, \qquad \psi = \psi_B \quad \text{given on } \partial R. \tag{5–39}$$

Here the S_i, for $1 \le i \le \beta_1(R)$, generate $H_2(R, \partial R)$, each S_i is built out of faces of tetrahedra which comprise the finite element mesh, and the jumps I_i are determined by Ampère's Law. Using barycentric coordinates $\{\lambda_i\}$, $1 \le i \le 4$, and interpolating ψ linearly on the vertices of the kth tetrahedron,

$$\psi^k = \sum_{n=1}^{4} \lambda_n \psi_n^k,$$

where ψ_n^k represents the discretization of ψ. Then the assembled finite element problem is

$$F(\psi) = \frac{1}{2} \sum_{l=1}^{m_3} \sum_{m=1}^{4} \sum_{n=1}^{4} \psi_m^l \mathcal{K}_{mn}^l \psi_n^l, \tag{5–40}$$

where the element stiffness matrix is now weighted by the element permeability μ. Define $\beta_1(R)$ indicator functions $\{\mathcal{I}_p\}$, $1 \leq p \leq \beta_1(R)$, as follows:

$$\mathcal{I}_p(m,l) = \begin{cases} 1 & \text{if } m\text{th node of } l\text{th tetrahedron} \in S_p^+, \\ 0 & \text{otherwise.} \end{cases}$$

Connecting the solution across tetrahedra and allowing for jumps in the scalar potential across cuts,

$$\psi_n^l = \sum_{i=1}^{m_0} C_{ni}^l u_i + \sum_{p=1}^{\beta_1(R)} I_p \mathcal{I}_p(n,l). \tag{5–41}$$

In (5–41) interelement jumps in the potential occur across cuts. Putting (5–41) into (5–40), we get

(5–42) $F(u_i, I_k) =$

$$\frac{1}{2} \sum_{l=1}^{m_3} \sum_{m=1}^{4} \sum_{n=1}^{4} \left(\sum_{i=1}^{m_0} C_{mi}^l u_i + \sum_{p=1}^{\beta_1} I_p \mathcal{I}_p(m,l) \right) \mathcal{K}_{mn}^l \left(\sum_{j=1}^{m_0} C_{nj}^l u_j + \sum_{q=1}^{\beta_1} I_q \mathcal{I}_q(n,l) \right).$$

Now write the global finite element stiffness matrix as

$$\mathcal{K}_{ij} = \left(\sum_{l=1}^{m_3} \sum_{m=1}^{4} \sum_{n=1}^{4} C_{nj}^l \mathcal{K}_{mn}^l C_{mi}^l \right).$$

Then separating constant, linear, and quadratic terms in u_i, and using the expression for the stiffness matrix, (5–42) is re-expressed as

$$F(u_i, I_k) = \frac{1}{2} \sum_{i=1}^{m_0} \sum_{j=1}^{m_0} u_i u_j \mathcal{K}_{ij} + \sum_{i=1}^{m_0} u_i \left(\sum_{p=1}^{\beta_1} I_p f_{pi} \right) + \text{ constants,}$$

where

$$f_{pi} = \sum_{l=1}^{m_3} \sum_{m=1}^{4} \sum_{n=1}^{4} C_{mi}^l \mathcal{K}_{mn}^l \mathcal{I}_p(n,l).$$

The discretized Euler–Lagrange equation for the functional (5–38) subject to (5–39) is

$$\mathcal{K} u = -\sum_{p=1}^{\beta_1} I_p f_p,$$

where the elements of $\mathcal{K}_{ij}$ are now weighted by μ, and f_p is the vector with entries $\{f_{pi}\}$.

The preceding formulation outlines how cuts are incorporated into the formulation of the finite element equations. We now proceed with the eddy current problem and analyze in more detail block partitions of the matrices which come into this problem. With expressions (5–36) and (5–37) in hand, we use the Poynting form (5–34) to get the finite element equations. The two terms of the discretized Poynting form are magnetic energy and eddy current power dissipation:

$$E_M = \tfrac{1}{2}\psi_d^T S_d^E \psi_d, \tag{5–43}$$

$$P_J = \chi_d^T S_d^P \chi_d, \tag{5–44}$$

where S_d^E and S_d^P are "unassembled finite element stiffness matrices" (see Chapter 4) for $\tilde{K}$ and $\partial\tilde{K} \cup S$, respectively. The magnetic permeability μ for each element is included in S_d^E and S_d^P includes the element permittivity. Putting the expressions for ψ_d and χ_d into E_M and P_J assembles the stiffness matrices, so that

$$E_M = \tfrac{1}{2}(\psi_c \;\; \chi_c \;\; I)^T\, S_c^E\, (\psi_c \;\; \chi_c \;\; I) \tag{5–45}$$

and

$$P_J = (\chi_c \;\; I)^T\, S_c^P\, (\chi_c\, I), \tag{5–46}$$

where the stiffness matrices are

$$S_c^E = (C_\psi^E \;\; C_\chi^E \;\; C_I^E)^T\, S_d^E\, (C_\psi^E \;\; C_\chi^E \;\; C_I^E)$$

and

$$S_c^P = (C_\chi^P \;\; C_I^P)^T\, S_d^P\, (C_\chi^P \;\; C_I^P).$$

By this formulation, S_c^E and S_c^P partition into blocks as follows:

$$S_c^E = \begin{pmatrix} E_{\psi\psi} & E_{\psi\chi} & E_{\psi I} \\ E_{\chi\psi} & E_{\chi\chi} & E_{\chi I} \\ E_{I\psi} & E_{I\chi} & E_{II} \end{pmatrix}, \tag{5–47}$$

$$S_c^P = \begin{pmatrix} P_{\chi\chi} & P_{\chi I} \\ P_{I\chi} & P_{II} \end{pmatrix}.$$

The variables ψ, χ, and I are subject to boundary conditions or are prescribed, so that the connected variables partition into free and prescribed components:

$$\psi_c = \begin{pmatrix} \psi_f \\ \psi_p \end{pmatrix}, \qquad \chi_c = \begin{pmatrix} \chi_f \\ \chi_p \end{pmatrix}, \qquad I = \begin{pmatrix} I_f \\ I_p \end{pmatrix}. \tag{5–48}$$

Thus there is a further partition of the stiffness matrices where each block in (5–47) divides into 2×2 subblocks:

(5–49)
$$S_c^E = \begin{pmatrix} E_{\psi_f\psi_f} & E_{\psi_f\psi_p} & E_{\psi_f\chi_f} & E_{\psi_f\chi_p} & E_{\psi_f I_f} & E_{\psi_f I_p} \\ E_{\psi_p\psi_f} & E_{\psi_p\psi_p} & E_{\psi_p\chi_f} & E_{\psi_p\chi_p} & E_{\psi_p I_f} & E_{\psi_p I_p} \\ E_{\chi_f\psi_f} & E_{\chi_f\psi_p} & E_{\chi_f\chi_f} & E_{\chi_f\chi_p} & E_{\chi_f I_f} & E_{\chi_f I_p} \\ E_{\chi_p\psi_f} & E_{\chi_p\psi_p} & E_{\chi_p\chi_f} & E_{\chi_p\chi_p} & E_{\chi_p I_f} & E_{\chi_p I_p} \\ E_{I_f\psi_f} & E_{I_f\psi_p} & E_{I_f\chi_f} & E_{I_f\chi_p} & E_{I_f I_f} & E_{I_f I_p} \\ E_{I_p\psi_f} & E_{I_p\psi_p} & E_{I_p\chi_f} & E_{I_p\chi_p} & E_{I_p I_f} & E_{I_p I_p} \end{pmatrix},$$

(5–50)
$$S_c^P = \begin{pmatrix} P_{\chi_f\chi_f} & P_{\chi_f\chi_p} & P_{\chi_f I_f} & P_{\chi_f I_p} \\ P_{\chi_p\chi_f} & P_{\chi_p\chi_p} & P_{\chi_p I_f} & P_{\chi_p I_p} \\ P_{I_f\chi_f} & P_{I_f\chi_p} & P_{I_f I_f} & P_{I_f I_p} \\ P_{I_p\chi_f} & P_{I_p\chi_p} & P_{I_p I_f} & P_{I_p I_p} \end{pmatrix}.$$

Since variations in the connected variables of (5–48) are only in the free portions, variations in the prescribed variables are zero, that is, $\delta u_p = 0$. The discretized equations for the problem come out of (5–35) which is rewritten below in terms of the free variables:

$$0 = \frac{d}{dt}\left(\frac{\delta E_M}{\delta \psi_f} + \frac{\delta E_M}{\delta \chi_f} + \frac{\delta E_M}{\delta I_f}\right) + \frac{\delta P_J}{\delta \chi_f} + \frac{\delta P_J}{\delta I_f}.$$

Putting (5–45), (5–46), (5–49), and (5–50) into the equation above, taking variations in the free variables, and separating free and prescribed variables gives a matrix equation of the form

$$(E_{ff}\ \ 2P_{ff})\begin{pmatrix} du_f/dt \\ u_f \end{pmatrix} = -(E_{fp}\ \ 2P_{fp})\begin{pmatrix} du_p/dt \\ u_p \end{pmatrix}$$

where

$$u_x = \begin{pmatrix} \psi_x \\ \chi_x \\ I_x \end{pmatrix}.$$

In particular, the finite element matrix equation is

$$\begin{pmatrix} E_{\psi_f\psi_f} & E_{\psi_f\chi_f} & E_{\psi_f I_f} \\ E_{\chi_f\psi_f} & E_{\chi_f\chi_f} & E_{\chi_f I_f} \\ E_{I_f\psi_f} & E_{I_f\chi_f} & E_{I_f I_f} \end{pmatrix}\begin{pmatrix} d\psi_f/dt \\ d\chi_f/dt \\ dI_f/dt \end{pmatrix} + 2\begin{pmatrix} P_{\chi_f\chi_f} & P_{\chi_f I_f} \\ P_{I_f\chi_f} & P_{I_f I_f} \end{pmatrix}\begin{pmatrix} \chi_f \\ I_f \end{pmatrix}$$
$$= -\begin{pmatrix} E_{\psi_f\psi_p} & E_{\psi_f\chi_p} & E_{\psi_f I_p} \\ E_{\chi_f\psi_p} & E_{\chi_f\chi_p} & E_{\chi_f I_p} \\ E_{I_f\psi_p} & E_{I_f\chi_p} & E_{I_f I_p} \end{pmatrix}\begin{pmatrix} d\psi_p/dt \\ d\chi_p/dt \\ dI_p/dt \end{pmatrix} - 2\begin{pmatrix} P_{\chi_f\chi_p} & P_{\chi_f I_p} \\ P_{I_f\chi_p} & P_{I_f I_p} \end{pmatrix}\begin{pmatrix} \chi_p \\ I_p \end{pmatrix}.$$

In this form, the equation can be handled by the combination of a time integration algorithm and a matrix equation solver. For nonlinear magnetic constitutive relations, Newton's method can be used at each time step of the time integration algorithm.

In summary, once cuts are computed a compact, orientable, embedded manifolds in the nonconducting region, a magnetic scalar potential is well-defined (single-valued) on the current-free region. Furthermore, the boundaries of the cuts are a set of cuts for a stream function describing eddy currents on current-carrying sheets or on the boundary of the nonconducting region. Here we have used these facts and a weak Galerkin form based on Poynting's theorem to give the discretized linear equations for computing fields and currents entirely by scalar potentials. The required input data are the finite element connection matrices of the mesh of the region with the current-carrying sheet(s), the submesh representing the sheet itself, the submeshes which represent cuts, and coordinates of mesh nodes. When these are specified as surfaces embedded in the region, the assembly process for the finite element equations is essentially the same as the assembly for Laplace's equation on the mesh, subject to the modifications required and described here.

Topology has the peculiarity that questions belonging to its domain may under certain circumstances be decidable even though the continua to which they are addressed may not be given exactly, but only vaguely, as is always the case in reality.

H. Weyl, *Philosophy of Mathematics and Natural Science*, 1949

6

An Algorithm to Make Cuts for Magnetic Scalar Potentials

6A. Introduction and Outline

In this chapter we consider a general finite element-based algorithm to make cuts for magnetic scalar potentials and investigate how the topological complexity of the three-dimensional region, which constitutes the domain of computation, affects the computational complexity of the algorithm. The algorithm is based on standard finite element theory with an added computation required to deal with topological constraints brought on by a scalar potential in a multiply connected region. The process of assembling the finite element matrices is also modified in the sense described at length in the previous chapter.

Regardless of the topology of the region, an algorithm can be implemented with $\mathcal{O}(m_0^3)$ time complexity and $\mathcal{O}(m_0^2)$ storage where m_0 denotes the number of vertices in the finite element discretization. However, in practice this is not useful since for large meshes the cost of finding cuts would become the dominant factor in the magnetic field computation. In order to make cuts worthwhile for problems such as nonlinear or time-varying magnetostatics, or in cases of complicated topology such as braided, knotted, or linked conductor configurations, an implementation of $\mathcal{O}(m_0^2)$ time complexity and $\mathcal{O}(m_0)$ storage is regarded as ideal. The obstruction to ideal complexity is related to the structure of the fundamental group This chapter describes an algorithm that can be implemented with $\mathcal{O}(m_0^2)$ time complexity and $\mathcal{O}(m_0^{4/3})$ storage complexity given no more topological data than that contained in the finite element connection matrix.

Electromagnetic Context and Numerical Motivation. The proper context for the algorithm of this chapter is in variational principles, the finite element method, and their connection to the topology of the domain of computation. Before seeing how topology enters the picture when considering a scalar potential, recall that magnetoquasisatics refers to the class of electromagnetics problems where the magnetic field is described by the following limiting case of Maxwell's equations:

$$\int_{\partial V} \boldsymbol{B} \cdot d\boldsymbol{S} = 0,$$

$$\int_{\partial S} \boldsymbol{H} \cdot dl = \int_{S} \boldsymbol{J} \cdot \boldsymbol{n} \, ds,$$

$$\int_{\partial S'} \boldsymbol{E} \cdot dl = -\frac{d}{dt} \int_{S'} \boldsymbol{B} \cdot \boldsymbol{n} \, ds.$$

Here ∂V refers to the boundary of a region V and ∂S is the boundary of a surface S. The displacement current $\partial \boldsymbol{D}/\partial t$ in Ampère's law is neglected and the current density vector $\boldsymbol{J}$ is assumed to be solenoidal. The magnetic flux density vector $\boldsymbol{B}$ is related to the magnetic field intensity $\boldsymbol{H}$ by $\boldsymbol{B} = \boldsymbol{B}(\boldsymbol{H})$, or for linear isotropic media, $\boldsymbol{B} = \mu \boldsymbol{H}$. The current density $\boldsymbol{J}$ in conductors is related to the electric field intensity vector $\boldsymbol{E}$ by Ohm's law: $\boldsymbol{J} = \sigma \boldsymbol{E}$. Let R denote a region which is the intersection of the region where it is desired to compute the magnetic field and where the current density $\boldsymbol{J}$ is zero. In R,

$$\operatorname{curl} \boldsymbol{H} = 0,$$

so that in terms of a scalar potential ψ,

$$\boldsymbol{H} = \operatorname{grad} \psi \tag{6–1}$$

in any contractible subset of R but in general ψ is globally multivalued as seen via Ampère's Law. This was illustrated in Figure 1.8 for a current-carrying trefoil knot c with current I and cut S. In that case, if $\boldsymbol{H} = \operatorname{grad} \psi$, Ampère's Law implies that a scalar potential ψ is multivalued as illustrated with loop c_1 where $\oint_{c_1} \boldsymbol{H} \cdot dl = I$ implies that ψ is multivalued when there is no cut. On the other hand, $\oint_{c_2} \boldsymbol{H} \cdot dl = I - I = 0$, even though c_2 is not contractible to a point implying that c_2 gives no information about ψ. With the cut in place as shown in the figure and a discontinuity I imposed on ψ from one side of the cut to the other, ψ is made to be single-valued on the cut complement. Note that Ampère's law does not require that c_2 intersect the cut.

It is common practice to sidestep the multivalued scalar potential by expressing the magnetic field as $\boldsymbol{H} = \boldsymbol{H}_p + \operatorname{grad} \tilde{\psi}$, where $\boldsymbol{H}_p$ is a particular solution for the field obtained, say, from the Biot–Savart Law. However, in this case, if $\boldsymbol{B} = \mu \boldsymbol{H}$ and $\mu \to \infty$, then $\boldsymbol{H} \to 0$ so that $\boldsymbol{H}_p = -\operatorname{grad} \tilde{\psi}$, leading to significant cancellation error for computation in regions where $\boldsymbol{H} \simeq 0$ while a "total" scalar potential as in (6–1) does not suffer from cancellation error. In addition, integration of the Biot–Savart integral destroys the sparsity of any discretization. The particular solution, $\boldsymbol{H}_p$, can be set to zero in a multiply connected region when the notion of a cut which makes the scalar potential single-valued is introduce. In practice use of cuts is viable if the software to generate cuts does not require input from the user. These reasons are an incentive to developing an algorithm for automatic computation of cuts.

Outline. With the preceding motivation in mind, we outline the main sections of this chapter. Section 6B introduces the essential pieces of (co)homology theory and differential forms required for defining the notion of cuts and for finding an algorithm to compute them. Section 6C presents a variational formulation which can be used in the context of the finite element method Section 6E fills in the piece missing from Section 6D, describing an algorithm for finding the topological constraints on the variational problem, and gives an analysis of the computational complexity of finding cuts. The overall process of computing cuts is summarized in algorithm 6.1 and the algorithm of Section 6E is summarized in algorithm 6.2. Two example problems are considered in order to illustrate the obstruction to $\mathcal{O}(n)$ complexity. Section 6F concludes with a summary of the main results, a review of geometric insights used to reduce the complexity of the algorithm, and suggestions for future work.

6B. Topological and Variational Context

Preceding chapters have generously set the context for the algebraic structures and duality theorems needed in order to establish a general cuts algorithm. However, the following points regarding the relevance of these tools add further motivation to the purpose of this chapter. One advantage of formulating cuts in terms of cohomology groups is that when a constructive proof of the existence of cuts is phrased in terms of the formalism of a certain homology theory, the proof gives way to a variational formulation for a cuts algorithm. The proof is sketched here while some more details are given in Section MA-I. Another advantage is that various homology theories give several ways to view cuts. In particular, when a finite element mesh is viewed as a simplicial complex as in Chapter 4, simplicial homology theory provides a framework for implementing an algorithm to make cuts and determines appropriate data structures. Finally, since the homology groups can be computed with coefficients in $\mathbb{Z}$ an implementation of the algorithm uses only integer arithmetic, thus avoiding introduction of rounding errors associated with real arithmetic. This implies that rounding error analysis is required only for the well-understood parts of the algorithm which depend on standard finite element theory.

Before considering the algorithm, we summarize the relevant groups from earlier chapters. Let R be the nonconducting region with boundary ∂R. The complement of R relative to $\mathbb{R}^3$, denoted by R^c, is the union of the problem "exterior" and the conducting region. Recall that Ampère's law is a statement about closed loops in R which link nonzero current. In terms of homology, $H_1(R;\mathbb{Z})$ can be viewed as the group of equivalence classes of closed loops in R which link closed paths in R^c which may be current paths. Two closed loops in R lie in the same equivalence class if together they comprise the boundary of a surface in R. As noted, the $\mathbb{Z}$ in $H_1(R;\mathbb{Z})$ expresses the fact that in this case there is no loss of information if one builds H_1 by taking only integer linear combinations of closed loops. As discussed in Section 3D, the homology groups for R are torsion-free so that integer coefficients are sufficient. The rank of $H_1(R)$, denoted by $\beta_1(R)$, is a characteristic parameter of R, significant because it describes the number of independent closed loops in R which link nonzero current. As such it will characterize the number of variational problems to be solved in the cuts algorithm.

The first cohomology group of R, denoted by $H^1(R;\mathbb{Z})$ can be regarded as the group of curl-free vector fields in R modulo vector fields which are the gradient of some function. As in the case of H_1, it is enough to take "forms with integer periods" meaning that integrating the field about a generator of H_1 gives an integer. The rank of $H^1(R;\mathbb{Z})$ is also $\beta_1(R)$. In three dimensions, $H_1(R)$ and $H^1(R)$ formalize Ampère's law in the sense that, respectively, they are algebraic structures describing linking of current and irrotational fields in R due to currents in R^c.

The second relative homology group $H_2(R,\partial R;\mathbb{Z})$, is the group of equivalence classes of surfaces in R with boundary in ∂R. Classes in $H_2(R,\partial R;\mathbb{Z})$ are surfaces with boundary in ∂R but which are not themselves boundaries of a volume in R. Its rank is the second Betti number, $\beta_2(R,\partial R)$. $H_2(R,\partial R;\mathbb{Z})$ is the quotient space of surfaces which are 2-cycles up to boundary in ∂R modulo surfaces which are boundaries (Figure 6.1). In three dimensions these turn out to be surfaces used for flux linkage calculations. This will be shown precisely, but we need to start with a definition to get to the algorithm.

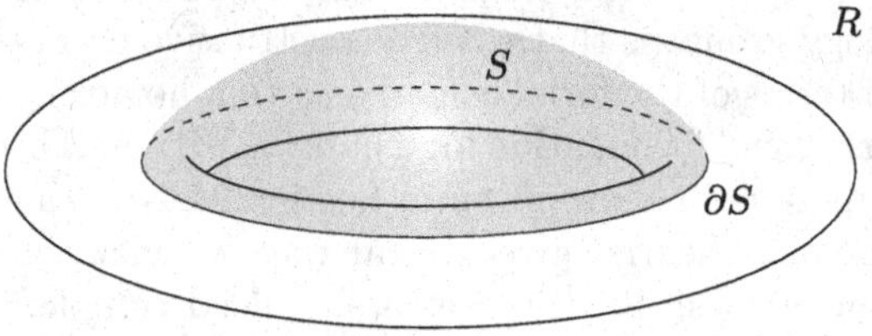

Figure 6.1. S is a surface in R with $\partial S \in \partial R$, but S is not the boundary of a volume in R. Also see Figure 1.8.

The reader should note that an essential requirement for a cut is that it must be a barrier to every loop which links a current in the sense of Ampère's law. For this to occur, the boundaries of the cuts must be on the boundary ∂R of the region in question. Thus we expect to have at least $\beta_1(R)$ cuts, one cut corresponding to each independent current. In fact there are $\beta_1(R)$ families of cuts where each family is an equivalence class of cuts associated with each current. This is a geometric and intuitive way of understanding what cuts are, but does not establish their existence or give a way of computing them. The duality theorems considered in Chapter 3 and the isomorphism described below provide the formalism required for showing existence and constructing an algorithm.

The Isomorphism $H^1(R;\mathbb{Z}) \simeq [R, S^1]$. The isomorphism discussed here allows us to restrict our attention to cuts which are embedded manifolds in R, and at the same time gives a way of computing these cuts through a variational problem.

The isomorphism establishes a relationship between $H^1(R;\mathbb{Z})$ and maps from R to the circle S^1, where S^1 is regarded as the unit circle in the complex plane. Let $[R, S^1]$ denote the space of maps $f : R \to S^1$ up to the equivalence relation given by homotopy. It is the case that

$$H^1(R;\mathbb{Z}) \simeq [R, S^1], \tag{6–2}$$

which says that the group of cohomology classes of R with integer periods is isomorphic to the group of homotopy classes of maps from R to S^1 [Kot87]. The reason for introducing (6–2) is twofold: First, there is no guarantee that homology classes can be represented by manifolds (surfaces). As discussed below, Equation (6–2) provides this guarantee. Second, since it associates each class in H^1 to a class of maps from R to S^1, a suitable choice of energy functional on $[R, S^1]$ will give a variational problem and an algorithm for computing cuts. Hence, this isomorphism and Poincaré–Lefschetz duality give way to an algorithm for cuts.

Concretely, choosing $\mu = d\theta/2\pi$, a closed, nonexact 1-form on S^1, then $f^*(\mu)$ is the "pullback" of μ via f so that one can regard $f^*(\mu)$ as monitoring the change in θ as one covers a family of cuts in R. Poincaré–Lefschetz duality and the preimage theorem [GP74] say that for a regular point p of f on S^1,

$$\int_R \omega \wedge f^*\left(\frac{d\theta}{2\pi}\right) = \int_{f^{-1}(p)} \omega \tag{6–3}$$

for any closed 2-form ω. Thus, given $f : R \to S^1$ where f winds around S^1, the pullback $f^*(\mu)$ is the Poincaré–Lefschetz dual to $f^{-1}(p)$ [Kot87].

In terms of vector calculus, given a map $f : R \to S^1$, Equation (6–3) can be re-written as

$$\frac{1}{2\pi i}\int_R G \cdot \operatorname{grad}(\ln f)\, dV = \int_{f^{-1}(p)} G \cdot \boldsymbol{n}\, dS \tag{6–4}$$

for any $G \in \mathcal{G}$. When $G = \mu \boldsymbol{H}$ and $\boldsymbol{H} = 1/(2\pi i)\ \operatorname{grad}(\ln f)$, each side of Equation (6–4) can be regarded as an expression for the energy of a system of (unit) currents in $\mathbb{R}^3 - R$. In particular, note that the right hand side is the integral over a cut of the magnetic flux due to a unit current.

Variational Aspects of an Algorithm for Cuts. This section considers a variational formulation for a cuts algorithm, the associated the Euler–Lagrange equation, and discusses the exact solutions of this variational problem while hinting at its topological flavor. Though not necessary for continuity to following sections, we employ the framework described above to find an explicit solution to the nonconvex variational problem and verify that the resulting function is single-valued.

The cuts algorithm consists of finding a solution to the variational problem of minimizing

$$F(f) = \int_V \operatorname{grad} \bar{f} \cdot \operatorname{grad} f\, dV, \tag{6–5}$$

subject to

$$\bar{f}f = 1 \quad \text{in } V, \tag{6–6}$$

and for $1 \le j \le \beta_1(V)$, the jth cut requires

$$\frac{1}{2\pi i}\oint_{c_k} \operatorname{grad}(\ln f) \cdot dl = \delta_{jk} \tag{6–7}$$

for $1 \le k \le \beta_1(V)$. Here f is a map from V to $\mathbb{C}$, and the c_l, for $1 \le l \le \beta_1(V)$, are curves representing a basis for $H_1(V;\mathbb{Z})$. Equation (6–6) shows that the solution to the above problem defines a map to the unit circle in the complex plane

$$f : V \longrightarrow S^1. \tag{6–8}$$

Taking the inverse image of a regular value on S^1 (i.e., a point p on S^1 such that the gradient of f is nonzero at every point in the preimage), we end up with a surface whose boundary lies in ∂V. This "cut surface" represents a relative homology class in $H_2(V, \partial V; \mathbb{Z})$ which is dual to the constraint represented by (6–7). This comes about because S^1 is an Eilenberg–MacLane space $K(\mathbb{Z}, 1)$ as discussed in detail in Section MA-I.

Our immediate objective is to make a distinction between how the variational problem (6–5)–(6–7) is handled numerically and analytically. First we note that one can handle the constraint set forth in (6–6) by letting

$$f = e^{2\pi i\psi} \tag{6–9}$$

where ψ is a real differentiable function which, by (6–7), must be multivalued. Substituting (6–9) into (6–5) gives

$$F(e^{2\pi i\psi}) = 4\pi^2 \int_V \operatorname{grad}\psi \cdot \operatorname{grad}\psi \, dV. \tag{6–10}$$

The starting point for an algorithm is the observation that the Euler–Lagrange equation corresponding to (6–10) is just Laplace's equation. Hence, in principle, an algorithm to find cuts is easily implemented once one can modify existing finite element code for solving Laplace's equation. Two subtleties which must be addressed are, first, interelement continuity conditions must be modified in order to respect (6–7) and second, from Equation (6–8), the inverse image of f can be obtained by considering the equipotentials of ψ modulo integers. Addressing these two subtleties, an algorithm to find cuts in any region can be implemented provided a "reasonable" finite element mesh exists, that is, a mesh on which Laplace's equation can be solved.

For a deeper understanding of situations where a complete set of cuts $\{S_i\}$, $1 \le i \le \beta_1(V)$, enable one to use a single-valued scalar potential in

$$\widetilde{V} = V - \bigcup_{i=1}^{\beta_1(V)} S_i,$$

but $\widetilde{V}$ is not simply connected, we need a better understanding of the solution of the variational problem (6–5)–(6–7). To this end, we will now handle the constraint (6–6) by a Lagrange multiplier which can be eliminated to obtain a "harmonic map equation" for f. When confronted with this nonlinear partial differential equation we will use what we have established regarding magnetic scalar potentials, the Biot–Savart law, and linking numbers, to produce an explicit solution.

If we append to the functional (6–5) a Lagrange multiplier term corresponding to the constraint (6–6), we end up with a variational problem for the functional

$$\widetilde{F}(f,\lambda) = \int_V \operatorname{grad} \bar{f} \cdot \operatorname{grad} f + \lambda(\bar{f} f - 1)\, dV, \tag{6–11}$$

subject to the constraint (6–7). When the first variation of this functional with respect to f is set to zero we obtain the weak Galerkin form

$$0 = \int_V \big((\operatorname{grad} \delta f) \cdot (\operatorname{grad} \bar{f}) + (\operatorname{grad} \delta \bar{f}) \cdot (\operatorname{grad} f) + \lambda(f\, \delta \bar{f} + \bar{f}\, \delta f)\big)\, dV.$$

If we eliminate the derivatives of the variation of f through integration by parts, we obtain

$$0 = \int_V \big(\delta f(-\operatorname{div}\operatorname{grad} \bar{f} + \bar{f}) + \delta \bar{f}(-\operatorname{div}\operatorname{grad} f + \lambda f)\big)\, dV + \int_{\partial V} \left(\delta f\, \frac{\partial \bar{f}}{\partial n} + \delta \bar{f}\, \frac{\partial f}{\partial n}\right) dS.$$

Writing $f = f_r + i f_i$ where f_r and f_i are real functions that can be varied independently, one finds that the vanishing of the above expression for all admissible δf implies

$$\begin{aligned} \operatorname{div}\operatorname{grad} f &= \lambda f \quad && \text{in } V,\\ \frac{\partial f}{\partial n} &= 0 && \text{on } \partial V. \end{aligned} \tag{6–12}$$

When the variation of the functional (6–11) with respect to λ is set to zero, we recover the constraint (6–6). We begin to eliminate the Lagrange multiplier from Equation (6–12) by first taking the Laplacian of the constraint (6–6) to obtain

$$0 = \operatorname{div}\operatorname{grad}(\bar{f} f) = (\operatorname{div}\operatorname{grad} \bar{f}) f + 2 \operatorname{grad} \bar{f} \cdot \operatorname{grad} f + \bar{f} \operatorname{div}\operatorname{grad} f$$

or

$$\Re(\bar{f}\ \operatorname{div}\operatorname{grad} f) = -|\operatorname{grad} f|^2, \tag{6–13}$$

where $\Re(\cdot)$ denotes the real part of $(\cdot)$. Multiplying Equation (6–12) by $\bar{f}$ and using Equations (6–6) and (6–13) we can solve for λ:

$$-|\operatorname{grad} f|^2 = \Re(\bar{f}\ \operatorname{div}\operatorname{grad} f) = \lambda \bar{f} f = \lambda$$

, and rephrase (6–12) as

$$\begin{aligned} \operatorname{div}\operatorname{grad} f &= -|\operatorname{grad} f|^2 f \quad && \text{in } V,\\ \frac{\partial f}{\partial n} &= 0 && \text{on } \partial V. \end{aligned} \tag{6–14}$$

Equations (6–14) and (6–7) provide a set of equations for the single-valued function defined in the discussion leading to Equation (6–8). At first sight, the solution to these equations in the exterior of current carrying wires is not obvious. If we perform the substitution given by Equation (6–9) then (6–14) reduces to a boundary value problem involving Laplace's equation and a multivalued function. In the algorithm for computing cuts it was necessary to deal with this multivaluedness in the context of interelement constraints — appealing to physical intuition would have lead to a circular argument where the magnetic scalar potential would be needed to define the cuts for the magnetic scalar potential! In the present case we want to develop our intuition about cuts and seek explicit expressions for the cuts. Hence we are free to use the equipotentials of the multivalued scalar potential tied to any easy-to-use constitutive law as equivalent cuts.

In order to find the function f satisfying (6–6) and (6–14) we first find a set of integer-valued currents in the exterior of V which insure that the corresponding scalar potential satisfies (6–7). Suppose this is accomplished by imposing $\beta_1(V)$ integer valued currents $\{n_i\}$ on $\beta_1(V)$ closed curves $\{c'_i\}$ in the exterior of V. The scalar potential is given by the line integral of the magnetic field found through the use of the Biot–Savart Law:

$$\psi(p) - \psi(p_0) = \frac{1}{4\pi} \int_{p_0}^{p} \sum_{i=1}^{\beta_1(V)} n_i \oint_{c'_i} \frac{[(\boldsymbol{r} - \boldsymbol{r}') \times d\boldsymbol{r}] \cdot d\boldsymbol{r}'}{|\boldsymbol{r} - \boldsymbol{r}'|^3}. \tag{6–15}$$

Note that we can modify this ψ so that it satisfies a Neumann boundary condition on ∂V by adding a single valued harmonic function which vanishes at p_0. From Equation (6–9), the desired f is easily seen to be

$$f(\boldsymbol{r}) = f_0 \exp\left(\frac{i}{2} \int_{p_0}^{p} \sum_{l=1}^{\beta_1(V)} n_l \oint_{c'_l} \frac{((\boldsymbol{r} - \boldsymbol{r}') \times d\boldsymbol{r}) \cdot d\boldsymbol{r}'}{|\boldsymbol{r} - \boldsymbol{r}'|^3} \right), \tag{6–16}$$

where the multiplicative factor

$$f_0 = e^{2\pi i \, \psi(p_0)} \tag{6–17}$$

is an arbitrary complex number of unit length. If we go from p_0 to p_0 along a closed loop c the expression for the linking number (3–13) shows us that the expression for f changes by

$$\exp\left(2\pi i \sum_{l=1}^{\beta_1(V)} n_l \, \text{Link}(c, c'_l) \right),$$

which is equal to 1. This shows that a concrete understanding of the linking number makes the single-valuedness of f manifest and finally, up to boundary conditions, Equation (6–16) is indeed the solution to the variational problem defined by (6–5)–(6–7) and the boundary value problem defined by (6–14) and (6–7). Lefschetz duality and the theory underlying the algorithm ensure that cuts take care of current linkage, but nowhere has anything been done to make $V - \bigcup S_i$ simply connected.

	Node (X_0 & F_0)	Edge (X_1 & F_1)
X/X_0^s	9	7.5
F/F_0^s	7.5–7.6	9.7–10.7

Table 6.1. Relative number (X) of nonzero entries in the stiffness matrix and number of floating point operations (F) per CG iteration for node- and edge-based vector interpolation compared to scalar potential (X_0^s and F_0^s) on large tetrahedral meshes.

Computational Overhead Required for Cuts. In order to evaluate the utility of having cuts, one must also compare computational overhead with the expected acceleration in solution time. Given a finite element mesh it is possible to compare the solution time complexities of the associated finite element matrix equations for scalar potential and vector methods. Since semidirect matrix solvers are commonly used for large problems, we argue in terms of a conjugate gradient (CG) solver iteration for the finite element matrix equations. Let F_0^s denote the number of floating point operations (FLOPs) per conjugate gradient solver iteration for nodal interpolation of a scalar potential on a tetrahedral mesh. Similarly, let F_0 and F_1 respectively denote the number of FLOPs per CG iteration for node- and edge-based vector interpolation [Kot91]. Finally, let X_0^s, X_0, and X_1 denote the number of nonzero entries in the stiffness matrix for nodal interpolation of a scalar potential, and node- and edge-based vector interpolation, respectively. Assuming similar distributions of eigenvalues in the matrices, the convergence of CG for each matrix is the same and the ratio of the complexity of the CG iterations is a reflection of the ratio of computer run times. Table 6.1 summarizes how vector formulations compare to a scalar potential formulation based on analysis of simplices in a mesh discussed in Section 4C. The top line of the table is derived from Equations (4–17), (4–21), and (4–23), then using (4–18), forming ratios, and letting $m_0 \to \infty$ in the ratios.

If a scalar potential can be computed, it provides a substantial speed-up, but the overhead is that of computing cuts. Hence, cuts may be most useful in the context of time-varying or nonlinear problems where cuts are computed only once but iterative solutions are required for the field.

Algorithm design must begin with the choice of an algebraic framework for the computation, and for reasons of computability, this is the most critical choice. The often-made assumption that cuts must render the region simply connected forces one to work with a structure called the first homotopy group for which basic questions related to this group are not known to be algorithmically decidable. In practical terms, homotopy-based algorithms are limited to problems reducible to a planar problem. Thus, their success depends on the fact that 2-d surfaces are completely classified (up to homeomorphism) by their Euler characteristic and numbers of connected components and boundary components. The (co)homology arguments presented here lead to a general definition of cuts and an algorithm for computing them by linear algebra techniques, but when using sparsity of the matrices to make the computation efficient, homotopy emerges as an important and useful tool.

6C. Variational Formulation of the Cuts Problem

On the basis of the tools introduced above, the computation of cuts can be formulated as a novel use of finite elements subject to two constraints imposed by the topology of R. The idea is to come up with a variational problem for finding minimum energy maps f from classes in $[R, S^1]$. Hence, a principle for finding cuts is to compute a collection of maps, $f_q : R \to S^1$, $1 \le q \le \beta_1(R)$ which correspond to a basis of the first cohomology group with integer coefficients $H^1(R;\mathbb{Z})$ and by duality, to a basis for $H_2(R, \partial R; \mathbb{Z})$. Any map in the homotopy class can be used, but picking harmonic maps reduces the problem to quadratic functionals tractable by the finite element method. Furthermore, the level surfaces of these maps are nicer than in the generic case.

As a variational problem, finding cuts can be rephrased in the following manner. "Computing maps" means finding the minima to $\beta_1(R)$ "energy" functionals

$$F(f_q) = \int_R \operatorname{grad} \bar{f}_q \cdot \operatorname{grad} f_q \, dR, \quad 1 \le q \le \beta_1(R), \tag{6–18}$$

subject to two constraints:

$$\bar{f}_q f_q = 1 \quad \text{in } R \tag{6–19}$$

and, for the jth cut, $1 \le j \le \beta_1(R)$,

$$\frac{1}{2\pi i} \oint_{z_k} \operatorname{grad}(\ln f_j) \cdot dl = P_{jk}, \tag{6–20}$$

where $\{z_q\}$, $1 \le q \le \beta_1(R)$, defines a set of generators of $H_1(R)$ and P_{jk} is the period of the jth 1-form on z_k is an entry of a nonsingular period matrix P. Intuitively, one might require the period matrix to be the identity matrix, but this is overly restrictive for a practical implementation of the algorithm. In fact, as discussed in Section 6E, direct computation of a basis for $H_1(R)$ is impractical but an equivalent criterion can be used to satisfy the constraint imposed by $H_1(R)$.

The solution to each map in the variational problem is unique since the "angle" of each f_q is a (multivalued) harmonic function which is uniquely specified by Equation (6–20) [GK95]. When the functionals are minimized, a set of cuts is computed by the formula

$$S_q = f_q^{-1}(p_q), \tag{6–21}$$

where p_q is any regular value of f_q, $1 \le q \le \beta_1(R)$. Note that S_q is the Poincaré–Lefschetz dual to $d(\ln f/2\pi i)$, as seen in equation 6–3 [Kot87].

On any contractible subset of R, constraint (6–19) is satisfied by letting

$$f_q = \exp(2\pi i \varphi), \quad 1 \le q \le \beta_1, \tag{6–22}$$

where φ is some real, locally single-valued, but globally multivalued, differentiable function. Choosing f_q this way, the Euler–Lagrange equation of (6–18) is Laplace's equation [GK95].

Equation (6–22) is satisfied on open, contractible subsets U_i and their intersections. When the U_i form a cover of R, the global problem is assembled by considering the combinatorics of intersections $\bigcap U_i$ as noted in [BT82, §13] and also described in [GR65] in the context of the Cousin problem in complex analysis. For a computer implementation using standard data structures, it is sufficient to take a (tetrahedral) discretization of R and consider what happens across faces of the tetrahedral elements. Using the normalized angle of f_q [Kot89a],

$$\theta_q = (\ln f_q / 2\pi i) \mod 1, \tag{6–23}$$

on an element and interpolating θ_q linearly over each element, we must consider that (6–20) prevents θ_q from being globally well-defined.

Section 6E addresses how constraints (6–20) are handled without making explicit reference to a set of curves representing a basis for $H_1(R)$. However, we begin by considering constraint (6–19) and the finite element-based part of the algorithm in the next section. Since each of the β_1 problems is treated in the same way, in the next section we drop the subscript denoting the qth variational problem in order to simplify notation.

6D. The Connection Between Finite Elements and Cuts

Here we consider the variational problems for (any) one of the $\beta_1(R)$ maps f_i representing a class in $H_2(R, \partial R)$ (by duality) and how the variational problem is handled via the finite element method. While the principles behind computing cuts are not dependent on the type of discretization used, this section is set in the context of first-order nodal variables on a tetrahedral finite element mesh as the data structures are simple.

The Role of Finite Elements in a Cuts Algorithm. Consider a tetrahedral discretization of R, denoted by K, with m_3 elements, m_0 nodes. The ith tetrahedron in K is denoted by σ_i^3. Recalling Equations (6–22) and (6–23), let θ be discretized on each element by the set θ_j^i, $1 \leq i \leq m_3$, $1 \leq j \leq 4$ for each one of the $\beta_1(R)$ variational problems. Here the subscript refers to the jth node of the ith tetrahedron and the θ_j^i on individual elements are defined modulo integers since we seek a map into the circle. Furthermore, constraints (6–20) require that there be discontinuities in θ_j^i between pairs of adjacent elements. This is not a problem since the finite element connection process is modified accordingly as described below. To make a bridge to the finite element method, we also let u_k, $1 \leq k \leq m_0$ denote a potential discretized on nodes of the mesh [Kot89a].

The usual finite element connection matrix is defined as

$$C_{jk}^i = \begin{cases} 1 & \text{if global node } k \text{ is the } j\text{th local node in } \sigma_i^3, \\ 0 & \text{otherwise.} \end{cases}$$

The modified connection process is the marriage, via the finite element connection matrix, of variables defined locally on the nodes of individual elements (θ_j^i) to variables defined on the global node set of the mesh (u_k). Consequently, variables defined element-by-element are said to be on the unassembled mesh while those defined on the global nodes are on the assembled mesh.

The global constraints (6–20) are handled via nodal discontinuities on the unassembled mesh, $\mathcal{J}_j^i$, $1 \leq i \leq m_3, 1 \leq j \leq 4$ (Figure 6.2). The jumps are

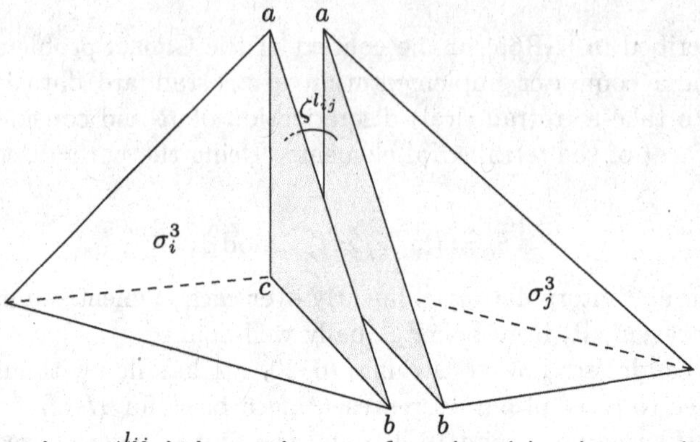

Figure 6.2. Jump $\zeta_k^{l_{ij}}$ is imposed across face shared by elements i and j and $\mathcal{J}_a^i - \mathcal{J}_a^j = \zeta^{l_{ij}}$.

specified so that for a given global node, $\mathcal{J}_j^i$ is an integer-valued jump relative to a $\mathcal{J}_l^k$ which is set to zero. From the perspective of a global node, there is a set of $\mathcal{J}_j^i$ associated with the global node (one $\mathcal{J}_j^i$ per element in which the global node is a local node), and one $\mathcal{J}_j^i$ per set can be set to zero. Then θ_j^i, which is defined on the unassembled mesh, is written

$$\theta_j^i = \sum_{k=1}^{m_0} C_{jk}^i u_k + \mathcal{J}_j^i, \quad 1 \le i \le m_3,\ 1 \le j \le 4. \tag{6–24}$$

For each of the $\beta_1(R)$ variational problems, there is one set of variables $\mathcal{J}_j^i$.

To finish specifying θ_j^i, the relationship between sets $\mathcal{J}_j^i$ and the homology class of the corresponding cut is needed. Recall that the sets of local nodal jumps $\mathcal{J}_j^i$ are defined on the unassembled finite elements problem. Since the discontinuity in θ must be consistent across the face of a tetrahedron, we introduce a set of discontinuities across faces, ζ^l, $1 \le l \le m_2$ where m_2 is the number of faces in the mesh. These are illustrated in Figure 6.2. Since there are β_1 variational problems there is a set $\{\zeta_1^l, \dots, \zeta_{\beta_1}^l\}$, and for the kth cut, given ζ_k^i, the remaining $\mathcal{J}_j^i$ can be found by the back substitution

$$\mathcal{J}_m^i - \mathcal{J}_n^j = \zeta_k^{l_{ij}}, \tag{6–25}$$

when elements i and j share face l_{ij}. The "topological computation" relating face jumps ζ_k to the relative homology class of the kth cut is discussed in Section 6E. At this point, if one is interested in using a scalar potential, but not in making the cuts visible to the user as a diagnostic tool, the set of face jumps $\{\zeta_1, \dots, \zeta_{\beta_1}\}$ is a set of cuts. However, note that these cuts are not embedded manifolds, but nonetheless represent a basis for $H_2(R, \partial R; \mathbb{Z})$. We will see in Section 6E that the face jumps can be identified with a class in the simplicial homology group related to $H_2(R, \partial R)$.

On the basis of the variational problem defined in equations (6–18)–(6–20), "finite element analysis" can be used to solve for each of $\beta_1(R)$ potentials, u_k. On each connected component of R, begin by setting one (arbitrary) variable in u_k to zero. Then define barycentric coordinates λ_i^l, $1 \le i \le 4$, on the lth tetrahedron, σ_l^3, and build the element stiffness matrix $\mathcal{K}_{mn}^l$:

$$\mathcal{K}_{mn}^l = \int_{\sigma_l^3} \operatorname{grad} \lambda_m^l \cdot \operatorname{grad} \lambda_n^l \, dV.$$

Discretizing the normalized angle θ (6–23) on σ_l^3 by

$$\theta^l = \sum_{i=1}^{4} \lambda_i^l \theta_i^l \tag{6–26}$$

and substituting into the functional (6–18) gives

$$F(\theta) = 4\pi^2 \sum_{l=1}^{m_3} \sum_{m=1}^{4} \sum_{n=1}^{4} \theta_m^l \theta_n^l \mathcal{K}_{mn}^l. \tag{6–27}$$

Using Equation (6–24) to "assemble the mesh" gives a quadratic form in u_k. The minimum of the quadratic form is obtained by differentiating with respect to the u_k, resulting in the matrix equation

$$\sum_{i=1}^{m_0} \mathcal{K}_{ij} u_i = -f_j. \tag{6–28}$$

Here $\{\mathcal{K}_{ij}\}$ forms the usual stiffness matrix,

$$\mathcal{K}_{ij} = \left(\sum_{l=1}^{m_3} \sum_{m=1}^{4} \sum_{n=1}^{4} C_{nj}^l \mathcal{K}_{mn}^l C_{mi}^l \right), \tag{6–29}$$

and, by Equation (6–25), the source term

$$f_j = \left(\sum_{l=1}^{m_3} \sum_{m=1}^{4} \sum_{n=1}^{4} \mathcal{J}_n^l \mathcal{K}_{mn}^l C_{mj}^l \right) \tag{6–30}$$

is related to the homology class of a relative cycle in $H_2(R, \partial R)$ by means of $\{\zeta_j^i\}$. Thus, with the exception of computing $\{\zeta_j^i\}$ and forming source term (6–30), the algorithm is readily implemented in any finite element analysis program. This gives the maps from R to S^1. To find the cut, recall Poincaré–Lefschetz duality and Equation (6–21). For each connected component of R,

$$(\theta')_j^i = C_{jk}^i (u_k + c) \mod 1, \tag{6–31}$$

where c is a constant so that $\theta' = 0$ is a regular value of f. After solving $\beta_1(R)$ variational problems, we proceed element by element to find and plot $f_q^{-1}(\theta' = 0)$, $1 \le q \le \beta_1(R)$, to obtain a set of cuts. This is done in an unambiguous way if the mesh is fine enough to ensure that, over an element, θ_j^i does not go more than one third of the way around the circle.

In order to use the cut for a scalar potential computation, the cut must be specified in terms of internal faces of the mesh, much as sets ζ_i are defined. For this we define β_1 sets S_i of faces obtained by perturbing a level set of the harmonic map onto internal faces of the finite element mesh. On a tetrahedral mesh this is done by simply choosing the element face which is on the side of the level set indicated by the normal (gradient direction) when the level set passes through the element. This is illustrated in Figure 6.3.

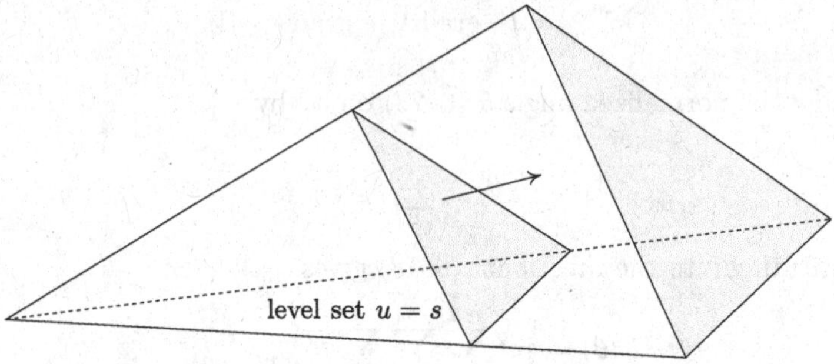

Figure 6.3. Level set of harmonic map is perturbed in gradient direction onto a face of the tetrahedral element. The face is selected merely by virtue of the fact that the potentials $u(v_i \geq s)$ where v_i is the ith vertex. In other words, the direction of $\operatorname{grad} u$ defines a normal to the cut, but it is unnecessary to actually compute the gradient. Cases in which the level set is perturbed onto an edge or vertex of the tetrahedron can be ignored since the level set will perturb onto the face of a neighboring tetrahedron.

Besides computing the $\{\zeta_1, \ldots, \zeta_{\beta_1}\}$ and incorporating the S_q as subcomplexes of the mesh, the computation of cuts makes use of standard software found in finite element software packages. Although the sign of a floating point calculation is required at one step, the "topological" part of the code is otherwise implemented with integer arithmetic and is therefore immune to rounding errors.

The following "algorithm" shows how cuts software fits into the typical finite element analysis process. The second and third steps are not "standard" finite element software, but are implemented with integer arithmetic and thus avoid introducing rounding error.

ALGORITHM 6.1. Finite elements and cuts

(1) **Tetrahedral mesh generation and refinement**
(2) **Extraction of simplicial complex:** Employ the data structures from Chapter 4 and generate the data needed for computing interelement constraints ζ_i.
(3) **Topological processing:** Compute interelement constraints $\{\zeta_1, \ldots, \zeta_{\beta_1}\}$ defined in Equation (6–25) and described in section 6E and algorithm 6.2.
(4) **Finite element solution:** Use (6–29) and (6–30) to form (6–28) for each of β_1 variational problems, and solve them by the finite element method.
(5) **Obtain cuts:** Level sets of the harmonic maps computed in the last step are cuts. When perturbed onto the 2-skeleton of the mesh to define β_1 surfaces S_i, they are the data needed to do a magnetic scalar potential calculation.

6E. Computation of 1-Cocycle Basis

Now we consider the computation of the interelement constraints which come about as a result of constraint (6–20) on the variational problem and were defined in Equation (6–25). The computation must be in terms of the finite element discretization K, and as indicated at the beginning of Section 6B, simplicial (co)homology is the algebraic framework. Here we use the results of companion chapter [GK01a] which describes the simplicial complex, and how it generates the finite element data structures required for computing cuts when taking the view that K and its simplicial complex are the same thing.

Chapter 4 discusses the duality between the boundary homomorphism on a simplicial complex K and its dual complex DK and shows that the duality appears naturally in the data structures describing K. This duality is a discrete version of Poincaré–Lefschetz duality introduced in Chapter 3, but stated at the level of the simplicial chain complex (the algebraic structure which is most suited to describing finite element meshes). The dual complex is also the most appropriate structure for describing the computation of the topological constraints for the variational problem. Below we give the basic definitions needed to formulate the computation of the topological constraints.

Definitions. For a tetrahedral mesh K, denote the nodes, edges, faces, and tetrahedra as *0-*, *1-*, *2-*, and *3-simplexes*, respectively, though in general it is possible to have an n-dimensional simplicial mesh. The total number of p-simplexes in a mesh is denoted by m_p. Each set of p-simplexes forms a linear space with (for the present purpose) coefficients in $\mathbb{Z}$, $C_p(K;\mathbb{Z})$, called the *p-chain group.* There is a boundary homomorphism $\partial_p : C_p(K;\mathbb{Z}) \to C_{p-1}(K;\mathbb{Z})$ which takes p-simplexes to $(p-1)$-simplexes such that the composition of two successive transformations is zero:

$$\partial_i \partial_{i+1} = 0, \quad 1 \le i \le n. \tag{6–32}$$

In other words, $\operatorname{im} \partial_{i+1} \subseteq \ker \partial_i$, and the sequences in (6–32) are summarized in the *simplicial chain complex*:

$$0 \longrightarrow C_n(K) \xrightarrow{\partial_n} \cdots \xrightarrow{\partial_p} C_{p-1}(K) \xrightarrow{\partial_{p-1}} \cdots \longrightarrow C_1(K) \xrightarrow{\partial_1} C_0(K) \longrightarrow 0. \tag{6–33}$$

As in the continuum case, this allows us to define homology (and cohomology) groups.

The adjoint operator of the boundary homomorphism is the *coboundary operator* $\partial^T_{p+1} : C^p(K;\mathbb{Z}) \to C^{p+1}(K;\mathbb{Z})$ where $C^p(K;\mathbb{Z})$ is the simplicial cochain group of functionals on p-chains; formally, $C^p(K;\mathbb{Z}) = \hom(C_p(K),\mathbb{Z})$. The cochain $c^p \in C^p$ satisfies the relation

$$\langle c^p, \partial_{p+1} c_{p+1} \rangle = \langle \partial^T_{p+1} c^p, c_{p+1} \rangle \tag{6–34}$$

for any $c_{p+1} \in C_{p+1}$. This is a discrete version of Stokes' theorem (compare Equations (3–2) and (3–3)) and serves as a definition of ∂^T_{p+1}. $\partial^T_i \partial^T_{i+1} = 0$ so that there is a *cochain complex*:

$$0 \longleftarrow C^n(K) \xleftarrow{\partial^T_n} \cdots \xleftarrow{\partial^T_p} C^{p-1}(K) \xleftarrow{\partial^T_{p-1}} \cdots \longleftarrow C^1(K) \xleftarrow{\partial^T_1} C^0(K) \longleftarrow 0.$$

Simplicial homology groups are quotient groups $H_p(K) = \ker \partial_p / \operatorname{im} \partial_{p+1}$ and the pth Betti number β_p is the rank of H_p; the *simplicial cohomology groups* are defined by $H^p(K) = \ker \partial_{p+1}^T / \operatorname{im} \partial_p^T$. An element of $H^p(K)$ is the coset

$$\zeta + B^p, \tag{6–35}$$

where $B^p(K) = \operatorname{im} \partial_p^T$ is the *p-coboundary subgroup* of $C^p(K)$ and $\zeta \in \ker \partial_{p+1}^T$ is a *p-cocycle*. The ranks of the homology and cohomology groups are related: $\operatorname{Rank}(H_p) = \operatorname{Rank}(H^p)$.

By identifying nonboundary p-simplexes in K with $(n-p)$-cells (so that when $n = 3$, 3-simplexes become nodes, 2-simplexes become edges, etc.), a formal dual of K, called the dual complex DK, can be formed directly from the connection matrix describing K. (The boundary is excluded since there are no 3-simplexes in ∂K.) The dual DK is not simplicial but cellular, and the number of p-cells of DK is denoted by $\breve{m}_p$. DK is a cellular complex in the sense of (6–33). Poincaré duality amounts to saying that the coboundary operators of the simplicial complex are dual to the boundary homomorphisms of the cellular complex, denoted by $\breve{\partial}_p$ in the sense that

$$\partial_{p+1}^T = \breve{\partial}_{n-p} \tag{6–36}$$

[GK01a]. Thus, for a 3-dimensional complex,

$$0 \longrightarrow C_3(DK) \overset{\breve{\partial}_3=\partial_1^T}{\longrightarrow} C_2(DK) \overset{\breve{\partial}_2=\partial_2^T}{\longrightarrow} C_1(DK) \overset{\breve{\partial}_1=\partial_3^T}{\longrightarrow} C_0(DK) \longrightarrow 0 \tag{6–37}$$

and

$$0 \longleftarrow C^3(DK) \overset{\breve{\partial}_3^T}{\longleftarrow} C^2(DK) \overset{\breve{\partial}_2^T}{\longleftarrow} C^1(DK) \overset{\breve{\partial}_1^T}{\longleftarrow} C^0(DK) \longleftarrow 0. \tag{6–38}$$

The (co)homology of DK is isomorphic to the (co)homology of K in complementary dimensions. In other words, Poincaré duality on the (co)chain level gives us the Poincaré duality of Section 6B.

Formulation of a 1*-cocycle generator set.* The duality between boundary and coboundary operators as set forth above is useful for formulating the outstanding problem of computing $\{\zeta_1, \dots, \zeta_{\beta_1}\}$ introduced in equation (6–25). These variables were introduced in order to handle interelement topological contraints (6–20) of the variational problem, but Equation (6–20) cannot be applied directly since a set of generators for $H_1(R)$ is generally not known beforehand and is hard to compute. On the other hand, (6–20) simply gives the periods of 1-cocycles integrated on a homology basis, so that it is enough to know a basis for the nontrivial (i.e. noncoboundary) 1-cocyles. Here, the ζ_i are described by $\beta_1(R)$ 1-cocycles which are generators for classes in $H^1(DK;\mathbb{Z})$, and by duality represent sets of faces having nonzero jumps in backsubstitution Equation (6–25). The advantage of formulating the problem in terms of $H^1(DK;\mathbb{Z})$ is that it immediately yields a matrix equation, and the 1-cocycles form $\beta_1(R)$ source terms for the right-hand side of Equation (6–28).

In general, equivalence classes in $H^1(DK;\mathbb{Z})$, the first cohomology group of the dual complex with integer coefficients, can be represented by integer-valued 1-forms which are functionals on 1-cells of the dual mesh. However, it is only possible (and necessary) to compute, for each equivalence class, a generating 1-cocycle defined by two properties described below.

First, to obtain a basis for noncobounding 1-cocycles a basis for the 1-coboundary subgroup $B^1(DK;\mathbb{Z})$ must be fixed by constructing the image of a 0-coboundary map $\breve{\partial}_1^T : C^0 \to C^1$. This not only fixes how $B^1(DK;\mathbb{Z})$ is represented, but allows the computation of 1-cocycles which represent closed, nonexact 1-forms.

Second, by definition, the 1-cocycle must also satisfy the condition that on the boundary of each 2-cell in DK, $\breve{\partial}_2(\text{2-cell}) = \varepsilon_1 e_1 + \varepsilon_2 e_2 + \cdots + \varepsilon_n e_n$,

$$\langle \zeta_j, \breve{\partial}_2(\text{2-cell})\rangle = \sum_{i=1}^{n} \varepsilon_i \zeta_j^i(e_i) = 0, \tag{6–39}$$

where $\varepsilon_i = \pm 1$ denotes the orientation of the ith 1-cell on the boundary of the 2-cell (Figure 6.4). The condition must be satisfied on any simply connected subset of the mesh, but the data readily available from the finite element connection matrix relates to 2-cells. Since Poincaré duality for complexes K and DK (6–36) says that $\breve{\partial}_2$ and ∂_2^T are identified, the coboundary operator ∂_2^T is the incidence matrix of 2- and 1-cells in DK and contains the data for the 1-cocycle conditions over all of DK. Equations (6–34) and (6–39) together say that, for any 1-cocycle ζ,

$$\partial_2^T \zeta = 0 \tag{6–40}$$

(on DK). Once a basis for the 1-coboundary subgroup has been fixed, a set of nontrivial 1-cocycles is found by computing a basis of the nullspace of the operator ∂_2^T. Equation (6–40) is an exceedingly underdetermined system, but as shown below, fixing a basis for B^1 induces a block partition of the matrix, and reduces the computation to a block whose nullspace rank is precisely $\beta_1(R)$.

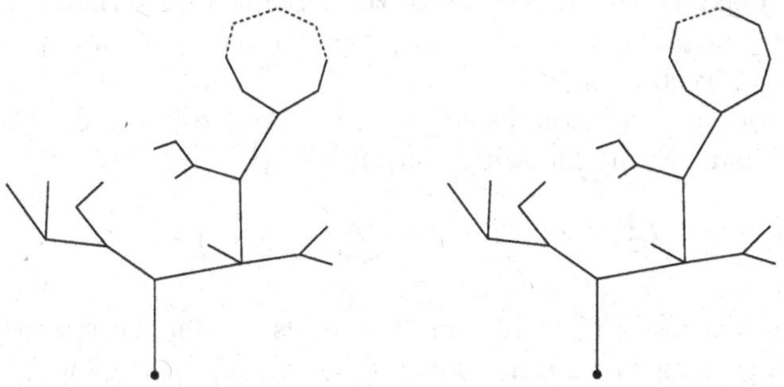

Figure 6.4. 1-cells on maximal spanning tree (solid) and 2-cells in DK. On the right, the 2-cell cocycle condition has four free variables (those not on the tree) while on the left, the condition can be satisfied (trivially) in terms of variables on the tree.

In summary there are two conditions which must be satisfied in order to find a set of 1-cocycles which are not coboundaries:

(1) A basis for B^1 must be fixed by considering the image of a map

$$\breve{\partial}_1^T : C^0 \longrightarrow C^1.$$

(2) The 1-cocycles must independently satisfy the 1-cocycle condition on each 2-cell of the mesh.

Structure of the Matrix Equation for Computing the 1-Cocycle Generators. The strategy outlined above amounts to constructing bases for $\operatorname{im}\breve{\partial}_1^T$ and $\ker\breve{\partial}_2^T$ (subject to $\operatorname{im}\breve{\partial}_1^T = 0$) in the complex (6–38). When $\operatorname{im}\breve{\partial}_1^T$ is annihilated, the surviving piece of $\ker\breve{\partial}_2^T$ gives a basis for 1-cohomology generators. In this section we give a method for constructing the required bases while retaining the sparsity of $\breve{\partial}_2^T$ and show how the construction yields a natural block partition of the matrix.

The arguments of this section have the same motivation as techniques of electrical circuit analysis. The rank argument of this section is a formalization of a familiar equation which relates the number of free variables n_{free} to the numbers of Kirchhoff current law (node) and Kirchhoff voltage law (loop) equations (n_{KCL} and n_{KVL}):

$$n_{\text{KCL}} + n_{\text{KVL}} - \beta_0 = n_{\text{free}},$$

where $\beta_0 = \operatorname{Rank}(H_0)$ is the number of connected components of the mesh or circuit.

To fix $B^1(DK)$, we construct a map $\breve{\partial}_1^T$ satisfying the Stokes Equation (6–34), which for this case says $\langle c^0, \partial e\rangle = \langle \breve{\partial}^T c^0, e\rangle$, where $e \in C_1(DK)$, $c^0 \in C^0(DK)$. Defining c^0 on vertices of DK and building a maximal tree on the 1-skeleton of DK fixes a basis for $B^1 = \operatorname{im}(\breve{\partial}_1^T)$ up to a constant on a single vertex on each connected component of DK by associating each vertex (functional) with an edge (functional) on the 1-skeleton. There are $m_0 - \beta_0$ 1-cells on the maximal tree, the same as the rank of $\operatorname{im}\breve{\partial}_1^T$. Since the coboundary subgroup is annihilated in the equivalence relation for cohomology, the variables of ζ_i corresponding to edges on the maximal tree can be set to zero. Below we see that this reduces the number of free variables enough to permit computation of an appropriate set of independent nullvectors of $\breve{\partial}_2^T$.

The reduction of free variables for each 1-cocycle solution ζ_i obtained by the maximal tree induces the following partition on $\breve{\partial}_2^T$:

$$\breve{\partial}_2^T \zeta_i = \Big(\underbrace{\;T\;}_{\breve{m}_0-\beta_0} \;\Big|\; \underbrace{\;U\;}_{\breve{m}_1-\breve{m}_0+\beta_0}\Big)\begin{pmatrix} 0_T \\ \zeta_U \end{pmatrix} = 0, \tag{6–41}$$

where columns of block U correspond to 1-cells not on the tree while columns of block T correspond to 1-cells on the tree. Variables in ζ_i which correspond to 1-cells on the tree are zero, so that there are $\breve{m}_1 - \breve{m}_0 + \beta_0(DK)$ free variables remaining for any nontrivial 1-cocycle (or nullspace) solution to the matrix equation. The following shows that the dimension of the nullspace of block U is $\beta_1(R)$.

The rank of $\breve{\partial}_2^T$ can be found by a standard argument which considers the ranks of the kernel and image of the boundary homomorphism in the cellular complex (6–37) and ranks of the corresponding homology groups. Since $\breve{\partial}_2$ is a linear map,

$$\dim\operatorname{im}\breve{\partial}_2 = \dim C_2 - \dim\ker\breve{\partial}_2 = \breve{m}_2 - \dim\ker\breve{\partial}_2. \tag{6–42}$$

In terms of the rank of the second homology group,

(6–43) $$\dim \ker \breve{\partial}_2 = \beta_2 + \dim \operatorname{im} \breve{\partial}_3 = \beta_2 + \breve{m}_3 - \beta_3,$$

where $\dim \breve{\partial}_3 = \breve{m}_3 - \beta_3$ follows from (6–37). In this case, since K is the triangulation of a connected 3-manifold with boundary, $\beta_3 = 0$. In any case, (6–43) and (6–42) give the rank of ∂_2^T:

$$\dim \operatorname{im} \breve{\partial}_2 = \breve{m}_2 - \breve{m}_3 - \beta_2 + \beta_3.$$

In terms of cocycle conditions, this result can be interpreted as counting the number of linearly independent cocycle conditions in rows of $\breve{\partial}_2$. Considering the set of cocycle conditions on a 3-cell, there is one linearly dependent cocycle condition, giving $\breve{m}_3 - \beta_3$ extra cocycle conditions in ∂_2^T. There is one linearly dependent equation among each set of cocycle conditions describing "cavities" of the region, giving another β_2 linearly dependent equations.

Consequently, the dimension of the nullspace of block U, $\mathcal{N}(U)$, in the partition of Equation (6–41) is

$$\begin{aligned}\dim \mathcal{N}(U) &= (\breve{m}_1 - \breve{m}_0 + \beta_0) - (\breve{m}_2 - \breve{m}_3 - \beta_2 + \beta_3)\\ &= -\chi(DK) + \beta_2 + \beta_0 - \beta_3 = \beta_1,\end{aligned}$$

since the Euler characteristic satisfies

$$\chi(DK) = \sum_{i=0}^{n} (-1)^i \beta_i = \sum_{i=0}^{n} (-1)^i \breve{m}_i.$$

Accounting for $\breve{m}_3 + \beta_2 - \beta_3$ linearly dependent cocycle conditions, the following partition of U into blocks of linearly independent (U_i) and linearly dependent (U_d) equations is a useful picture to keep in mind for the rank argument:

$$U = \left(\begin{array}{c} U_i \\ \hline U_d \end{array} \right) \begin{array}{l} \}\ \breve{m}_2 - \breve{m}_3 - \beta_2 + \beta_3 \\ \}\ \breve{m}_3 + \beta_2 - \beta_3 \end{array}.$$

In practice, the linear dependence of rows in U can be exploited when finding a diagonalization of U so that the nullspace basis $\{\zeta_1, \ldots, \zeta_{\beta_1}\}$ is relatively sparse.

Sparsity of ∂_2^T and U. Recall that nonboundary 2-simplexes in K are mapped to 1-cells in DK and nonboundary 1-simplexes in K are mapped to 2-cells in DK. In K, the boundary of every 2-simplex has three 1-simplexes so that in DK each 1-cell is in at most three 2-cells. The inequality comes about because ∂K does not enter into the contruction of DK; in particular, a 2-simplex with some of its boundary in ∂K corresponds to a 1-cell which is an edge in fewer than three 2-cells. Consequently, columns of ∂_2^T have at most three nonzero entries, or $3\breve{m}_1$ is an upper bound on the number of nonzero entries in ∂_2^T.

A lower (upper) bound on the difference between $3\breve{m}_1$ and the number of nonzero entries is given by $(b-2)n_1$ where n_1 is the number of 1-simplexes in ∂K and b is an upper (lower) bound on the number of 2-simplexes which meet at a boundary 1-simplex. In the estimate we take two less than b since the two faces meeting at a boundary edge do not have entries in U.

Block Partition and Sparsity of the Matrix Equation. At this point we are free to choose any method for finding a basis for the nullspace of U. Typical methods for matrices with integer coefficients are the Smith and Hermite normal form algorithms [Coh93]. Since ∂_2^T is an incidence matrix with nonzero entries ± 1, problems such as pivot selection can be avoided, but they also destroy the sparsity of ∂_2^T and their time complexity is $\mathcal{O}(\breve{m}_0^3)$. This indicates that the combinatorial structure of the matrix is more important than its numerical structure. The literature on computing sparse nullspace bases of real matrices is applicable here [CEG86, PF90].

U can be block partitioned into a form which preserves most of its sparsity. The partition is based on the observation that a 2-cell Equation (6–39) which has only one free variable after fixing $\operatorname{im} \breve{\partial}_1^T$ is satisfied trivially—variables for such 1-cells do not contribute to the 1-cocycles and can be set to zero in ζ_i. In terms of the maximal tree, this case corresponds to Figure 6.4. In matrix ∂_2^T, this elimination amounts to forward substitution of variables on the tree, forming a lower triangular block in U and eliminating variables which are not essential to the description of the null basis while avoiding zero fill-in. When the process of forward substitution halts (as it must if the null space basis we seek is nontrivial), the remaining free variables and cocycle conditions contain a full description of the complex on a substantially smaller set of generators and relations. This results in the following block partition of the matrix equation, where U_{11} is the lower triangular block resulting from the forward substitution:

$$\partial_2^T = \left(\begin{array}{c|cc} T & U_{11} & 0 \\ & U_{21} & U_{22} \end{array} \right).$$

Block T corresponds to a maximal tree on the 1-cells of DK and, variables associated with T are zero in the nullspace basis. Forward substitution of nullspace basis variables on T gives the lower triangular block U_{11} so that the nullspace basis vectors have the form

$$\zeta_i = \begin{pmatrix} 0_T \\ 0_{U_{11}} \\ \zeta_{U_{22},i} \end{pmatrix}.$$

As with ∂_2^T, block U_{22} has at most three nonzero entries per column since no operations involve zero fill-in. Figure 6.6 shows examples of U_{22} for two interesting cases. The first example is the Borromean rings and the second example is the trefoil knot, both shown in Figure 6.5.

At this point it is best to admit that there are two ideas from topology which have strong ties to the present construction. One of them is Poincaré's algorithm for computing the generators and relations of the fundamental group of a complex [Sti93]. This construction is similar with the added constraint of preserving sparsity of the equations and reduction of the Poincaré data into block U_{22} of ∂_2^T. Another relevant notion is that of the spine of a manifold [Thu97].

ALGORITHM 6.2. Algorithm for 1-cocycle generator set

(1) **Initialize:** Set $\{\zeta_1, \ldots, \zeta_{\beta_1}\}$ to be zero.

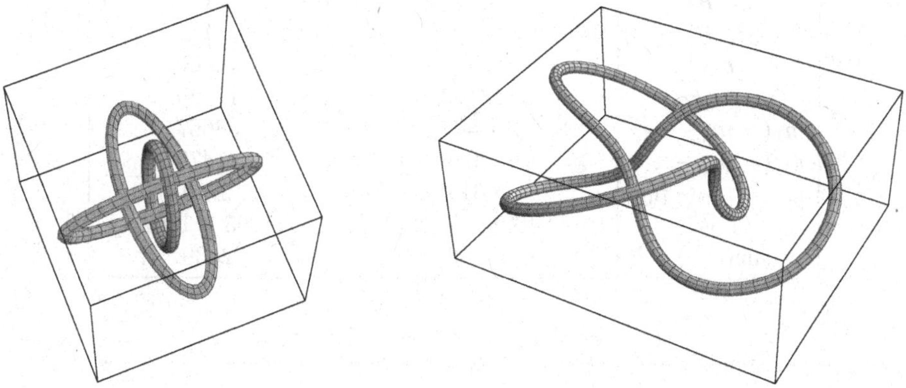

Figure 6.5. Borromean rings and trefoil knot. The Borromean rings are three rings which pairwise have zero linking number but are inseparable.

(2) **Maximal Tree:** Construct maximal spanning tree $\mathcal{T}$ on 1-skeleton of DK.
(3) **Partition:** Set $\zeta_i|_{\mathcal{T}} = 0$ and partition ∂_2^T as in (6–41).
(4) **Forward Substitution:** Forward substitute variables $\zeta_i|_{\mathcal{T}}$ (for all ζ_i these are the same variables) through U iteratively until the process halts.
(5) U_{22} **Nullbasis:** Compute nullspace basis of U_{22} by a sparse null basis technique or by computing the Smith normal form.

The size of U_{22}. The process of partitioning ∂_2^T effectively retracts all the information about the topology of the mesh onto a 2-subcomplex of the mesh. The tree gives a retraction onto the 2-skeleton of $K - \partial K$, and the reduction by forward substitution is a retraction onto a 2-subcomplex $\tilde{K}$ represented by U_{22}. In the dual mesh, the retraction, $D\tilde{K}$, has the same "homotopy type" as DK and hence the same (co)homology groups. For a sufficiently good maximal spanning tree (one that is, in some sense, short and fat), the number of 1-cells in U_{22} is the number of faces (of K) on S' a set of cuts plus additional surfaces which make any noncontractible loop on a cut contractible. Let N be some measure of the number of degrees of freedom per unit length in DK so that m_0 is $\mathcal{O}(N^3)$. Note that $\breve{m}_1$ is linearly related to m_0. Let k be the number of 1-cells in U_{22}, namely the number of free variables remaining in the reduced matrix. As the mesh is refined, k is on the order of the area of S', that is $\mathcal{O}(N^2)$ or $\mathcal{O}(m_0^{2/3})$. The complexity of an algorithm to compute the nullspace basis is $\mathcal{O}(m_0^2) + \mathcal{O}(k^3)$ in time and $\mathcal{O}(m_0) + \mathcal{O}(k^2)$ in storage, where k is $\mathcal{O}(m_0^{2/3})$, so the time complexity becomes $\mathcal{O}(m_0^2)$ and space complexity is $\mathcal{O}(m_0^{4/3})$. The overall time requirement for computing cuts is that of finding $\{\zeta_1, \ldots, \zeta_{\beta_1}\}$ for each cut and β_1 solutions of Laplace's equation to find the nodal potential described in Section 6C.

	a. Borromean Rings	b. Trefoil Knot
$\beta_0(R)$	1	1
$\beta_1(R)$	3	1
m_0	9665	3933
$m_3(=\breve{m}_0)$	48463	19929
$\breve{m}_1(=m_2-n_2)$	93877	38667
$\breve{m}_2(=m_1-n_1)$	52029	21479
$\breve{m}_3(=m_0-n_0)$	6614	2740
U_{22}	4008×2888	1393×1007
$\mathrm{nz}(U_{22})$	8156	2839

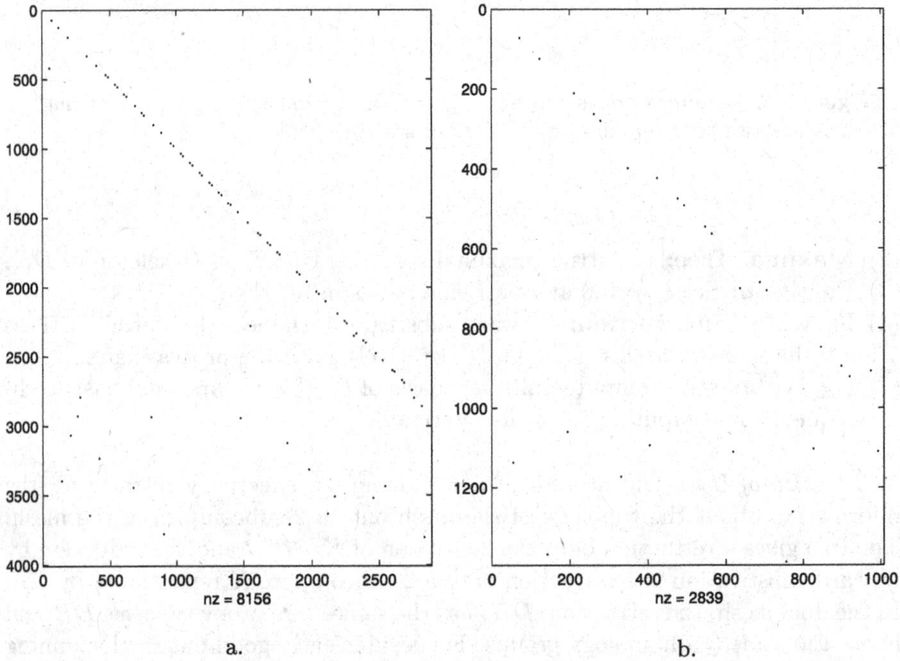

Figure 6.6. U_{22} for two cases of interesting topology: (a) complement of Borromean rings (three unlinked but inseparable rings) and (b) complement of a trefoil knot.

6F. Summary and Conclusions

While Ampère's law gives intuition about the role and nature of cuts, it sheds no light on their construction and computation. On the other hand, the algebraic structures of (co)homology theory are adequate for formulation of an algorithm for finding cuts on finite element meshes which are orientable, embedded submanifolds of the nonconducting region. The algorithm fits naturally into finite element theory.

Starting with the connection matrix, cuts can always be found in $\mathcal{O}(m_0^3)$ time and $\mathcal{O}(m_0^2)$ storage. However, complexity can be improved to $\mathcal{O}(m_0^2)$ time and $\mathcal{O}(m_0^{4/3})$ storage by the algorithm outlined above. Moreover, the algorithm discussed in Section 6E preserves sparsity in the finite element matrices and thus does not adversely affect complexity in subsequent computation of a scalar potential with cuts. The speed of the algorithm can be further improved if one starts with a coarse mesh or information about the fundamental group π_1 and its commutators [Sti93]. It is clear that in the context of adaptive mesh refinement, cuts should be computed on a coarse mesh and then refined with the mesh since even the most coarse mesh contains all the information required for the topological computation. On the other hand, since the topological computation involves only integer arithmetic, computation on a fine mesh does not introduce rounding error.

The program itself is the only complete description of what the program will do.

P. J. Davis

Do you know of Grassmann's Ausdehnungslehre? *Spottiswood spoke of it in Dublin as something above and beyond quaternions. I have not seen it, but Sir William Hamilton of Edinburgh used to say that the greater the extension the smaller the intention.*

James Clerk Maxwell, in a letter to P. G. Tait.

7

A Paradigm Problem

7A. The Paradigm Problem

The purpose of this chapter is to show how the formalism of differential forms reduces a broad class of problems in computational electromagnetics to a common form. For this class of problems, the differential complexes and orthogonal decompositions associated with differential forms make questions of existence and uniqueness of solution simple to answer in a complete way which exposes the role played by relative homology groups. When this class of problems is formulated variationally, the orthogonal decomposition theorem developed in Section MA-M generalizes certain well known interrelationships between gauge transformations and conservation laws (see [Ton68]) to include global conditions between dual cohomology groups. The orthogonal decomposition theorem can then be used to construct an alternate variational principle whose unique extremal always exists and can be used to obtain a posteriori measures of problem solvability, that is to verify if any conservation law was violated in the statement of the problem. A diagrammatic representation of the problem along the lines of [Ton72b] will be given and the role of homology groups will be reconsidered in this context. This of course will be of interest to people working in the area complementary variational principles.

In addition to the usual literature cited in the bibliography, the work of Tonti [Ton68, Ton69, Ton72b, Ton72a, Ton77], Sibner and Sibner [SS70, SS79, SS81] and [Kot82] have been particularly useful in developing the ideas presented in this chapter. From the view of computational electromagnetics, the beauty of formulating a paradigm variational problem in terms of differential forms is that the finite element method and Whitney form interpolation yield a discretization which faithfully reproduces all the essential features of the continuum problem. Although this point of view was advocated two decades ago [Kot84], general acceptance by the engineering community has come as a result of a lot of hard work [Bos98]. See Hiptmair [Hip02] for a recent survey of interpolation for electromagnetic field problems based on differential forms.

The notation used in this chapter is quite distinct from notation used in previous chapters. Readers not familiar with Hodge theory on manifolds with boundary can consult Section MA-J (page 243) and subsequent ones, where the notation in developed.

The paradigm problem of this chapter will now be considered. Let M be a compact orientable n-dimensional Riemannian manifold with boundary. In the paradigm problem to be considered, the field is described by two differential forms

$$\beta \in C^p(M), \qquad \eta \in C^{n-p}(M),$$

which are related to another differential form

$$\lambda \in C^{n-p+1}(M),$$

which describes the sources in the problem. These differential forms are required to satisfy the following key pair of equations:

$$\boxed{\int_{\partial c_{p+1}} \beta = 0}$$

$$\boxed{\int_{\partial c_{n-p+1}} \eta = \int_{c_{n-p+1}} \lambda}$$

for all $c_{p+1} \in C_{p+1}(M)$ and $c_{n-p+1} \in C_{n-p+1}(M)$. If S is a set of $n-1$-dimensional interface surfaces where β may be discontinuous, the first integral equation implies that

$$d\beta = 0 \quad \text{on } M - S.$$

Also one can define an orientation on S so that there is a plus side and a minus side and

$$t\beta^+ = t\beta^- \quad \text{as } S \text{ is traversed.}$$

Here t is the pullback i^* of a differential form by the inclusion map

$$i : S \to M.$$

The $+$ and $-$ superscripts denote limiting values as the orientable codimension-one surface S is approached from within M.

It is natural to inquire whether there exists a potential

$$\alpha \in C^{p-1}(M)$$

such that

$$\beta = d\alpha.$$

In other words, the first integral equation shows that β is a closed form and one would like to know whether it is exact. The answer, of course, is given by the de Rham isomorphism, that is, β is exact if all of its periods vanish on a basis of the homology group $H_p(M)$. In addition to the structure above , the paradigm problem to be considered has a constitutive relation relating β and η. Further consideration will not be given to this constitutive relation until the next section.

Although various boundary conditions can be imposed on β and η so that a boundary value problem can be defined, to simplify the presentation it is assumed in accordance with the general philosophy adopted here that

$$\partial M = S_1 \cup S_2, \qquad t\beta = 0 \quad \text{on } S_1, \qquad t\eta = 0 \quad \text{on } S_2,$$

where $S_1 \cap S_2$ is an $(n-2)$-dimensional manifold whose connected components represent intersections between symmetry planes and connected components of the boundary of an original problem where symmetries were not exploited.

It is important, before going on, to list the specific problems that occur as special cases of this general problem. They are:

(1) Electrodynamics in four dimensions.
(2) Electrostatics in three dimensions.
(3) Magnetostatics in three dimensions.
(4) Currents in three-dimensional conducting bodies where displacement currents are neglected.
(5) Low frequency steady or eddy current problems where currents are confined to surfaces that are modeled as two-dimensional manifolds. In this case, the local sources or "excitation" is the time variation of the magnetic field transverse to the surface.
(6) Magnetostatics problems that are two-dimensional in nature because of rotational or translational symmetry in a given three-dimensional problem.
(7) Electrostatics problems that are two-dimensional in nature because of rotational or translational symmetry in a given three-dimensional problem.

Note that the last two problems have not been discussed so far because of their "topologically uninteresting" properties. They are included here for completeness, and a word of caution is in order. For two-dimensional problems which arise from axially symmetric three-dimensional problems it is important to remember that the metric tensor on M is not the one inherited from $\mathbb{R}^3$ but rather is a function of the distance from the axis of symmetry.

Tables 7.1 and 7.2 summarize the correspondence between the paradigm problem defined in terms of differential forms and the various cases listed above. Table 7.2 lists examples considered so far which are useful for sorting out topological or other details.

7B. The Constitutive Relation and Variational Formulation

In order to define a constitutive relation between β and η, consider a mapping

$$C : C^p_c(M) \to C^p_c(M)$$

that, when restricted to a point of M, becomes a transformation mapping one differential form into another. In addition, given arbitrary $\omega, \omega_1, \omega_2 \in C^p_c(M)$ and a positive definite Riemannian structure on M that induces a positive definite inner product $\langle \cdot , \cdot \rangle_p$ on $C^p_c(M)$, the following two properties are required of the mapping C:

Inst.	n, p	α	β	η	λ	$t\beta$ (= 0 on S_1)	$t\eta$ (= 0 on S_2)
1	4, 2	$\boldsymbol{A}, \phi$	$\boldsymbol{E}, \boldsymbol{B}$	$\boldsymbol{D}, \boldsymbol{H}$	$\boldsymbol{J}, \rho$	$\boldsymbol{n}\times\boldsymbol{E}$, $\boldsymbol{B}\cdot\boldsymbol{n}$	$\boldsymbol{n}\times\boldsymbol{H}$, $\boldsymbol{D}\cdot\boldsymbol{n}$
2	3, 1	ϕ	$\boldsymbol{E}$	$\boldsymbol{D}$	ρ	$\boldsymbol{n}\times\boldsymbol{E}$	$\boldsymbol{D}\cdot\boldsymbol{n}$
3	3, 2	$\boldsymbol{A}$	$\boldsymbol{B}$	$\boldsymbol{H}$	$\boldsymbol{J}$	$\boldsymbol{B}\cdot\boldsymbol{n}$	$\boldsymbol{n}\times\boldsymbol{H}$
4	3, 2	$\boldsymbol{H}$	$\boldsymbol{J}$	$\boldsymbol{E}$	$-\partial\boldsymbol{B}/\partial t$	$\boldsymbol{J}\cdot\boldsymbol{n}$	$\boldsymbol{n}\times\boldsymbol{E}$
5	2, 1	ψ	$\boldsymbol{n}'\times\boldsymbol{J}$	$\boldsymbol{E}$	$-(\partial\boldsymbol{B}/\partial t)\cdot\boldsymbol{n}'$	J_n	E_t
6	2, 1	A_n	$\boldsymbol{n}'\times\boldsymbol{B}$	$\boldsymbol{H}$	$\boldsymbol{J}\cdot\boldsymbol{n}'$	B_n	H_t
7	2, 1	ϕ	$\boldsymbol{E}$	$\boldsymbol{n}'\times\boldsymbol{D}$	ρ/length	E_t	D_n

Table 7.1. Instances of the paradigm problem (keyed to the list on page 185). Note that two-dimensional problems are assumed to be embedded in a three-dimensional space, where $\boldsymbol{n}'$ is the unit normal to M.

Instance of paradigm problem	Relevant examples
1	7.10
2	1.6, 1.7, 1.14, 3.2, 3.6, 7.2, 7.11, 7.13
3	1.9, 1.15, 3.3, 3.6, 3.4, 7.2, 7.11, 7.13
4	1.9, 2.3, 3.6, 7.2, 7.11, 7.13
5	1.5, 1.8, 1.13, 3.5, 3.6, 7.12, 7.14
6	1.8, 7.12, 7.14
7	1.8, 7.12, 7.14

Table 7.2. Instances of the paradigm problem cross-referenced to examples in this book. Refer to the list of examples (page 273) for page numbers.

(1) Strict monotonicity:

$$\langle C(\omega_1) - C(\omega_2), \omega_1 - \omega_2\rangle \geq 0,$$

with equality if and only if $\omega_1 = \omega_2$.

(2) Symmetry: defining the linear functional

$$f_\omega(\omega_1) = \langle C(\omega), \omega_1\rangle_p$$

and denoting its Gâteaux variation by

$$f'_\omega(\omega_1, \omega_2) = \langle C'_\omega(\omega_2), \omega_1\rangle_p,$$

it is required that this function is a symmetric bilinear function of ω_1 and ω_2. That is,

$$\langle C'_\omega(\omega_1), \omega_2\rangle_p = \langle C'_\omega(\omega_2), \omega_1\rangle_p.$$

The first of these two conditions ensures the invertibility of C. When there is a pseudo-Riemannian structure on the manifold the inner product $\langle\cdot,\cdot\rangle_p$ is indefinite as is the case in four-dimensional versions of electromagnetics and the appropriate reformulation of condition 1 is found in [Ton72a] pp. 351–352. The second of the two conditions above will imply that there exists a variational principle for the problem at hand. See [Vai64] or [Ton69] for a thorough discussion. This being said, let the constitutive relation between β and η be expressed in terms of the Hodge star operator

$$\eta = *C(\beta).$$

Defining a potential. The next step in formulating the paradigm problem variationally, is to relate $\beta \in Z^p_c(M - S_1)$ to a potential α. In all of the special cases of the paradigm problem shown in Table 7.1, with the exception of the fifth case involving current flow on sheets, the physics of the problem shows that it is reasonable to assume that

$$\beta = d\alpha,$$

since $M \subset \mathbb{R}^n$ and the preceding equation is true for $\mathbb{R}^n$. In the fifth case involving currents on sheets, one can use the techniques developed in Example 2.4 (page 85) to express the current density vector in terms of a stream function which has jump discontinuities on a set of curves representing generators of $H_1(M, S_2)$. The values of these jump discontinuities are related to the time rate of change of magnetic flux through "holes" and "handles" and are prescribed as a principal condition in any variational formulation. Keeping this in mind, it is assumed that

$$\beta = d\alpha \quad \text{for some } \alpha \in C^{p-1}(M)$$

in the paradigm problem. The next thing to do in formulating a variational principle, where $\beta \in Z^p_c(M - S_1)$ is imposed as a principal condition, is to figure out a way of imposing the condition

$$t\beta = 0 \quad \text{on } S_1$$

in terms of a vector potential α. In general, since the exterior derivative commutes with pullbacks, the observation that

$$t\alpha = 0 \implies 0 = dt\alpha = td\alpha = t\beta \quad \text{on } S_1$$

does not mean that it is advisable to make the pullback of α to S_1 vanish. To see why this is so, consider the following portion of the long exact homology sequence for the pair (M, S_1):

$$\cdots \xrightarrow{\delta_{p+1}} H_p(S_1) \xrightarrow{\tilde{\imath}_p} H_p(M) \xrightarrow{\tilde{\jmath}_p} H_p(M, S_1) \longrightarrow$$
$$\xrightarrow{\delta_p} H_{p-1}(S_1) \xrightarrow{\tilde{\imath}_{p-1}} H_{p-1}(M) \xrightarrow{\tilde{\jmath}_{p-1}} H_{p-1}(M, S_1) \longrightarrow \cdots$$

The three-step procedure for finding homology generators introduced in Chapter 1 (page 38) gives

$$H_p(M, S_1) \simeq \delta_p^{-1}\left(\ker(\tilde{\imath}_{p-1})\right) \oplus \tilde{\jmath}_p\left(\frac{H_p(M)}{\tilde{\imath}_p\left(H_p(S_1)\right)}\right).$$

The above arguments concerning the existence of a potential α deal with the periods of β on generators of $H_p(M)$ and hence the generators of $H_p(M, S_1)$ corresponding to $\text{im}(\tilde{\jmath}_p)$. It remains to consider how the periods of β on generators of $H_p(M, S_1)$ corresponding to

$$\delta_p^{-1}(\ker(\tilde{\imath}_{p-1}))$$

depend on the tangential components of α on S_1. Let $z_p \in Z_p(M, S_1)$ represent a nonzero homology class in

$$\delta_p^{-1}(\ker(\tilde{\imath}_{p-1})),$$

and consider the calculation of the period of β on this homology class:

$$\int_{z_p} \beta = \int_{z_p} d\alpha = \int_{\partial z_p} t\alpha = 0 \quad \text{if } t\alpha = 0 \text{ on } S_1.$$

Hence, unless the periods of β vanish on

$$\delta_p^{-1}(\ker(\tilde{\imath}_{p-1})),$$

there is no hope of making the tangential components of α vanish on S_1. Instead, one must find a way of prescribing $t\alpha$ on S_1 such that

$$dt\alpha = td\alpha = t\beta = 0 \quad \text{on } S_1$$

and the periods of β on generators of

$$\delta_p^{-1}(\ker(\tilde{\imath}_{p-1}))$$

are prescribed. This is simple in the case where $p = 1$, since a scalar potential or stream function is forced to be a constant on each connected component of S_1 if its exterior derivative vanishes there. For vector potentials ($p = 2$) the problem is trickier, since the tangential components of the vector potential on S_1 are related to some scalar function which has jump discontinuities on curves representing generators of $H_1(S_2, \partial S_2)$. This situation should present no difficulties since it has been considered in examples 1.14, 1.15, and 2.3.

A First Variational Formulation. As a prelude to the variational formulation of the paradigm problem, one has

$$\eta = *C(\beta) \text{ in } M, \qquad \beta = d\alpha \text{ in } M, \qquad t\alpha \text{ specified on } S_1.$$

The last two conditions are used to ensure that $\beta \in Z_c^p(M - S_1)$ and the periods of β on $\delta_p^{-1}(\ker(\tilde{\imath}_{p-1}))$ are prescribed in some definite way. One is now required to find a variational principle which would have

$$d\eta = \lambda \text{ in } M, \qquad t\eta = 0 \text{ on } S_2$$

as the Euler–Lagrange equation and natural boundary condition respectively. A variational principle for this problem is a functional

$$F : C^{p-1}(M) \to \mathbb{R}$$

which is stationary at the $p-1$ form α and satisfies the requirements above. To define a variational principle, consider a family of $(p-1)$-forms parametrized differentiably by s, that is, a curve

$$\gamma : [0, 1] \to C^{p-1}(M),$$

with $\gamma(0) = \alpha_0$ an initial state and $\gamma(1) = \alpha$ an extremal; in order to respect the principal boundary condition on S_1, we fix

$$t\gamma(s) = t\alpha \quad \text{for all } s \in [0,1].$$

No other constraints are placed on γ, so that

$$\left.\frac{\partial \gamma}{\partial s}\right|_{s=1} = \tilde{\alpha},$$

the variation of the extremal, can be an arbitrary element of the space $C^p_c(M-S_1)$ of admissible variations. Next, the symmetry condition that

$$\langle C'_\omega(\omega_1), \omega_2\rangle_p = \langle C'_\omega(\omega_2), \omega_1\rangle_p \tag{7–1}$$

for all $\omega, \omega_1, \omega_2 \in C^p(M)$ ensures that the value of the functional F defined by

$$F(\alpha) = F(\alpha_0) + \int_0^1 \left(\left\langle C\,(d\gamma(s)),\, d\frac{\partial \gamma(s)}{\partial s} \right\rangle_p + (-1)^r \left\langle *\lambda, \frac{\partial \gamma(s)}{\partial s} \right\rangle_{p-1} \right) ds, \tag{7–2}$$

where $r = (n-p+1)(p-1) + (p-1)$, is independent of the path in $C^{p-1}(M)$ joining α_0 and α. That is, the value of the right-hand side of the equation above does not depend on the way in which $\gamma(s)$ goes from α_0 to α as s goes form zero to one (see [Ton72a]). For a general view of this formulation of variational functionals the reader is referred to [Ton69] and [Vai64]. The functional (7–2) follows directly from the equations defining the paradigm problem. In Vainberg's picture, symmetry condition (7–1) is analogous to the vanishing of the curl in function space, and this in turn ensures that a variational principle exists.

To verify that the extremal of the functional above has the required properties, recall that an extremal of the functional and the variation of the extremal are assumed to be

$$\gamma(1) = \alpha, \qquad \left.\frac{\partial \gamma(s)}{\partial s}\right|_{s=1} = \tilde{\alpha}.$$

This implies that variations of the extremal can be considered by looking at $\gamma(1-\varepsilon)$ for ε sufficiently small, and that the condition for the functional to be stationary at α is:

$$\left.\frac{\partial F}{\partial \varepsilon}(\gamma(1-\varepsilon))\right|_{\varepsilon=0} = 0.$$

Using the definition of the inner product, one can rewrite the functional as

$$F(\alpha) = F(\alpha_0) + \int_0^1 \left(\int_M d\frac{\partial \gamma(s)}{\partial s} \wedge *C(d\alpha) + (-1)^{p-1} \int_M \frac{\partial \gamma(s)}{\partial s} \wedge \lambda \right) ds.$$

Using this expression it is seen that the functional is stationary at α when

$$\begin{aligned} 0 &= \frac{\partial}{\partial \varepsilon} \int_0^{1-\varepsilon} \left(\int_M d\frac{\partial \gamma(s)}{\partial s} \wedge *C\,(d\gamma(s)) + (-1)^{p-1} \int_M \frac{\partial \gamma(s)}{\partial s} \wedge \lambda \right) ds \Bigg|_{\varepsilon=0} \\ &= -\int_M d\frac{\partial \gamma(1)}{\partial s} \wedge *C\,(d\gamma(1)) - (-1)^{p-1} \int_M \frac{\partial \gamma(1)}{\partial s} \wedge \lambda \\ &= -\int_M d\tilde{\alpha} \wedge *C(d\alpha) - (-1)^{p-1} \int_M \tilde{\alpha} \wedge \lambda \end{aligned}$$

for all $\tilde{\alpha} \in C_c^{p-1}(M - S_1)$. The integration by parts formula, which was obtained as a corollary to Stokes' theorem, shows that:

$$\int_M d\tilde{\alpha} \wedge *C(d\alpha) = \int_{\partial M} t\tilde{\alpha} \wedge t\,(*C(d\alpha)) - (-1)^{p-1} \int_M \tilde{\alpha} \wedge d * C(d\alpha).$$

Combining these two equations, we see that the functional is stationary at α if

$$0 = (-1)^{p-1} \int_M \tilde{\alpha} \wedge (d * C(d\alpha) - \lambda) - \int_{\partial M} t\tilde{\alpha} \wedge t\,(*C(d\alpha))$$

for all $\tilde{\alpha} \in C_c^{p-1}(M - S_1)$. This of course means that

$$d * C(d\alpha) = \lambda \quad \text{in } M$$

is the Euler–Lagrange equation and

$$t\,(*C(d\alpha)) = 0 \quad \text{on } \partial M - S_1 = S_2$$

the natural boundary condition. Noting that

$$\eta = *C(\beta), \qquad \beta = d\alpha$$

the Euler–Lagrange equation and the natural boundary conditions state that the functional is stationary when

$$d\eta = \lambda \quad \text{in } M,$$
$$t\eta = 0 \quad \text{on } S_2.$$

Thus it is seen that the paradigm problem is amenable to a variational formulation.

Interface conditions revisited. Before moving to the questions of existence and uniqueness of extremal, it is useful to mention how the interface conditions associated with the two integral laws of the paradigm problem are handled in the variational formulation, since this aspect has been ignored in the above calculations. Interface conditions are considered when the function C is discontinuous along some $(n-1)$-dimensional manifold S. In the variational formulation it is assumed that the potential α is continuous everywhere in M and differentiable in $M - S$. One can define an orientation locally on S and hence a plus side and a minus side. In this case, if superscripts refer to a limiting value of a differential form from a particular side of S, then

$$t\beta^+ = t\beta^- \quad \text{on } S$$

is the interface condition associated with the integral law

$$\int_{\partial c_{p+1}} \beta = 0 \quad \text{for all } c_{p+1} \in C_{p+1}(M).$$

That this interface condition results as a consequence of the continuity requirements imposed on α is seen from the following argument. Since α is continuous in M one has

$$t\alpha^+ = t\alpha^- \quad \text{on } S.$$

Both sides of the above equation are differentiable with respect to the directions tangent to S because α is assumed differentiable in $M-S$. The exterior derivative in $C^*(S)$ involves only these tangential directions, so

$$dt\alpha^+ = dt\alpha^- \quad \text{on } S$$

but

$$td\alpha^+ = td\alpha^- \quad \text{on } S$$

or

$$t\beta^+ = t\beta^- \quad \text{on } S,$$

since exterior differentiation commutes with pullbacks. Similarly, when λ has bounded coefficients, the interface condition

$$t\eta^+ = t\eta^- \quad \text{on } S$$

is associated with the integral law

$$\int_{\partial c_{n-p+1}} \eta = \int_{c_{n-p+1}} \lambda.$$

To see how this interface condition comes out of the variational formulation, notice that

$$d{*}C(d\alpha)$$

need not exist on S. Hence, if there are interfaces, then taking the variation of the functional one must use the integration by parts formula in $M - S$. When this is done, one obtains the same answer as before plus the term

$$-\int_S t\tilde{\alpha} \wedge t\left({*}C(d\alpha^+) - {*}C(d\alpha^-)\right).$$

The arbitrariness of $t\tilde{\alpha}$ on S implies that

$$t * C(d\alpha^+) = t * C(d\alpha^-).$$

With the identifications

$$\eta = {*}C(\beta), \qquad \beta = d\alpha$$

one has the desired result. This completes the discussion of the constitutive relation and the variational principle.

7C. Gauge Transformations and Conservation Laws

The objective of this section is to develop a feeling for how nonunique the solution of the paradigm problem can be and to show how this nonuniqueness is related to the compatibility conditions which must be satisfied in order for a solution to the paradigm problem to exist. The approach taken in this section is basically due to [Ton68], however, it is more general than Tonti's in that the role of homology groups is considered. Every effort is made to avoid using the words local and global because the mathematical usage of the words local and global does not coincide with the meanings attributed to these words by physicists working in field theory.

For the paradigm problem being considered, we define a gauge transformation as a transformation on the potential α which leaves the following quantities untouched:

$$\beta = d\alpha \quad \text{in } M,$$
$$t\alpha \quad \text{on } S_1.$$

The gauge transformation is assumed to have the form

$$\alpha \to \alpha + \alpha_G \quad \text{in } M,$$

where $\alpha_G \in Z_c^{p-1}(M - S_1)$. It is obvious that α_G cannot lie in any bigger space since, by definition

$$Z_c^{p-1}(M - S_1) = \{\omega \mid \omega \in C_c^{p-1}(M - S_1),\ d\omega = 0 \text{ in } M\}.$$

By the orthogonal decomposition discussed in Section MA-M, it is known that

$$Z_c^{p-1}(M - S_1) = B_c^{p-1}(M - S_1) \oplus \mathcal{H}^{p-1}(M, S_1),$$

where

$$\mathcal{H}^{p-1}(M, S_1) = \{\omega \mid \omega \in Z_c^{p-1}(M - S_1),\ \ n\omega = 0 \text{ on } S_2,\ \delta\omega = 0 \text{ in } M\}$$

and

$$\beta_{p-1}(M, S_1) = \dim \mathcal{H}^{p-1}(M, S_1).$$

This orthogonal decomposition enables one to characterize the space of the gauge transformations. In scalar potential problems, that is, cases 2, 5, 6, 7 in Table 7.1, p is equal to 1 and $\alpha_G \in \mathcal{H}^0(M, S_1)$ since $B_c^0(M - S_1)$ is the space containing only the zero vector. This situation is trivial to interpret since α_G is equal to some constant in each connected component of M which does not contain a subset of S_1. In problems where $p = 2$, that is, cases 1, 3, 4 in Table 7.1, one has

$$\alpha_G \in B_c^1(M - S_1) \oplus \mathcal{H}^1(M, S_1).$$

Thus it is expected that the gauge transformation can be described by a scalar function which vanishes on S_1 and $\beta_1(M, S_1)$ other degrees of freedom. The case where n is equal to three is treated explicitly in [Kot82, Section 4.2.2].

Since the gauge transformation is supposed to leave the differential form β invariant, one would hope that the gauge transformation would also leave the stationary value of the functional invariant. To formalize this intuition, suppose α is an extremal and let

$$\gamma : [0, 1] \to C_c^{p-1}(M - S_1),$$

where

$$\gamma(s) = \alpha + s\alpha_G$$

and

$$\frac{\partial \gamma(s)}{\partial s} = \alpha_G \in Z_c^{p-1}(M - S_1) \quad \text{for all } s \in [0, 1].$$

In this case, recalling the definition of the variational functional for the paradigm problem, one has

$$\begin{aligned}
F(\alpha+\alpha_G) - F(\alpha) \\
&= F(\gamma(1)) - F(\gamma(0)) \\
&= \int_0^1 \left(\int_M d\left(\frac{\partial\gamma(s)}{\partial s}\right) \wedge *C\left(d\gamma(s)\right) + (-1)^{p-1}\int_M \left(\frac{\partial\gamma(s)}{\partial s}\right)\wedge\lambda \right) ds \\
&= (-1)^{p-1}\int_0^1 \left(\int_M \alpha_G\wedge\lambda\right) ds \quad \text{since } d\left(\frac{\partial\gamma(s)}{\partial s}\right) = 0 \\
&= (-1)^{p-1}\int_M \alpha_G\wedge\lambda.
\end{aligned}$$

Thus the gauge transformation leaves the value of the functional invariant if and only if

$$\int_M \alpha_G\wedge\lambda = 0 \quad \text{for all } \alpha_G \in Z_c^{p-1}(M-S_1).$$

This condition can be rewritten as

$$\langle \alpha_G, *\lambda\rangle_{p-1} = 0 \quad \text{for all } \alpha_G \in Z_c^{p-1}(M-S_1).$$

However, from the orthogonal decomposition theorem developed in Section MA-M, it is known that

$$\left(Z_c^{p-1}(M-S_1)\right)^{\perp} = \widetilde{B}_{p-1}(M,S_2) = *B_c^{n-p+1}(M-S_2);$$

hence $*\lambda \in *B_c^{n-p+1}(M-S_2)$ or $\lambda \in B_c^{n-p+1}(M-S_2)$. This condition is precisely the compatibility condition that ensures that the equations

$$\begin{aligned} d\eta &= \lambda \qquad \text{in } M, \\ t\eta &= 0 \qquad \text{on } S_2 \end{aligned}$$

are solvable for η. Thus the Euler–Lagrange equation and the natural boundary conditions can be satisfied only when the stationary value of the functional is invariant under any gauge transformation.

The compatibility condition on λ is not amenable to direct verification in its present form. However, since

$$Z_c^{n-p+1}(M-S_2) \simeq B_c^{n-p+1}(M-S_2) \oplus H_c^{n-p+1}(M-S_2),$$

one sees that the compatibility condition can by verified by checking

$$\left.\begin{aligned} d\lambda = 0 \quad &\text{in } M \\ t\lambda = 0 \quad &\text{on } S_2 \end{aligned}\right\} \Rightarrow \lambda \in Z_c^{n-p+1}(M-S_2)$$

and then verifying that the periods of λ vanish on a set of generators of

$$H_{n-p+1}(M,S_2).$$

This, in particular confirms the results given in [Kot82] which were considered in Example 3.4. This method of verifying the compatibility condition on λ also shows that the duality theorem

$$H_c^{p-1}(M-S_1) \simeq H_c^{n-p+1}(M-S_2)$$

plays a crucial role in interrelating degrees of freedom in the gauge transformation and degrees of freedom in λ constrained by the compatibility condition. It is

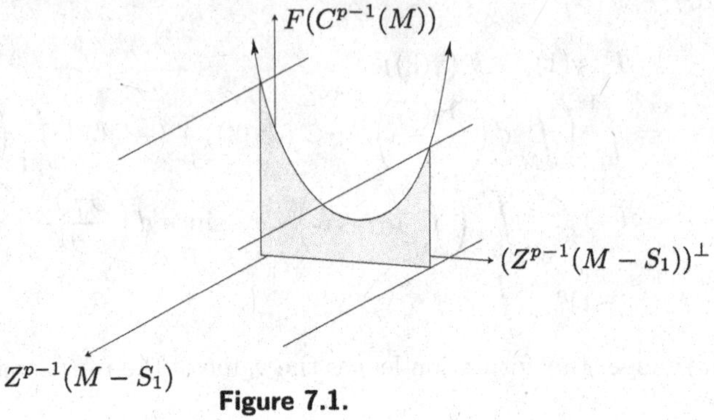

Figure 7.1.

worth mentioning that

$$\lambda \notin B_c^{n-p+1}(M-S_2)$$

implies that the value of the functional is not invariant under every gauge transformation and that the Euler–Lagrange equation or the natural boundary conditions cannot be satisfied. In this case the functional has no extremum and it is useful to have a geometrical picture of the situation. Consider the diagram given in Figure 7.1. The graph of the functional in the "plane" spanned by $F(C^{p-1}(M))$ and $(Z_c^{p-1}(M-S_1))^\perp$ is upward convex whenever the Riemannian structure on M is positive definite. This comes about as a result of the strict monotonicity assumption on the constitutive relation which is a valid assumption to make in all of the cases of the paradigm problem listed in Table 7.1 with the exception of electromagnetism in four dimensions. For simplicity, in the remainder of this section the discussion will focus on the case of convex functionals.

The Role of Convexity. When the functional is invariant under gauge transformations, moving in the direction of $Z_c^{p-1}(M-S_1)$ does not change the value of the functional so that the graph looks like an infinitely long level trough which is convex upward in the "plane" $F(C^{p-1}(M))$–$(Z_c^{p-1}(M-S_1))^\perp$. However, when the functional is not invariant under gauge transformations, that is, $\lambda \notin B_c^{n-p+1}(M-S_2)$ the trough is tilted and the functional has no stationary point. In this case the graph in the $F(C^{p-1}(M))$–$(Z_c^{p-1}(M-s_1))^\perp$ "plane" remains the same but the slope in the $Z_c^{p-1}(M-S_1)$ direction has a nonzero value depending on the value of the projection

$$\frac{\int_M \alpha_G \wedge \lambda}{\sqrt{\langle \alpha_G, \alpha_G \rangle_{p-1}}}.$$

Thus, the interplay between gauge conditions and conservation laws arises from the above projection and gives a geometrical picture as to what happens when conservation laws are violated.

It has been seen that the compatibility condition $\lambda \in B_c^{n-p+1}(M - S_2)$ is necessary for the functional to have a minimum. In the case of a linear constitutive relation the Euler–Lagrange equation is a linear operator equation, so if the spaces in question are chosen so that the range of the operator is closed, the condition

$$\langle \alpha_G, *\lambda \rangle_{p-1} = 0 \quad \text{for all } \alpha_G \in Z_c^{p-1}(M - S_1)$$

is sufficient to ensure the solvability of the Euler–Lagrange equation [Ton68], since the Fredholm alternative is applicable in this case. In the case of a nonlinear strictly monotone constitutive relation, the resulting convex functional may fail to have an extremum even if the above orthogonality condition holds. The extra condition which is required is

$$\frac{\langle C(\omega), \omega \rangle_p}{\sqrt{\langle \omega, \omega \rangle_p}} \to \infty \quad \text{as } \langle \omega, \omega \rangle_p \to \infty$$

for all $\omega \in C_c^p(M)$. The reason why this condition is necessary is best understood in terms of an example.

Example 7.1 A convex function without a minimum. Let

$$f(\xi) = \sqrt{1 + \xi^2} - l\xi, \; l, \xi \in \mathbb{R}^1.$$

It is readily verified that

$$f'(\xi) = \xi(1 + \xi^2)^{-1/2} - l,$$
$$f''(\xi) = (1 + \xi^2)^{-3/2}.$$

Since the second derivative is always positive, the function is convex for all values of ξ. At a minimum value we must have

$$f'(\xi) = 0 = \xi(1 + \xi^2)^{-1/2} - l$$

or

$$\xi = l(1 + \xi^2)^{1/2} \Rightarrow \xi = \frac{l}{\sqrt{1 - l^2}}.$$

Thus, the convex function f has no minimum if $|l| > 1$. To see how this example relates to the above condition, make the identifications

$$f(\xi) = \int_0^\xi C(\tau)\, d\tau - l\xi, \qquad (C(\tau), \tau) = \tau C(\tau),$$

so that

$$C(\xi) = f'(\xi) + l = \frac{\xi}{\sqrt{1 + \xi^2}};$$

in this case

$$\lim_{|\xi| \to \infty} \frac{C(\xi)\xi}{|\xi|} = \frac{\xi^2}{\sqrt{1 + \xi^2}|\xi|} = \frac{|\xi|}{\sqrt{1 + \xi^2}} = 1 < \infty,$$

so that the extra condition imposed on the constitutive relation is violated. □

Example 7.1 shows that in the paradigm problem being considered, if

$$\lim_{\sqrt{\langle\omega,\omega\rangle_p}\to\infty} \frac{\langle C(\omega),\omega\rangle_p}{\sqrt{\langle\omega,\omega\rangle_p}} < \infty$$

then one expects that for some $\lambda \in B_c^{n-p+1}(M - S_2)$ with sufficiently large norm, the functional of the paradigm problem may fail to have a minimum. The interpretation of this extra condition in terms of the trough picture is as follows. Suppose $\lambda \in B_c^{n-p+1}(M - S_2)$ and consider the graph of the functional in the plane defined by

$$F\big((Z_c^{p-1}(M - S_1))^{\perp}\big) \quad \text{and} \quad (Z_c^{p-1}(M - S_1))^{\perp}$$

as a function of the norm of λ as shown in Figure 7.2. This diagram illustrates how the minimum value of the functional may tend to minus infinity as the norm of λ increases and the condition

$$\lim_{\|\omega\|_p\to\infty} \frac{\langle C(\omega),\omega\rangle}{\|\omega\|_p} = \infty$$

is violated. Thus, when thinking of the graph of $F(\alpha)$ as a trough, one sees that the trough is tilted in the $Z_c^{p-1}(M - S_1)$ direction when λ violates some conservation law, and the trough "rolls over" when the above condition is not satisfied and λ is chosen in a suitable way.

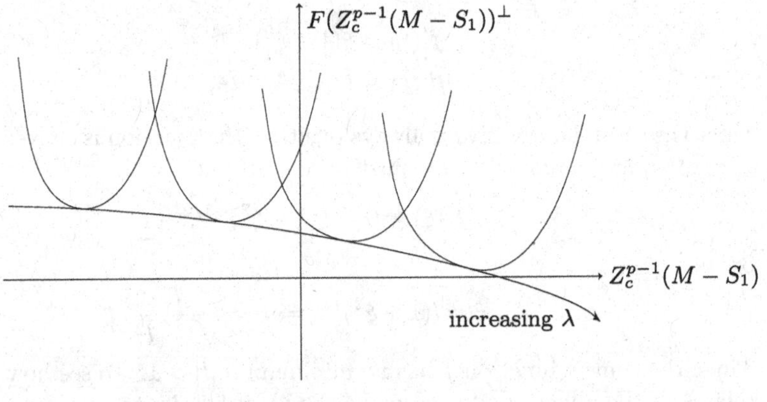

Figure 7.2.

For numerical work, one would like a variational principle whose extremum always exists and is unique. The variational principle for the paradigm problem has a unique solution if and only if the space $Z_c^{p-1}(M-S_1)$ which is homologous to $B_c^{p-1}(M - S_1) \oplus H_c^{p-1}(M - S_1)$ contains only the null vector. By the above direct sum decomposition this happens in practical problems where $p = 1$ (so that $B_c^0(M - S_1) = 0$) and there is a Dirichlet condition imposed on some part of the boundary of each connected component of M, (so that $H_c^0(M - S_1) = 0$). When the extremal of the functional is nonunique, the usual algorithms for minimizing convex functionals can be generalized to the case where the extremum of the functional is nonunique. For example, Newton's method as described by [Lue69],

Section 10.4 can be generalized as in [Alt55]. However, in such cases it is usually easier to reformulate the variational principle for the paradigm problem in such a way that there always exist a unique solution. There are two basic approaches to this problem which will be considered next.

7D. Modified Variational Principles

The purpose of this section is to formulate variational principles for the paradigm problem for which the potential α is unique. Such variational principles will have interesting consequences for conservation laws since a unique solution for the potential α implies that there is no gauge transformation which in turn implies that there is no conservation law which is naturally associated with the functional.

The first approach to the problem is to note that once the principal boundary conditions have been imposed on S_1 the space of admissible variations of the extremal is $C_c^{p-1}(M - S_1)$ and the space of gauge transformations is $Z_c^{p-1}(M - S_1)$. Hence, if the space of admissible variations of the functional and the domain of the functional is restricted to

$$\left(Z_c^{p-1}(M - S_1)\right)^{\perp} \cap C_c^{p-1}(M - S_1),$$

the functional's previous minimum can still be attained but the solution is now unique. By the orthogonal decomposition developed in Section MA-M, one has

$$(Z_c^{p-1}(M - S_1))^{\perp} = \widetilde{B}_{p-1}(M, S_2)$$

hence the space of admissible variations becomes

$$\begin{aligned}&\widetilde{B}_{p-1}(M, S_2) \cap C_c^{p-1}(M - S_1) = \\ &\qquad \left\{\tilde{\alpha} \mid t\tilde{\alpha} = 0 \text{ on } S_1,\ \tilde{\alpha} = \delta_p\omega \text{ in } M \text{ for some } \omega \in C_c^p(M) \text{ with } n\omega = 0 \text{ on } S_2\right\}.\end{aligned}$$

This procedure raises an interesting question. By the observations of [Ton68] one knows that the number of degrees of freedom in the gauge transformation is equal to the number of degrees of freedom by which the source, described by λ, is constrained by a conservation law. Hence in this case where the domain of the functional is constrained, so that the extremal is unique, one expects that the variational principle is completely insensitive to violations of the conservation law $\lambda \in B_c^{n-p+1}(M - S_2)$. To see why this is so, consider the unique decomposition

$$\lambda = \lambda_{\text{cons}} + \lambda_{\text{nonc}},$$

where

$$\lambda \in C_c^{n-p+1}(M), \quad \lambda_{\text{cons}} \in B_c^{n-p+1}(M - S_2), \quad \lambda_{\text{nonc}} \in (B_c^{n-p+1}(M - S_2))^{\perp}.$$

What is required is to show that the extremal of the functional is independent of the way in which λ_{nonc} is prescribed. Considering the functional of the paradigm problem as a function of λ, when α is restricted as above, one has

$$\begin{aligned}F_\lambda(\alpha) - F_{\lambda_{\text{cons}}}(\alpha) &= (-1)^{p-1}\int_M \alpha \wedge \lambda_{\text{nonc}} \\ &= (-1)^{(p-1)-(n-p+1)(p-1)}\int_M \lambda_{\text{nonc}} \wedge \alpha \\ &= (-1)^{(p-1)}\int_M \lambda_{\text{nonc}} \wedge **\alpha = (-1)^{(p-1)}\langle \lambda_{\text{nonc}}, *\alpha\rangle_{n-p+1}.\end{aligned}$$

However, while

$$*\alpha \in *\left((Z_c^{p-1}(M-S_1))^\perp\right) = *\widetilde{B}_{p-1}(M,S_2) = B_c^{n-p+1}(M-S_2),$$

we also have

$$\lambda_{\text{nonc}} \in \left(B_c^{n-p+1}(M-S_2)\right)^\perp,$$

and combining these two results, we get

$$\langle \lambda_{\text{nonc}}, *\alpha\rangle_{n-p+1} = 0,$$

so that

$$F_\lambda(\alpha) = F_{\lambda_{\text{cons}}}(\alpha).$$

Thus, by restricting the class of admissible variations of the functional's extremal, one obtains a variational formulation whose unique extremal is insensitive to violations of the compatibility condition $\lambda \in B^{n-p+1}(M-S_1)$. This approach to the problem is useful in the context of direct variational methods such as the Ritz method or the finite element method only when it is possible to find basis functions which ensure that

$$\alpha \in \left(Z_c^{p-1}(M-S_1)\right)^\perp = \widetilde{B}_{p-1}(M,S_2).$$

The second method for obtaining a variational formulation of the paradigm problem in which the extremal is unique is inspired by [Kot82] chapter 5. In this method, which at first sight resembles the "penalty function method" (see [Lue69], sect. 10.11), the domain of the functional before principal boundary conditions are imposed is $C_c^{p-1}(M)$. The method involves finding a functional $F^\perp(\alpha)$ whose graph looks like a trough perpendicular to the trough of $F(\alpha)$:

In this scheme the functional

$$G(\alpha) = F(\alpha) + F^\perp(\alpha)$$

has a unique minimum which lies above the $(Z_c^{p-1}(M-S_1))^\perp$ "axis" whenever the trough associated with $F(\alpha)$ is not tilted. That is if $F^\perp(\alpha)$ is designed so that its minimum is the $(Z_c^{p-1}(M-S_1))^\perp$ "axis" then the minimum of $G(\alpha)$ should lie above the $(Z_c^{p-1}(M-S_1))^\perp$ "axis" whenever $\lambda \in B_c^{n-p+1}(M-S_2)$. It is also desired that the contrapositives of these statements are also true in the following sense. If $\lambda \notin B_c^{n-p+1}(M-S_2)$ so that the trough associated with $F(\alpha)$ is "tilted" then the distance of the extremum of the functional $G(\alpha)$ to the $G(\alpha) - (Z_c^{p-1}(M-S_1))^\perp$ plane measures, in some sense, the value of the projection

$$\|\alpha_G\|_{p-1}^{-1} \max_{\alpha_G \in Z_c^{p-1}(M-S_1)} \int_M \alpha_G \wedge \lambda.$$

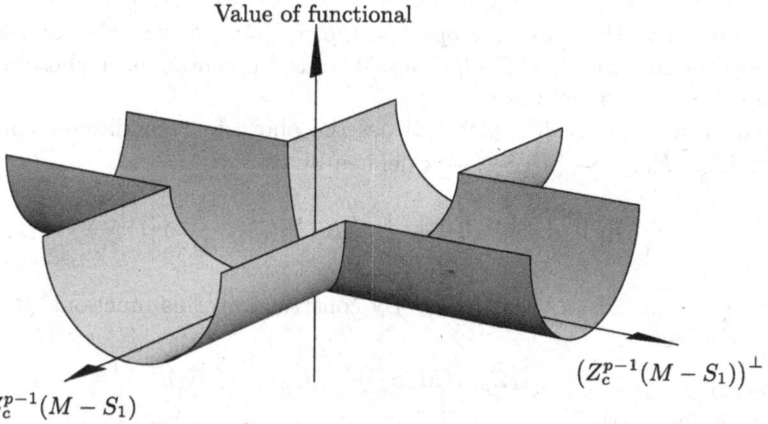

Figure 7.3. The troughs associated with $F(\alpha)$ (on the $Z_c^{p-1}(M-S_1)$ axis) and $F^\perp(\alpha)$.

Having this picture in mind, the first thing to do is construct a functional with the properties desired of $F^\perp(\alpha)$. To find a functional which is definite on $Z_c^{p-1}(M-S_1)$ and level on $(Z_c^{p-1}(M-S_1))^\perp$, one notes that the orthogonal decomposition of Section MA-M gives

$$\left(Z_c^{p-1}(M-S_1)\right)^\perp = \widetilde{B}_{p-1}(M,S_2), \qquad Z_c^{p-1}(M-S_1) = \left(\widetilde{B}_{p-1}(M,S_2)\right)^\perp.$$

Hence, one actually wants a functional $F^\perp(\alpha)$ which is level on $\widetilde{B}_{p-1}(M,S_2)$ and convex on $(\widetilde{B}_{p-1}(M,S_2))^\perp$.

As a prelude to the construction of $F^\perp(\alpha)$, let K be a map

$$K : C^{p-2}(M) \to C^{p-2}(M)$$

satisfying the same conditions associated with the constitutive mapping. That is, for ω, ω_1, $\omega_2 \in C^{p-2}(M)$ the following three properties are assumed to hold.

(1) Strict monotonicity:

$$\langle K(\omega_1) - K(\omega_2),\ \omega_1 - \omega_2\rangle_{p-2} \geq 0,$$

with equality if and only if $\omega_1 = \omega_2$.

(2) Symmetry: defining the functional

$$f_\omega(\omega_1) = \langle K(\omega), \omega_1\rangle_{p-2}$$

and denoting the Gâteaux variation of this functional by

$$f'_\omega(\omega_1,\omega_2) = \langle K'_\omega(\omega_2), \omega_1\rangle_{p-2},$$

it is required that this function is a symmetric bilinear function of ω_1 and ω_2. That is

$$\langle K'_\omega(\omega_1), \omega_2\rangle_{p-2} = \langle K'_\omega(\omega_2), \omega_1\rangle_{p-2}.$$

(3) Asymptotic property:

$$\lim_{\|\omega\|_{p-2}\to\infty} \frac{\langle K(\omega),\omega\rangle_{p-2}}{\|\omega\|_{p-2}} = \infty.$$

In addition to these usual properties the mapping K will also be assumed to satisfy the condition $K(0) = 0$, where 0 is the differential form whose coefficients vanish relative to any basis.

Given a mapping K which satisfies the above four conditions, consider the functional $F_0 : \widetilde{C}_{p-1}(M, S_2) \to \mathbb{R}$ defined by

$$F_0(\gamma(1)) = F_0(\gamma(0)) + \int_0^1 \left\langle K(\delta\gamma(s)), \frac{d}{ds}(\delta\gamma(s)) \right\rangle_{p-2} ds$$

for any $\gamma : [0,1] \to \widetilde{C}_{p-1}(M, S_2)$. By construction, this functional is convex in the subspace

$$\left(\widetilde{Z}_{p-1}(M, S_2)\right)^{\perp} \cap \widetilde{C}_{p-1}(M, S_2)$$

and "level" in the subspace $\widetilde{Z}_{p-1}(M, S_2)$. Furthermore, $F_0(\alpha) \geq F_0(0)$ with equality if and only if $\alpha \in \widetilde{Z}_{p-1}(M, S_2)$.

At this stage, the construction of $F^{\perp}(\alpha)$ is actually simple. Considering the orthogonal decomposition of Section MA-M, the following diagrams are readily seen to be true:

$$\begin{array}{ccccccc} B_c^{p-1}(M-S_1) & \subset & Z_c^{p-1}(M-S_1) & = & B_c^{p-1}(M-S_1) & \oplus & \mathcal{H}^{p-1}(M,S_1) \\ \| & & \| & & \| & & \| \\ \left(\widetilde{Z}_{p-1}(M,S_2)\right)^{\perp} & \subset & \left(\widetilde{B}_{p-1}(M,S_2)\right)^{\perp} & = & \left(\widetilde{Z}_{p-1}(M,S_2)\right)^{\perp} & \oplus & \mathcal{H}^{p-1}(M,S_1) \end{array} \tag{7–3}$$

and

$$\begin{array}{ccccccc} \widetilde{B}_{p-1}(M,S_2) & \subset & \widetilde{Z}_{p-1}(M,S_2) & = & \widetilde{B}_{p-1}(M,S_2) & \oplus & \mathcal{H}^{p-1}(M,S_1) \\ \| & & \| & & \| & & \| \\ (Z_c^{p-1}(M-S_1))^{\perp} & \subset & (B_c^{p-1}(M-S_1))^{\perp} & = & (Z_c^{p-1}(M-S_1))^{\perp} & \oplus & \mathcal{H}^{p-1}(M,S_1) \end{array} \tag{7–4}$$

where

$$*\mathcal{H}^{n-p+1}(M, S_2) = \mathcal{H}^{p-1}(M, S_1) = Z_c^{p-1}(M - S_1) \cap \widetilde{Z}_{p-1}(M, S_2).$$

Looking at these diagrams, we see that in (7–3) $F^{\perp}$ is supposed to be convex on the spaces listed in the second column while F_0 is convex in the spaces listed in the first column. Similarly, in (7–4), $F^{\perp}$ is invariant with respect to variations in the spaces listed in the first column while F_0 is invariant with respect to variations in the spaces listed in the second column. Thus, observing the direct sum decompositions in the third column of (7–3) and (7–4) it is obvious that the functional F_0 meets all of the specifications of $F^{\perp}$ except on the space $\mathcal{H}^{p-1}(M, S_1)$. More specifically, the functional $F^{\perp}$ is required to constrain the periods of a $p-1$ form in $\mathcal{H}^{p-1}(M, S_1)$ while the functional F_0 does not. To fix this discrepancy, let z_i, with $1 \leq i \leq \beta_{n-p+1}(M, S_2)$, be a set of generators of $H_{n-p+1}(M, S_2)$, and let k_i, with $1 \leq i \leq \beta_{n-p+1}(M, S_2)$, be a set of positive constants. For a curve

$$\gamma : [0,1] \to C_c^{p-1}(M),$$

the functionals

$$F_i : C_c^{p-1}(M) \to \mathbb{R}, \quad 1 \le i \le \beta_1(M, S_1),$$

defined by

$$\begin{aligned} F_i\Big(\gamma(1)\Big) &= F_i\Big(\gamma(0)\Big) + k_i \int_0^1 \left(\int_{z_i} *\gamma(s)\right)\left(\int_{z_i} *\frac{\partial\gamma(s)}{\partial s}\right) ds \\ &= F_i\Big(\gamma(0)\Big) + \frac{k_i}{2}\left(\left(\int_{z_i} *\gamma(1)\right)^2 - \left(\int_{z_i} *\gamma(0)\right)^2\right) \end{aligned}$$

have the property

$$F_i(\alpha) \ge F_i(0),$$

with equality if and only if the integral of $*\alpha$ over z_i vanishes. Next, consider the "candidate" functional

$$F_{\text{cand}}^{\perp}(\alpha) = \sum_{i=0}^{\beta_{n-p+1}(M,S_2)} F_i(\alpha).$$

Immediately, from the definitions of the F_i, one has

$$F_{\text{cand}}^{\perp}(\alpha) - F_{\text{cand}}^{\perp}(0) = \sum_{i=0}^{\beta_{n-p+1}(M,S_2)} (F_i(\alpha) - F_i(0)) \ge 0,$$

with equality if and only if

$$F_i(\alpha) = F_i(0), \quad 0 \le i \le \beta_{n-p+1}(M, S_2).$$

This last condition is equivalent to

$$\alpha \in \widetilde{Z}_{p-1}(M, S_2) \quad \text{and} \quad \int_{z_i} *\alpha = 0 \quad \text{for } 1 \le i \le \beta_{n-p+1}(M, S_2),$$

or

$$*\alpha \in *\widetilde{Z}_{p-1}(M, S_2) = Z_c^{n-p+1}(M - S_2), \qquad \int_{z_i} *\alpha = 0 \quad \text{for } 1 \le i \le \beta_{n-p+1}(M, S_2).$$

By the relative de Rham isomorphism, this is equivalent to

$$*\alpha \in B_c^{n-p+1}(M - S_2)$$

which in turn is equivalent to

$$\alpha \in *B_c^{n-p+1}(M - S_2) = \widetilde{B}_{p-1}(M, S_2) = \left(Z_c^{p-1}(M - S_1)\right)^{\perp}.$$

Hence, in summary,

$$F_{\text{cand}}^{\perp}(\alpha) - F_{\text{cand}}^{\perp}(0) \ge 0,$$

with equality if and only if

$$\alpha \in \left(Z_c^{p-1}(M - S_1)\right)^{\perp}.$$

Furthermore, by construction, $F_{\text{cand}}^{\perp}$ is convex when its domain is restricted to the space $Z_c^{p-1}(M - S_1)$. Thus interpreting $F_{\text{cand}}^{\perp}$ as a trough, it is seen that it satisfies the requirements of $F^{\perp}$ and hence can be used to define $F^{\perp}$. Hence, for a curve

$$\gamma : [0, 1] \to \widetilde{C}_{p-1}(M, S_2),$$

one can define the functional

$$F^{\perp} : \widetilde{C}_{p-1}(M, S_2) \to \mathbb{R}$$

as follows:

$$\begin{aligned} F^{\perp}(\gamma(1)) - F^{\perp}(\gamma(0)) &= F^{\perp}_{\text{cand}}(\gamma(1)) - F^{\perp}_{\text{cand}}(\gamma(0)) \\ &= \sum_{i=0}^{\beta_{n-p+1}(M,S_2)} \left(F_i(\gamma(1)) - F_i(\gamma(0))\right) \\ &= \int_0^1 \left\langle K(\delta\gamma(s)), \delta\left(\frac{\partial\gamma(s)}{\partial s}\right)\right\rangle_{p-2} ds \\ &\quad + \sum_{i=1}^{\beta_{n-p+1}(M,S_2)} k_i \int_0^1 \left(\int_{z_i} *\gamma(s)\right)\left(\int_{z_i} *\frac{\partial\gamma(s)}{\partial s}\right) ds. \end{aligned}$$

Finally one can complete the quest for a variational formulation of the paradigm problem in which the extremal of the functional is unique by letting

$$G(\alpha) - G(0) = (F(\alpha) - F(0)) + \left(F^{\perp}(\alpha) - F^{\perp}(0)\right),$$

while respecting the principal boundary conditions

$$t\alpha \text{ prescribed on } S_1, \qquad n\alpha = 0 \text{ on } S_2.$$

Explicit Variation of the Modified Variational Principle. To define $G(\alpha)$ more explicitly, consider a continuous differentiable curve

$$\gamma : [0,1] \to \widetilde{C}_{p-1}(M, S_2)$$

with

$$\begin{aligned} \gamma(0) &= \alpha_0 \text{ an initial state,} \\ \gamma(1) &= \alpha_1 \text{ an extremal;} \end{aligned}$$

and, in order to respect the principal boundary conditions,

$$\begin{aligned} t\alpha_0 = t\alpha &= t\gamma(s) && \text{on } S_1, \\ n\alpha_0 = n\alpha &= n\gamma(s) = 0 && \text{on } S_2, \end{aligned}$$

for all $s \in [0,1]$. No other constraints are placed on γ, so that

$$\frac{\partial\gamma}{\partial s} \in \widetilde{C}_{p-1}(M, S_2) \cap C_c^{p-1}(M - S_1) \quad \text{for all } s \in [0,1]$$

and the variation of the extremal

$$\left.\frac{\partial\gamma}{\partial s}\right|_{s=1} = \tilde{\alpha}$$

can be chosen to be any admissible variation where the space of admissible variations is $\widetilde{C}_{p-1}(M, S_2) \cap C_c^{p-1}(M - S_1)$. Thus writing out the functional $G(\alpha)$

explicitly one has

$$G(\alpha) - G(\alpha_0) = G(\gamma(1)) - G(\gamma(0))$$

$$= \int_0^1 \left(\left\langle C(d\gamma(s)), \frac{\partial}{\partial s} d\gamma(s) \right\rangle_p + (-1)^r \left\langle *\lambda, \frac{\partial \gamma(s)}{\partial s} \right\rangle_{p-1} + \left\langle K(\delta\gamma(s)), \frac{\partial}{\partial s}(\delta\gamma(s)) \right\rangle_{p-2} \right) ds$$

$$+ \sum_{i=1}^{\beta_{n-p+1}(M,S_2)} k_i \int_0^1 \left(\int_{z_i} *\gamma(s) \right) \left(\int_{z_i} * \frac{\partial \gamma(s)}{\partial s} \right) ds,$$

where $r = (n-p+1)(p-1) + (p-1)$. To investigate the stationary point of the functional one recalls that

$$\alpha = \gamma(1) \in C_c^{p-1}(M - S_1), \qquad \tilde{\alpha} = \left. \frac{\partial \gamma}{\partial s} \right|_{s=1} \in C_c^{p-1}(M - S_1) \cap \widetilde{C}_{p-1}(M, S_2),$$

and insists that

$$\left. \frac{\partial}{\partial \varepsilon} G(\gamma(1-\varepsilon)) \right|_{\varepsilon=0} = 0$$

for all admissible $\tilde{\alpha}$. Doing this shows that the following identity must be true for all α:

$$\text{(7–5)} \quad 0 = \langle C(d\alpha), d\tilde{\alpha} \rangle_p + (-1)^r \langle *\lambda, \tilde{\alpha} \rangle_{p-1} + \langle K(\delta\alpha), \delta\tilde{\alpha} \rangle_{p-2} + \sum_{i=1}^{\beta_{n-p+1}(M,S_2)} \left(\int_{z_i} *\alpha \right) \left(\int_{z_i} *\tilde{\alpha} \right).$$

It is in general not possible to integrate by parts to obtain an Euler–Lagrange equation in the usual sense because of the integral terms which constrain the integrals of $*\alpha$ on a set of generators of $H_{n-p+1}(M, S_2)$. Furthermore, in the present case it is not necessary to derive an Euler–Lagrange equation since the functional is designed to be extremized by direct variational methods. What is necessary to verify is the geometric picture developed when thinking about the troughs associated with the graphs of the functionals $F(\alpha)$ and $F^{\perp}(\alpha)$. That is, it must be verified that when λ obeys the conservation law

$$\lambda \in B_c^{n-p+1}(M - S_2)$$

the extremal of $G(\alpha)$ provides a physically meaningful solution to the paradigm problem and the projection of the extremal into $Z_c^{p-1}(M - S_1)$ vanishes. Alternatively, when the conservation law is violated one hopes that the extremal of $G(\alpha)$ can be interpreted as providing a "least squares" solution to the nearest physically meaningful problem where the conservation law is not violated and that the projection of the extremal into $Z_c^{p-1}(M - S_1)$ measures in some sense the extent by which the conservation law is violated. Hence let λ be prescribed in some way which does not necessarily respect a conservation law and consider the orthogonal decomposition

$$\lambda = \lambda_{\text{cons}} + \lambda_{\text{nonc}},$$

where

$$\lambda_{\rm cons} \in B_c^{n-p+1}(M - S_2)$$
$$\lambda_{\rm nonc} \in \left(B_c^{n-p+1}(M - S_2)\right)^{\perp} = \widetilde{Z}_{n-p+1}(M, S_1).$$

From this orthogonal decomposition, it follows immediately that

$$*\lambda = *\lambda_{\rm cons} + *\lambda_{\rm nonc},$$

where

$$*\lambda_{\rm cons} = *B_c^{n-p+1}(M - S_2) = \widetilde{B}_{p-1}(M, S_2),$$
$$*\lambda_{\rm nonc} = *\widetilde{Z}_{n-p+1}(M, S_1) = Z_c^{p-1}(M, S_1).$$

Similarly, for $\alpha \in \widetilde{C}_{p-1}(M, S_2)$ one has the orthogonal decomposition

$$\widetilde{C}_{p-1}(M, S_2) = \widetilde{B}_{p-1}(M, S_2) \oplus \left(Z_c^{p-1}(M - S_1) \cap \widetilde{C}_{p-1}(M, S_2)\right),$$

and α can be expressed as

$$\alpha = \alpha_0 + \alpha_G,$$

where

$$\alpha_0 \in \widetilde{B}_{p-1}(M, S_2),$$
$$\alpha_G \in Z_c^{p-1}(M - S_1) \cap \widetilde{C}_{p-1}(M, S_2).$$

Finally, for $\tilde{\alpha} \in \widetilde{C}_{p-1}(M, S_2) \cap C_c^{p-1}(M - S_1)$ one has the orthogonal decomposition

$$\widetilde{C}_{p-1}(M, S_2) \cap C_c^{p-1}(M - S_1)$$
$$= \left(\widetilde{B}_{p-1}(M, S_2) \cap C_c^{p-1}(M - S_1)\right) \oplus \left(Z_c^{p-1}(M - S_1) \cap \widetilde{C}_{p-1}(M, S_2)\right),$$

and $\tilde{\alpha}$ can be expressed as

$$\tilde{\alpha} = \tilde{\alpha}_0 + \tilde{\alpha}_G,$$

where

$$\tilde{\alpha}_0 \in \widetilde{B}_{p-1}(M, S_2) \cap C_c^{p-1}(M - S_1),$$
$$\tilde{\alpha}_G \in Z_c^{p-1}(M - S_1) \cap \widetilde{C}_{p-1}(M, S_2).$$

Before returning to the condition that ensures that the functional G is stationary at α, note that expressing α_0 and $\tilde{\alpha}_0$ as

$$\alpha_0 = \delta\theta, \qquad \tilde{\alpha}_0 = \delta\tilde{\theta}$$

it becomes apparent that

$$\int_{z_i} *\alpha_0 = \int_{z_i} *\delta\theta = (-1)^p \int_{z_i} d * \theta = 0, \qquad 1 \le i \le \beta_{n-p+1}(M - S_2),$$
$$\int_{z_i} *\tilde{\alpha}_0 = \int_{z_i} *\delta\tilde{\theta} = (-1)^p \int_{z_i} d * \tilde{\theta} = 0, \qquad 1 \le i \le \beta_{n-p+1}(M - S_2),$$

since the integral of a coboundary on a cycle vanishes. Next recall the identity (7–5) which must be satisfied for all $\tilde{\alpha} \in C_c^{p-1}(M - S_1) \cap \widetilde{C}_{p-1}(M, S_2)$ in order for the functional G to be stationary at α:

$$0 = \langle C(d\alpha), d\tilde{\alpha}\rangle_p + \langle K(\delta\alpha), \delta\tilde{\alpha}\rangle_{p-2} + (-1)^r \langle *\lambda, \tilde{\alpha}\rangle_{p-1} + \sum_{i=1}^{\beta_{n-p+1}(M,S_2)} k_i\left(\int_{z_i} *\alpha\right)\left(\int_{z_i} *\tilde{\alpha}\right).$$

Substituting the earlier orthogonal decompositions into this identity and recalling the definitions of the spaces involved gives

$$0 = \langle C(d\alpha_0), d\tilde{\alpha}_0\rangle_p + \big\langle K(\delta\alpha_G), \delta\tilde{\alpha}_G\big\rangle_{p-2} + (-1)^r \big\langle *\lambda_{\text{cons}} + *\lambda_{\text{nonc}}, \tilde{\alpha}_0 + \tilde{\alpha}_G\big\rangle_{p-1} \cdot + \sum_{i=1}^{\beta_{n-p+1}(M,S_2)} k_i\left(\int_{z_i} *\alpha_G\right)\left(\int_{z_i} *\tilde{\alpha}_G\right).$$

Keeping in mind that the spaces $Z_c^{p-1}(M - S_1)$ and $\widetilde{B}_{p-1}(M, S_2)$ are mutually orthogonal, the inner product involving the source term and the variation of the extremal can be simplified to yield

$$\begin{aligned} 0 = \big\langle C(d\alpha_0), d\tilde{\alpha}_0\big\rangle_p &+ (-1)^r \big\langle *\lambda_{\text{cons}}, \tilde{\alpha}_0\big\rangle_{p-1} \\ &+ \big\langle K(\delta\alpha_G), \delta\tilde{\alpha}_G\big\rangle_{p-2} + (-1)^r \big\langle *\lambda_{\text{nonc}}, \tilde{\alpha}_G\big\rangle_{p-1} \\ &+ \sum_{i=1}^{\beta_{n-p+1}(M,S_2)} k_i\left(\int_{z_i} *\alpha_G\right)\left(\int_{z_i} *\tilde{\alpha}_G\right). \end{aligned}$$

It is obvious by the independence of $\tilde{\alpha}_0$ and and $\tilde{\alpha}_G$ that the above condition is equivalent to the following two conditions:

$$0 = \langle C(d\alpha_0), d\tilde{\alpha}_0\rangle_p + (-1)^r \langle *\lambda_{\text{cons}}, \tilde{\alpha}_0\rangle_{p-1}, \tag{7–6}$$

for all $\tilde{\alpha}_0 \in \widetilde{B}_{p-1}(M, S_2) \cap C_c^{p-1}(M - S_1)$, and

$$0 = \langle K(\delta\alpha_G), \delta\tilde{\alpha}_G\rangle_{p-2} + (-1)^r \langle *\lambda_{\text{nonc}}, \tilde{\alpha}_G\rangle_{p-1} + \sum_{i=1}^{\beta_{n-p+1}(M,S_2)} \left(\int_{z_i} \alpha_G\right)\left(\int_{z_i} \tilde{\alpha}_G\right), \tag{7–7}$$

for all $\tilde{\alpha} \in Z_c^{p-1}(M - S_1) \cap \widetilde{C}_{p-1}(M, S_2)$. To deduce the properties of the extremal $\alpha = \alpha_0 + \alpha_G$ of the functional G, one can look at the consequences of the above two identities. This can be done in two steps as follows.

Consequences of condition (7–6). Condition (7–6) is precisely the criterion for the original functional F to be stationary at $\alpha_0 \in (Z_c^{p-1}(M - S_1))^\perp$ and where the source is λ_{cons}. Previous calculations show that the above identity implies

$$\begin{aligned} d * C(d\alpha_0) &= \lambda_{\text{cons}} &&\text{in } M, \\ t * C(d\alpha_0) &= 0 &&\text{in } S_2, \end{aligned}$$

so that the potential that makes $G(\alpha)$ stationary gives a solution to the paradigm problem where λ is replaced by $\lambda_{\rm cons}$. By the definition of $\lambda_{\rm cons}$ it follows that

$$\min_{\xi \in B_f^{n-p+1}(M-S_2)} \| \lambda - \xi \|_{n-p+1} = \| \lambda - \lambda_{\rm cons} \|_{n-p+1} .$$

Hence one can say that the extremal of G provides a solution to the nearest physically meaningful paradigm problem.

Consequences of condition (7–7). Noticing that $\tilde{\alpha}_G$ and α_G both belong to the space $Z_c^{p-1}(M - S_1) \cap \widetilde{C}_{p-1}(M, S_2)$ and that $\tilde{\alpha}_G$ is arbitrary, one can let $\tilde{\alpha}_G$ be equal to α_G, so that the identity (7–7) becomes

$$-(-1)^r \langle *\lambda_{\rm nonc}, \alpha_G \rangle_{p-1} = \langle K(\delta\alpha_G), \delta\alpha_G \rangle_{p-2} + \sum_{i=1}^{\beta_{n-p+1}(M,S_2)} k_i \left(\int_{z_i} *\alpha_G \right)^2 \geq 0,$$

with equality if and only if $\alpha_G \in \widetilde{B}_{p-1}(M, S_2)$. Since α_G is an element of $Z_c^{p-1}(M - S_1)$ it is seen that the expression becomes an equality if and only if $\alpha_G = 0$. Thus it is apparent that

$$*\lambda_{\rm nonc} = 0 \quad \text{implies} \quad \alpha_G = 0$$

and that

$$\alpha_G \neq 0 \quad \text{implies} \quad *\lambda_{\rm nonc} \neq 0.$$

To prove the converses of these statements it is necessary to show that it is possible to find a $\tilde{\alpha}_G$ such that if $\lambda_{\rm nonc} \neq 0$ then

$$\langle *\lambda_{\rm nonc}, \tilde{\alpha}_G \rangle_{p-1} \neq 0$$

and that α_G is such an $\tilde{\alpha}_G$. Unfortunately,

$$*\lambda_{\rm nonc} \in Z_c^{p-1}(M - S_1)$$

and

$$\tilde{\alpha}_G \in Z_c^{p-1}(M - S_1) \cap \widetilde{C}_{p-1}(M, S_2);$$

hence $\tilde{\alpha}_G$ can be selected to reflect the projection of $\lambda_{\rm nonc}$ in $Z_c^{p-1}(M - S_1) \cap \widetilde{C}_{p-1}(M, S_2)$ and nothing more. Note, however, that if one imposes with complete certainty $t\lambda = 0$ on S_2, then $\lambda \in C_c^{n-p+1}(M - S_2)$ and hence

$$*\lambda_{\rm nonc} \in *\widetilde{Z}_{n-p+1}(M, S_1) \cap *C_c^{n-p+1}(M - S_2) = Z_c^{p-1}(M - S_1) \cap \widetilde{C}_{p-1}(M, S_2).$$

In this case α_G, $\tilde{\alpha}_G$ and $*\lambda_{\rm nonc}$ all belong to the space

$$Z_c^{p-1}(M - S_1) \cap \widetilde{C}_{p-1}(M, S_2)$$

and it is always possible to find an $\tilde{\alpha}_G$ such that $*\lambda_{\rm nonc} \neq 0$ implies

$$\langle *\lambda_{\rm nonc}, \tilde{\alpha}_G \rangle_{p-1} \neq 0.$$

However, by Equation (7–7) this implies that $\alpha_G \neq 0$ and since Equation (7–7) is valid for all possible $\tilde{\alpha}_G$, one can set $\tilde{\alpha}_G$ equal to α_G to obtain

$$-(-1)^r \langle *\lambda_{\rm nonc}, \alpha_G \rangle_{p-1} = \langle K(\delta\alpha_G), \delta\alpha_G \rangle_{p-2} + \sum_{i=1}^{\beta_{n-p+1}(M,S_2)} k_i \left(\int_{z_i} *\alpha_G \right)^2 > 0.$$

Hence $\lambda_{\text{nonc}} \neq 0$ implies $\alpha_G \neq 0$, or $\alpha_G = 0$ implies $\lambda_{\text{nonc}} = 0$, and it is proved that

$$\alpha_G = 0 \text{ if and only if } *\lambda_{\text{nonc}} = 0.$$

It is seen that the identities (7–6) and (7–7) adequately describe what happens in a neighborhood of the extremal α in $\widetilde{C}_{p-1}(M, S_2)$ when one thinks in terms of tilted troughs.

Two final points are in order. The first point is that the value of the functional $F^{\perp}$ evaluated at the extremal of G provides an *a posteriori* estimate of how large λ_{nonc} is. This is apparent from the trough picture. The second point is that when there is a pseudo-Riemannian structure on the manifold M, the expression $\langle \cdot, \cdot \rangle_k$ is no longer positive definite, hence the functionals considered are no longer convex and the trough picture is no longer valid. Although the orthogonal decomposition of Section MA-M is still a legitimate direct sum decomposition and when $\beta_{n-p+1}(M, S_2) = 0$ the functional G still provides an effective way of imposing the Lorentz gauge $\delta\alpha = 0$ whenever the conservation of charge ($d\lambda = 0$) is respected, it is not clear what the exact properties of G are. From the point of view of computational electromagnetics, there is little motivation for pursuing this question and so the case of pseudo-Riemannian structures is ignored.

7E. Tonti Diagrams

In this final section, Tonti diagrams and the associated framework for complementary variational principles will be considered. This work is well known to people in the field of computational electromagnetics and an overview of the literature in this context is given in the paper by Penmann and Fraser [PF84]. In this connection the author also found the seminar paper by [Cam83] most useful. The basis of the following discussion are the papers of Enzo Tonti [Ton72b, Ton72a] where certain short exact sequences associated with differential operators appearing in field equations are recognized as being a basic ingredient in formulating a common structure for a large class of physical theories. This work of Tonti fits hand in glove with the work of J. J. Kohn [Koh72] on differential complexes. The point of view taken here is that for the practical problems described by the paradigm problem being considered in this chapter, the interrelationship between the work of Tonti and Kohn is easily seen by considering the complexes associated with the exterior derivative and its adjoint on a Riemannian manifold with boundary. The idea of introducing complexes and various concepts from algebraic topology into Tonti diagrams is not new and is developed in the companion papers [Bra77] and [Ton77].

The main conclusion to be drawn from the present approach is that the differential complexes associated with the exterior derivative give, when applicable, a deeper insight into Tonti diagrams than is usually possible since the de Rham isomorphism enables one to give concrete and intuitive answers to questions involving the (co)homology of the differential complexes. More precisely, the usual development of Tonti diagrams involves differential complexes where the symbol sequence of the differential operators involved is exact while what is actually desired is that the (co)homology of the complex be trivial. That is, if the (co)homology of the differential complex is nontrivial, then reasoning with the exactness of the symbol sequence alone may lead to false conclusions concerning the existence and uniqueness of solutions to equations. To the best of the author's knowledge, the only differential complexes of practical use for which something concrete can be said about (co)homology, are the differential complexes associated with the exterior derivative since in this case the de Rham isomorphism applies.

To formulate a Tonti diagram for the paradigm problem, consider first the paradigm problem and suppose that $H_c^{p-1}(M - S_1)$ is trivial. In this case,

$$Z_c^{p-1}(M - S_1) = B_c^{p-1}(M - S_1)$$

and

$$\alpha \to \alpha + d\chi, \quad \chi \in C_c^{p-2}(M - S_1)$$

is a gauge transformation describing the nonuniqueness in the potential α. Next, when dealing with complementary variational principles, we must find an

$$\eta_{\text{part}} \in C_c^{n-p}(M - S_2) = *\widetilde{C}_p(M, S_2)$$

such that

$$d\eta_{\text{part}} = \lambda$$

and

$$\eta - \eta_{\text{part}} \in B_c^{n-p}(M - S_2) = *\widetilde{B}_p(M, S_2).$$

In this case the forms β and η are determined by reducing the problem to a boundary value problem for

$$\nu \in C_c^{n-p-1}(M - S_2) = *\widetilde{C}_{p+1}(M, S_2),$$

where ν is defined by

$$d\nu = \eta - \eta_{\text{part}}.$$

This boundary value problem for ν is deduced from the equations

$$\begin{aligned} d\beta &= 0 && \text{in } M \\ t\beta &= 0 && \text{on } S_1 \\ d\nu + \eta_{\text{part}} &= *C(\beta) && \text{in } M. \end{aligned}$$

From these equations the boundary value problem is seen to be

$$\begin{aligned} d(C^{-1}((-1)^{p(n-p)} * (\eta_{\text{part}} + d\nu))) &= 0 \quad \text{in } M \\ t(C^{-1}((-1)^{p(n-p)} * (\eta_{\text{part}} + d\nu))) &= 0 \quad \text{on } S_1 \\ t\nu &= 0 \quad \text{on } S_2. \end{aligned}$$

The variational formulation for this problem is obtained by considering a curve

$$\gamma : [0, 1] \to C_c^{n-p-1}(M - S_2)$$

and defining the functional for the complementary problem as follows:

$$J(\gamma(1)) - J(\gamma(0))$$

$$= -(-1)^{p(n-p)} \int_0^1 \left\langle C^{-1}((-1)^{p(n-p)} * (\eta_{\text{part}} + d\gamma(s))), *\left(\eta_{\text{part}} + d\frac{\partial \gamma(s)}{\partial s}\right)\right\rangle_p ds$$

$$= -\int_0^1 \int_M (C^{-1}((-1)^{p(n-p)} *(\eta_{\text{part}} + d\gamma(s)))) \wedge \left(\eta_{\text{part}} + d\frac{\partial \gamma(s)}{\partial s}\right) ds.$$

To verify that this is indeed the correct functional let

$$\gamma(1) = \nu, \qquad \left.\frac{\partial \gamma(s)}{\partial s}\right|_{s=1} = \tilde{\nu}$$

be the extremal and any variation of the extremal of the functional where the space of admissible variations is $C_c^{n-p-1}(M - S_2) = *\widetilde{C}_{p+1}(M, S_2)$. The functional is stationary when

$$\left.\frac{\partial J}{\partial \varepsilon}\gamma(1-\varepsilon)\right|_{\varepsilon=0} = 0$$

for all admissible variations of the extremal. This condition amounts to

$$0 = \int_M C^{-1}((-1)^{p(n-p)} * (\eta_{\text{part}} + d\gamma(1))) \wedge d\left(\left.\frac{\partial \gamma(s)}{\partial s}\right|_{s=1}\right)$$

or

$$0 = \int_M C^{-1}((-1)^{p(n-p)} * (\eta_{\text{part}} + d\nu)) \wedge d\tilde{\nu}$$

for all admissible $\tilde{\nu}$. Integrating this expression by parts and using the fact that

$$t\nu = 0 \quad \text{on } S_2,$$

one obtains

$$0 = \int_M d\left(C^{-1}\left((-1)^{p(n-p)} * (\eta_{\text{part}} + d\nu)\right)\right) \wedge \tilde{\nu}$$
$$- \int_{S_1} t\left(C^{-1}\left((-1)^{p(n-p)} * (\eta_{\text{part}} + d\nu)\right)\right) \wedge t\tilde{\nu},$$

from which it is apparent that the functional is the desired one since $\tilde{\nu}$ can be taken to be any admissible variation. In this formulation, the extremal of the functional J is unique up an to an element of $Z_c^{n-p-1}(M - S_2)$ and the nonuniqueness can be described by a gauge transformation

$$\nu \to \nu + \nu_G, \quad \text{where } \nu_G \in Z_c^{n-p-1}(M - S_2).$$

Hence whenever there is a Riemannian structure on M which induces the inner product $\langle \cdot, \cdot \rangle_k$ on k-forms, the functional J is convex on

$$(Z_c^{n-p-1}(M - S_2))^\perp = \widetilde{B}_{n-p-1}(M, S_1) = *B_c^{p+1}(M - S_1)$$

and level on the space

$$Z_c^{n-p-1}(M - S_2) = *\widetilde{Z}_{p+1}(M, S_2).$$

Just as the construction of a functional $F^{\perp}$ enabled one to modify the functional F in order to construct a variational formulation involving a functional

$$G(\alpha) = F(\alpha) + F^{\perp}(\alpha)$$

for which the resulting extremal is unique, one can construct a functional $J^{\perp}(\nu)$ such that

$$I(\nu) = J(\nu) + J^{\perp}(\nu)$$

is a functional whose unique extremal is also an extremal of the functional J. This, of course, happens when the functional $J^{\perp}$ is constructed so that it is convex on the space

$$Z_c^{n-p-1}(M - S_2) = *\widetilde{Z}_{p+1}(M, S_2)$$

and level on

$$(Z_c^{n-p-1}(M - S_2))^{\perp} = \widetilde{B}_{n-p-1}(M, S_1) = *B_c^{p+1}(M - S_1).$$

Thus, again, one is led to a situation involving two troughs as shown in Figure 7.3. With this picture in mind, the functional $J^{\perp}$ can be constructed in analogy with the construction of $F^{\perp}$. Consider first a mapping

$$K' : C^{n-p-2}(M) \to C^{n-p-2}(M)$$

which satisfies the same symmetry, monotonicity, and asymptotic properties required of the function K used in the construction of $F^{\perp}$. Define a functional $J^{\perp}$ as follows. Given

$$\gamma : [0,1] \to \widetilde{C}_{n-p-1}(M, S_1),$$

let

$$J^{\perp}(\gamma(1)) - J^{\perp}(\gamma(0)) = -\int_0^1 \left\langle K'^{-1}(\delta\gamma(s)), \delta\frac{\partial\gamma(s)}{\partial s}\right\rangle_{n-p-2} - \sum_{i=0}^{\beta_{p+1}(M,S_1)} l_i \int_0^1 \left(\int_{z_i} *\gamma(s)\right)\left(\int_{z_i} *\frac{\partial\gamma(s)}{\partial s}\right) ds,$$

where the z_i are associated with generators of the homology group $H_{p+1}(M, S_1)$ and the l_i are positive constants. The functional $J^{\perp}$ thus defined is convex on the space

$$Z_c^{n-p-1}(M - S_2) = \left(\widetilde{B}_{n-p-1}(M, S_1)\right)^{\perp}$$

and level on the space

$$\left(Z_c^{n-p-1}(M - S_2)\right)^{\perp} = \widetilde{B}_{n-p-1}(M, S_1).$$

This is easily seen, since the situations involving $F^{\perp}$ and $J^{\perp}$ become identical if one interchanges the symbols

$$J^{\perp} \leftrightarrow F^{\perp},$$
$$S_1 \leftrightarrow S_2,$$
$$K' \leftrightarrow K,$$
$$n - p \leftrightarrow p.$$

Hence the functional I defined as

$$I(\nu) = J(\nu) + J^{\perp}(\nu),$$

with domain $\widetilde{C}_{n-p-1}(M, S_1) \cap C_c^{n-p-1}(M - S_2)$, has a unique extremum.

Finally, in order to finish this prelude to the Tonti diagram, note that if

$$\beta_{n-p-1}(M, S_2) = 0$$

the nonuniqueness in the complementary potential ν, when the variational formulation involving the functional J is used, can be described by a gauge transformation

$$\nu \to \nu + d\theta,$$

where

$$\theta \in C_c^{n-p-2}(M - S_2) = *\widetilde{C}_{p+2}(M, S_2).$$

Furthermore, when considering the Tonti diagram it is convenient to assume that β may be related to some type of source ρ through the equation

$$d\beta = \rho,$$

where in the present case $\rho = 0$. Hence, in terms of the notation introduced so far, the above formulation of the complementary variational principle for the paradigm problem is summarized by the Tonti type diagram in Figure 7.4 and used extensively in [PF84]. For the present purpose, the diagram in Figure 7.4 presents a simplistic view of the paradigm problem since boundary conditions

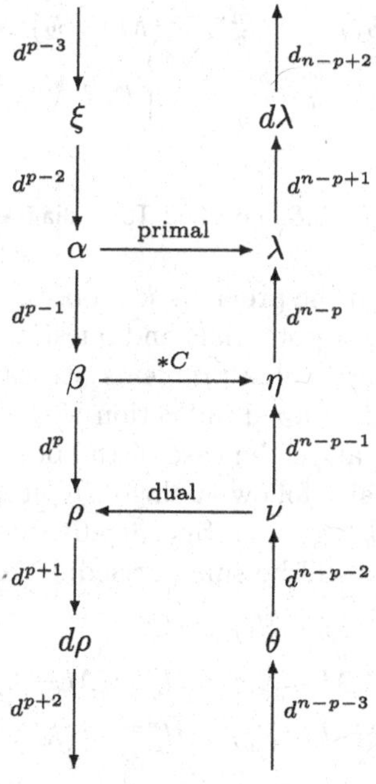

Figure 7.4. Typical Tonti diagram.

and domains of definition of operators have been ignored. Thus it is impossible to get a clear understanding of how homology groups come into play. To remedy this situation, one must realize that when boundary conditions are imposed, the left hand side of the above diagram is associated with the complex $C_c^*(M - S_1)$ while the right hand side of the diagram is associated with the complex $C_c^*(M - S_2) = *\widetilde{C}_*(M, S_2)$. Thus to be more explicit, the above diagram should be rewritten as shown in Figure 7.5. Once this structure has been

$$
\begin{array}{ccccc}
\Big\downarrow d^{p-3} & & \Big\uparrow d^{n-p+2} & & \Big\uparrow \delta_{p-2} \\
C_c^{p-2}(M - S_1) & & C_c^{n-p+2}(M - S_2) & \simeq & \widetilde{C}_{p-2}(M, S_2) \\
\Big\downarrow d^{p-2} & & \Big\uparrow d^{n-p+1} & & \Big\uparrow \delta_{p-1} \\
C_c^{p-1}(M - S_1) & & C_c^{n-p+1}(M - S_2) & \simeq & \widetilde{C}_{p-1}(M, S_2) \\
\Big\downarrow d^{p-1} & & \Big\uparrow d^{n-p} & & \Big\uparrow \delta_{p} \\
C_c^{p}(M - S_1) & \xrightarrow{*C} & C_c^{n-p}(M - S_2) & \simeq & \widetilde{C}_{p}(M, S_2) \\
\Big\downarrow d^{p} & & \Big\uparrow d^{n-p-1} & & \Big\uparrow \delta_{p+1} \\
C_c^{p+1}(M - S_1) & & C_c^{n-p-1}(M - S_2) & \simeq & \widetilde{C}_{p+1}(M, S_2) \\
\Big\downarrow d^{p+1} & & \Big\uparrow d^{n-p-2} & & \Big\uparrow \delta_{p+2} \\
C_c^{p+2}(M - S_1) & & C_c^{n-p-2}(M - S_2) & \simeq & \widetilde{C}_{p+2}(M, S_2) \\
\Big\downarrow d^{p+2} & & \Big\uparrow d^{n-p-3} & & \Big\uparrow \delta_{p+3}
\end{array}
$$

Figure 7.5. Fortified Tonti diagram.

identified, it is apparent from the previous sections of this chapter that questions of existence and uniqueness of potentials and questions of existence and uniqueness of solutions to boundary value problems are easily handled by using the orthogonal decomposition developed in Section MA-M. Though these questions have been considered in detail in the case of the potential α and the results for the complementary potential ν follow analogously, it is useful to outline the role played by various cohomology groups. Specifically, the role of the following pairs of groups and isomorphisms will be summarized:

$$H_c^{p-1}(M - S_1) \simeq H_c^{n-p+1}(M - S_2) \tag{7–8}$$

$$H_c^{p}(M - S_1) \simeq H_c^{n-p}(M - S_2) \tag{7–9}$$

$$H_c^{p+1}(M - S_1) \simeq H_c^{n-p-1}(M - S_2) \tag{7–10}$$

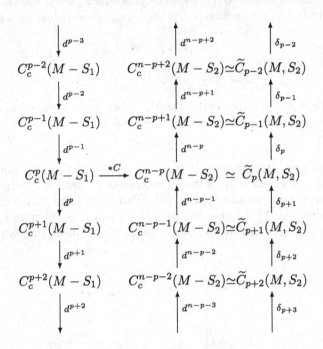

Consequences of (7–8). Once $t\alpha$ is prescribed on S_1, the group $H_c^{p-1}(M - S_1)$ was seen to describe the nonuniqueness of α in the paradigm problem which cannot be described by a gauge transformation of the form

$$\alpha \to \alpha + d\chi, \qquad \chi \in C_c^{p-2}(M - S_1).$$

In other words, the nonuniqueness of α is described by $Z_c^{p-1}(M - S_1)$ while the above gauge transformation involves $B_c^{p-1}(M - S_1)$, hence the difference is described by $H_c^{p-1}(M - S_1)$ since by definition

$$H_c^{p-1}(M - S_1) = Z_c^{p-1}(M - S_1)/B_c^{p-1}(M - S_1).$$

Dually, $H_c^{n-p+1}(M - S_2)$ was seen to be associated with the global conditions ensuring that $\lambda \in B_c^{n-p+1}(M - S_2)$ once it is known that $\lambda \in Z_c^{n-p+1}(M - S_2)$. Finally, the isomorphism

$$H_c^{p-1}(M - S_1) \simeq H_c^{n-p+1}(M - S_2)$$

expresses the duality between the global degrees of freedom in the nonuniqueness (gauge transformation) of α and the solvability condition (conservation law) involving λ. This isomorphism is exploited in the construction of the functional $F^{\perp}$ and its interpretation is best appreciated by using the de Rham isomorphism to reduce the above isomorphism to

$$H_{p-1}(M, S_1) \simeq H_{n-p+1}(M, S_2)$$

and to interpret this isomorphism in terms of the intersection numbers of the generators of these two homology groups as in Chapter 1.

Consequences of (7–9). The group $H_c^p(M - S_1)$ is associated with global conditions which ensure that $\beta \in B_c^p(M - S_1)$ once it is determined that $\beta \in Z_c^p(M - S_1)$. Furthermore, it gives insight into the conditions which α must satisfy on S_1 if $\beta = d\alpha$. Dually the group $H_c^{n-p}(M-S_2)$ is associated with global conditions which η_{part} must satisfy in order for there to be a $\nu \in C_c^{n-p-1}(M-S_2)$ such that

$$\begin{aligned} d\eta_{\text{part}} &= \lambda && \text{in } M, \\ d\nu &= \eta - \eta_{\text{part}} && \text{in } M. \end{aligned}$$

Thus the cohomology group $H_c^p(M - S_1)$ is used in formulating a primal variational principle, while the cohomology group $H_c^{n-p}(M - S_2)$ is used in formulating a dual variational principle and the isomorphism

$$H_c^p(M - S_1) \simeq H_c^{n-p}(M - S_2)$$

then expresses the fact that the number of global conditions is the same in both the original and complementary formulations. Note that for most problems, the periods of closed forms on the generators of

$$H_p(M, S_1), \quad H_{n-p}(M, S_2)$$

have the interpretation of a lumped parameter current, potential difference, or flux as was seen in Examples 1.14, 1.15, 2.3, and 2.4. Thus in these examples the isomorphism in homology has a direct interpretation.

Consequences of (7–10). Had β not been a closed form but rather tied to an equation of the form

$$d\beta = \rho,$$

then if $\rho \in Z_c^{p+1}(M - S_1)$ the group $H_c^{p+1}(M - S_1)$ is associated with the conditions which ensure that $\rho \in B_c^{p+1}(M - S_1)$. Thus the group $H_c^{p+1}(M - S_1)$ is associated with the global conditions which ensure the solvability of the equations for the extremal ν of the complementary variational principle. Dually, the group $H_c^{n-p-1}(M - S_2)$ describes the nonuniqueness in ν which cannot be described by a gauge transformation of the form

$$\nu \rightarrow \nu + d\theta, \qquad \theta \in C_c^{n-p-2}(M - S_2).$$

In other words, the nonuniqueness of ν is described by $Z_c^{n-p-1}(M - S_2)$ while the above gauge transformation involves $B_c^{n-p-1}(M - S_2)$ and the difference is characterized by $H_c^{n-p-1}(M - S_2)$. Finally the isomorphism

$$H_c^{p+1}(M - S_1) \simeq H_c^{n-p-1}(M - S_2)$$

expresses the duality between the global degrees of freedom in the nonuniqueness (gauge transformation) of ν and the solvability condition (conservation law) $\rho \in B_c^{p+1}(M - S_1)$. Thus the above isomorphism plays the same role in the complementary variational formulation as the isomorphism in (7–8) played in the primal variational formulation. This shows how the above isomorphism played a role in the construction of the functional $J^{\perp}$.

Summary. We have considered the role of homology and cohomology groups in the context of the Tonti diagram for the paradigm problem. The Tonti diagram for the paradigm problem includes as special cases electrostatics, magnetostatics and electromagnetics hence it unifies all of the cases considered in [PF84] and makes explicit the role of homology groups in this context. As mentioned in the introductory paragraph, the main virtue of the paradigm problem is that the Whitney form interpolation, introduced by Weil [Wei52], reproduces all of the features of the paradigm problem in a discrete setting. Specifically, Whitney forms provide a "chain homotopy" between the relative de Rham complex and its simplicial counterpart. By its algebraic structure, this chain homotopy preserves all considerations of homology and cohomology, that is, all circuit-theoretic information. Furthermore, the structure of the orthogonal decompositions is preserved in the discrete setting. Given the framework for quasistatic modeling developed in Section 2E, we see that the practical implications are far reaching. This chapter attempted to give a sketch of the mathematical coherence of the underlying ideas.

The paradox is now fully established that the utmost abstractions are the true weapons with which to control our thought of concrete fact.

A. N. Whitehead, *Science and the modern world*, 1925

In the year 1844 two remarkable events occurred, the publication by Hamilton of his discovery of quaternions, and the publication by Grassmann of his "Ausdehnungslehre". ... [I]t is regrettable but not surprising, that quaternions were hailed as a great discovery while Grassmann had to wait 23 years before his work received any recognition at all from professional mathematicians.

F. J. Dyson, [Dys72]

Mathematical Appendix: Manifolds, Differential Forms, Cohomology, Riemannian Structures

The systematic use of differential forms in electromagnetic theory started with the truly remarkable paper of Hargraves [Har08] in which the space-time covariant form of Maxwell's equations was deduced. Despite the efforts of great engineers such as Gabriel Kron (see [BLG70] for a bibliography) the use of differential forms in electrical engineering is, unfortunately, still quite rare. The reader is referred to the paper by Deschamps [Des81] for an introductory view of the subject. The purpose of this appendix is to summarize the properties of differential forms which are necessary for the development of cohomology theory in the context of manifolds without getting into the aspects which depend on metric notions. We also develop the aspects of the theory that both depend on the metric and are required for Chapter 7. Reference [Tei01] presents most of the topics in this chapter from the point of view of the numerical analyst interested in network models for Maxwell's equations.

There are several books which the authors found particularly invaluable. These are [War71, Chapters 4 and 6] for a proof of Stokes' theorem and the Hodge decomposition for a manifold without boundary, [Spi79, Chapters 8 and 11] for integration theory and cohomology theory in terms of differential forms, [BT82] for a quick route into cohomology and [Yan70] for results concerning manifolds with boundary. Finally, the papers by Duff, Spencer, Conner, and Friedrichs (see bibliography) are for basic intuitions about orthogonal decompositions on manifolds with boundary.

What remains to be developed is a systematic way of manipulating differential forms which involves only basic linear algebra and partial differentiation. Once the basic operations on differential forms have been defined, all of the properties of cohomology groups appear as in the first chapter.

MA-A. Differentiable Manifolds

In order to talk about differential forms, it is important first to have an acquaintance with the notion of a differentiable manifold. Roughly speaking, a differentiable manifold of dimension n can be described locally by n coordinates, that is, given any point p in an n-dimensional differentiable manifold M, one can find a neighborhood U of p which is homeomorphic to a subset of $\mathbb{R}^n$. More accurately, the one-to-one continuous mapping φ which takes U into a subset of $\mathbb{R}^n$ is differentiable a specified number of times. The reason why one is required to work in terms of open sets and not the whole manifold is because the simplest of n-dimensional manifolds are not homeomorphic to any subset of $\mathbb{R}^n$. The two-dimensional sphere S^2, for example, requires at least two such open sets for a cover.

More formally, an atlas is used in order to describe an "n-dimensional differentiable manifold M of class C^k". An atlas $\mathcal{A}$ is a collection of pairs (U_i, φ_i) called charts where U_i is an open set of M and φ_i is a one-to-one bijective map, differentiable of class C^k, mapping U_i into an open set of $\mathbb{R}^n$. In addition the charts in the atlas are assumed to satisfy:

(1) $\varphi_i \circ \varphi_j^{-1} : \varphi_j(U_i \cap U_j) \to \varphi_i(U_i \cap U_j)$ is a differentiable function of class C^k whenever $(U_i, \varphi_i), (U_j, \varphi_j) \in \mathcal{A}$ (see Figure MA-1). The functions $\varphi_i \circ \varphi_j^{-1}$ are called transition functions.

(2) $\bigcup U_i = M$.

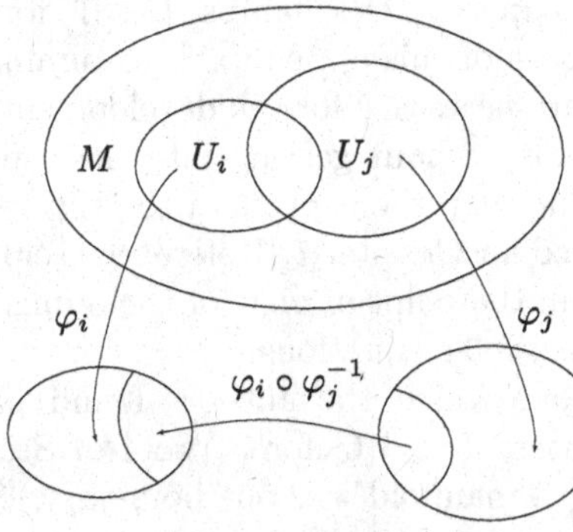

Figure MA-1.

Thus, referring back to the sphere we see that it is a 2-dimensional differentiable manifold of class C^∞ which can be described by an atlas which contains two charts. The actual definition of a differentiable manifold involves not only an atlas but an equivalence class of atlases where if $\mathcal{A}$ and $\mathcal{B}$ are atlases for a manifold M, then $\mathcal{A} \cup \mathcal{B}$ is also an atlas. That is, if

$$(U_i, \varphi_i) \in \mathcal{A} \quad \text{and} \quad (W_j, \psi_j) \in \mathcal{B},$$

then

$$\varphi_i \circ \psi_j^{-1} : \psi_j(U_i \cap W_j) \to \varphi_i(U_i \cap W_j)$$

is a continuous map which is just as smooth as φ_i or ψ_j. Thus a set M together with an equivalence class of atlases is called a differentiable manifold.

The local nature of the definition of a manifold is essential if one is not to constrain the global topology of the manifold. A fundamental property of differentiable manifolds is paracompactness, which enables one to construct partitions of unity [War71, pages 5–10]. The existence of a partition of unity is required to smoothly specify a geometrical object such as a vector field, differential form or Riemannian structure globally on a differentiable manifold by specifying the geometrical object locally in terms of the coordinate charts. Throughout this chapter it will be assumed that such geometrical objects are defined globally and most computations will be performed in local coordinate charts without regard to how the charts fit together globally. Another almost immediate consequence of the definition of a manifold is that once a notion of distance (Riemannian structure) is defined, the cohomology of the manifold is easily computed in terms of differential forms [BT82, §5]. Holding off on the questions of homology and cohomology the exposition will concentrate on the algebraic properties of differential forms.

MA-B. Tangent Vectors and the Dual Space of One-Forms

Suppose that in a neighborhood of a point p in an n-dimensional manifold M there are local coordinates

$$x^i, \quad 1 \le i \le n$$

The tangent space M_p at the point $p \in M$ is defined to be the linear span of all linear first-order differential operators. That is, if $X \in M_p$ then X can be represented as

$$X = \sum_{i=1}^{n} X^i(p) \frac{\partial}{\partial x^i}.$$

It is obvious that

$$\frac{\partial}{\partial x^i}, \quad 1 \le i \le n,$$

form a basis for M_p. The interpretation of the tangent space is obtained by considering

$$X(f)\Big|_p = \sum_{i=1}^{n} X^i(p) \frac{\partial f}{\partial x^i}\Big|_p.$$

The tangent vectors can be interpreted as giving directional derivatives of functions. The collection of all tangent spaces to a manifold is called the tangent bundle and is denoted by $T(M)$. Thus

$$T(M) = \bigcup_{p \in M} M_p.$$

A vector field on M is defined to be a smooth section of the tangent bundle, that is, if one writes a vector field X in terms of local coordinates in a neighborhood of p then

$$X = \sum_{i=1}^{n} X^i \frac{\partial}{\partial x^i},$$

where the X^i are smooth functions of the local coordinates.

Since a vector space has been defined it is natural to inquire about its dual space. An element of the dual space to M_p is a first-order differential form (or 1-form) evaluated at p, which is expressed as

$$\omega = \sum_{i=1}^{n} a_i(p)\, dx^i.$$

The dual space to the tangent space M will be denoted by M_p^*. In this scheme one identifies

$$dx^i, \quad 1 \le i \le n,$$

as a basis for M_p^* and the dual basis to

$$\frac{\partial}{\partial x^i}, \quad 1 \le i \le n.$$

Thus

$$dx^i \frac{\partial}{\partial x^j} = \delta^i_j$$

and

$$\begin{aligned} \omega(X)\Big|_p &= \sum_{i=1}^{n} a_i(p)\, dx^i \left(\sum_{j=1}^{n} X^j(p) \frac{\partial}{\partial x^j} \right) \\ &= \sum_{i=1}^{n} \sum_{j=1}^{n} a_i(p) X^j(p)\, dx^i \frac{\partial}{\partial x^j} = \sum_{i=1}^{n} a_i(p) X^i(p). \end{aligned}$$

Having defined the dual space to vectors as differential forms, one can also define the cotangent bundle to the manifold M as

$$T^*(M) = \bigcup_{p \in M} M_p^*.$$

In order to verify that $\omega(X)$ is really an invariant quantity, it is essential to know how ω and X behave under coordinate transformations. Suppose φ is a mapping between an m-dimensional manifold M' and an n-dimensional manifold M:

$$\varphi : M' \to M.$$

We now choose points $p' \in M'$, $p \in M$ such that $\varphi(p') = p$, and look at the actions induced by φ on 1-forms and vector fields:

$$\begin{aligned} \varphi^{\#} &: T^*(M) \to T^*(M'), \\ \varphi_{\#} &: T(M') \to T(M). \end{aligned}$$

These actions have the property that if $\omega \in T^*(M)$ and $X \in T(M')$, then

$$(\varphi^{\#}\omega)(X)\Big|_{p'} = \omega(\varphi_{\#}X)\Big|_p.$$

The form $\varphi^{\#}\omega$ is called the pullback of ω under φ. Let $(y^1,\dots,y^m)$ and $(x^1,\dots,x^n)$ be local coordinates around $p' \in M'$ and $p \in M$ respectively. In terms of these local coordinates there is a functional relationship

$$x^i = x^i(y^1,\dots,y^n), \quad 1 \le i \le n,$$

and the induced transformations $\varphi^{\#}$, $\varphi_{\#}$ transform basis vectors according to the rules

$$dx^j = \sum_{i=1}^{m} \frac{\partial x^j}{\partial y^i}\, dy^i,$$

$$\frac{\partial}{\partial y^j} = \sum_{i=1}^{n} \frac{\partial x^i}{\partial y^j}\frac{\partial}{\partial x^i}.$$

Hence, if

$$\omega = \sum_{i=1}^{n} a_i(p)\, dx^i,$$

$$X = \sum_{j=1}^{m} X^j\left(\varphi(p')\right) \frac{\partial}{\partial y^j},$$

then

$$\varphi^{\#}(\omega) = \sum_{j=1}^{m}\left(\sum_{i=1}^{n} a_i\left(\varphi(p')\right)\frac{\partial x^i}{\partial y^j}\right) dy^j,$$

$$\varphi_{\#}(X) = \sum_{i=1}^{n}\left(\sum_{i=1}^{m} X^j(p)\frac{\partial x^i}{\partial y^j}\right)\frac{\partial}{\partial x^i}$$

and

$$\begin{aligned}(\varphi^{\#}\omega)(X)\big|_{p'} &= \sum_{i=1}^{n}\sum_{j=1}^{m} a_i\left(\varphi(p')\right)\frac{\partial x^i}{\partial y^j}X^j\left(\varphi(p')\right)\\ &= \sum_{i=1}^{n}\sum_{j=1}^{m} a_i(p)\frac{\partial x^i}{\partial y^j}X^j(p) = \omega\left(\varphi_{\#}X\right)\big|_{p},\end{aligned}$$

which is the desired transformation. The transformation rule for the basis vectors dx^i $(1 \le i \le n)$ and $\partial/\partial y^j$ $(1 \le j \le m)$ is intuitive if these basis vectors are regarded as infinitesimals and partial derivatives and the usual rules of calculus are used. However the reader should avoid making any interpretation of the symbol d until the exterior derivative is defined. That is,

$$d(\text{something})$$

should not be interpreted as an infinitesimal.

One more remark is in order. Suppose M'' is another manifold and there is a transformation $\psi : M'' \to M'$. There is a composite transformation $\varphi \circ \psi$ which

makes the following diagram commutative

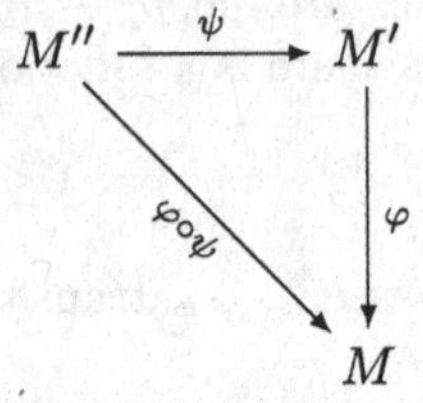

The chain rule for partial derivatives shows that the induced transformations on vector fields and 1-forms make the following diagrams commutative

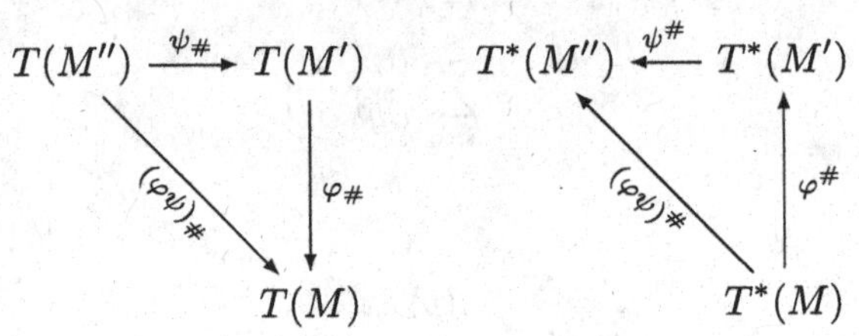

Hence

$$(\varphi\psi)_{\#} = \varphi_{\#}\psi_{\#},$$
$$(\varphi\psi)^{\#} = \psi^{\#}\varphi^{\#}.$$

Thus vector fields transform covariantly while 1-forms transform contravariantly and the whole scheme is invariant under transformations.

MA-C. Higher-Order Differential Forms and Exterior Algebra

The identification of 1-forms at a point p as elements of the dual space of M_p enables one to regard a differential 1-form at a point p as a linear functional on the tangent space M_p. Higher-order k-forms are a generalization of this idea. At a point $p \in M$, a differential k-form is defined to be an alternating k-linear functional on the tangent space M_p. That is, if ω is a k-form then

$$\omega\big|_p : \underbrace{M_p \times M_p \times \cdots \times M_p}_{k\ times} \to \mathbb{R},$$

which is linear in each argument and satisfies the following. If

$$X_1, X_2, \ldots, X_k \in M_p$$

then for any permutation π of k integers $(1, \ldots, k)$ we have

$$\omega\left(X_{\pi(1)}, X_{\pi(2)}, \ldots, X_{\pi(k)}\right) = \operatorname{sgn}(\pi)\omega(X_1, X_2, \ldots, X_k),$$

where

$$\operatorname{sgn}(\pi) = \begin{cases} 1 & \text{if } \pi \text{ is an even permutation,} \\ -1 & \text{if } \pi \text{ is an odd permutation.} \end{cases}$$

The set of k-forms at a point p form a vector space denoted by

$$\Lambda_k(M_p^*).$$

In addition, the following definitions are made:

$$\Lambda_k(M_p^*) = 0, \qquad k < 0,$$
$$\Lambda_1(M_p^*) = M_p^*,$$
$$\Lambda_0(M_p^*) = \text{values of functions evaluated at } p.$$

When thinking of alternating multilinear mappings, it is useful to remember the *alternation mapping* which sends any multilinear mapping into an alternating one:

$$\mathrm{Alt}\,(T(X_1, \ldots, X_k)) = \sum_{\pi \in S_k} \frac{\mathrm{sgn}(\pi)}{k!} T\left(X_{\pi(1)}, X_{\pi(2)}, \ldots, X_{\pi(k)}\right)$$

where S_k is the group of permutations of k objects. The alternation mapping has the same properties as the determinant function and from this fact one can deduce that

$$\Lambda_k(M_p^*) = 0 \quad \text{if } k > n.$$

The exterior algebra of M_p^* is defined as

$$\Lambda(M_p^*) = \bigoplus_{k=0}^{n} \Lambda_k(M_p^*).$$

By forming

$$\Lambda_k^*(M) = \bigcup_{p \in M} \Lambda_k(M_p^*)$$

and considering the k-forms whose coefficients are differentiable functions of coordinates, one has the exterior k-bundle of the manifold M. The set of all differential forms on a manifold M form the exterior algebra bundle of M which is defined as

$$\Lambda^*(M) = \bigcup_{p \in M} \Lambda(M_p^*) = \bigoplus_{k=0}^{n} \Lambda_k^*(M).$$

The term *exterior algebra* has been used several times without any mention of what this algebra is. There is a product

$$\wedge : \Lambda^*(M) \times \Lambda^*(M) \to \Lambda^*(M),$$

called the exterior product (or wedge product, or Grassmann product) that takes a k-form and an l-form and gives a $(k+l)$-form according to the following rule. If $\omega \in \Lambda_k^*(M)$, $\eta \in \Lambda_l^*(M)$, and $(X_1, X_2, \ldots, X_{k+l}) \in T(M)$, then

$$(\omega \wedge \eta)(X_1, X_2, \ldots, X_{k+l}) = \frac{1}{(k+l)!} \sum_{\pi \in S_{k+l}} \mathrm{sgn}(\pi) \omega\left(X_{\pi(1)}, \ldots, X_{\pi(k)}\right) \eta\left(X_{\pi(k+l)}, \ldots, X_{\pi(k+l)}\right).$$

This definition of wedge multiplication is not very useful for explicit calculations, being rather like the formal definition of a matrix determinant. For practical computations it is important to remember that wedge multiplication is:

(1) bilinear:

$$\omega \wedge (a_1\eta_1 + a_2\eta_2) = a_1(\omega \wedge \eta_1) + a_2(\omega \wedge \eta_2),$$
$$(a_1\eta_1 + a_2\eta_2) \wedge \omega = a_1(\eta_1 \wedge \omega) + a_2(\eta_2 \wedge \omega) \text{for } a_1, a_2 \in \mathbb{R};$$

(2) associative:

$$(\lambda \wedge \mu) \wedge \eta = \lambda \wedge (\mu \wedge \eta);$$

(3) graded commutative:

$$\omega \wedge \eta = (-1)^{kl} \eta \wedge \omega \quad \text{for } \omega \in \Lambda_k^*(M), \eta \in \Lambda_l^*(M).$$

Before considering some examples of wedge multiplication it is worth considering what differential forms look like at a point $p \in M$ where $(x^1, x^2, \ldots, x^n)$ are local coordinates. Let

$$dx^i, \qquad 1 \leq i \leq n$$

be a basis for M_p^*. By taking repeated wedge products in all possible ways, $\Lambda_k(M_p^*)$ is seen to be spanned by expressions of the form

$$dx^{i_1} \wedge dx^{i_2} \wedge \cdots \wedge dx^{i_k}.$$

Furthermore for $\pi \in S_k$ one has

$$dx^{i_1} \wedge dx^{i_2} \wedge \cdots \wedge dx^{i_k} = \operatorname{sgn}(\pi) dx^{i_{\pi(1)}} \wedge \cdots \wedge dx^{i_{\pi(k)}}$$

and since $dx^i, 1 \leq i \leq n$ span $\Lambda_1(M_p^*)$, one sees that $\Lambda_k(M_p^*)$ has a basis of the form

$$dx^{i_1} \wedge dx^{i_2} \wedge \cdots \wedge dx^{i_k}, \qquad 1 \leq i_1 < \cdots < i_k \leq n.$$

Therefore, for $k > 0, \omega \in \Lambda_k(M_p^*)$ looks like:

$$\omega = \sum_{i_1 < i_2 < \cdots < i_k} a_{i_1, i_2, \ldots i_k}(p)\, dx^{i_1} \wedge dx^{i_2} \wedge \cdots \wedge dx^{i_k}$$

and

$$\dim \Lambda_k(M_p^*) = \frac{n!}{(n-k)!k!} = \binom{n}{k}.$$

By the binomial theorem it is trivial to calculate the dimension of the exterior algebra of M_p^*

$$\dim\left(\Lambda(M_p^*)\right) = \sum_{k=0}^{n} \dim\left(\Lambda_k(M_p^*)\right) = \sum_{k=0}^{n} \binom{n}{k}$$
$$= (1+1)^n = 2^n.$$

At this point it is useful to consider an example.

Example 7.2 Wedge multiplication in three dimensions. Consider the cotangent bundle to a three-dimensional manifold embedded in $\mathbb{R}^m$ for some $m \geq 3$. Let $p \in M$ and dx^1, dx^2, dx^3 be a basis of $\Lambda_1(M_p^*) = M_p^*$. If

$$\omega_1, \omega_2, \omega_3 \in \Lambda_1(M_p^*) \quad \text{and} \quad \eta \in \Lambda_2(M_p^*),$$

where

$$\begin{aligned}\omega_1 &= A_1\,dx^1 + A_2\,dx^2 + A_3\,dx^3,\\ \omega_2 &= B_1\,dx^1 + B_2\,dx^2 + B_3\,dx^3,\\ \omega_3 &= C_1\,dx^1 + C_2\,dx^2 + C_3\,dx^3,\\ \eta &= P_1\,dx^2 \wedge dx^3 + P_2\,dx^3 \wedge dx^1 + P_3\,dx^1 \wedge dx^2,\end{aligned}$$

one obtains, using the rules for wedge multiplication,

$$\begin{aligned}\eta\wedge\omega_3 &= (P_1\,dx^2\wedge dx^3+P_2\,dx^3\wedge dx^1+P_3\,dx^1\wedge dx^2)\wedge(C_1\,dx^1+C_2\,dx^2+C_3\,dx^3)\\ &= (P_1C_1+P_2C_2+P_3C_3)\,dx^1\wedge dx^2\wedge dx^3,\end{aligned}$$

$$\begin{aligned}\omega_1\wedge\omega_2 &= (A_1\,dx^1\wedge A_2\,dx^2\wedge A_3\,dx^3)\wedge(B_1\,dx^1+B_2\,dx^2+B_3\,dx^3)\\ &= (A_2B_3-A_3B_2)\,dx_2\wedge dx_3+(A_3B_1-A_1B_3)dx^3\wedge dx^1\\ &\qquad\qquad +(A_1B_2-A_2B_1)\,dx^1\wedge dx^2.\end{aligned}$$

Identifying $\omega_1 \wedge \omega_2$ with η, these two formulas give

$$\begin{aligned}&\omega_1 \wedge \omega_2 \wedge \omega_3\\ &= \big((A_2B_3 - A_3B_2)C_1 + (A_3B_1 - A_1B_3)C_2 + (A_1B_2 - A_2B_1)C_3\big)\,dx^1 \wedge dx^2 \wedge dx^3.\end{aligned}$$

Hence, the scalar product, vector product and scalar triple product of vector calculus arise in the wedge multiplication of forms of various degrees on a three-dimensional manifold. □

MA-D. Behavior of Differential Forms Under Mappings

In the previous section many of the properties of differential forms were seen to be properties of alternating multilinear functionals over a vector space. The following fact is also a consequence of the definition of alternating multilinear functionals. Suppose there is a linear transformation on M_p then there is an induced exterior algebra homomorphism. That is, suppose there is a homomorphism

$$f_{\#} : M'_p \to M_p$$

that induces

$$f^{\#} : \Lambda(M^*_p) \to \Lambda(M'^*_p).$$

If

$$\omega \to f^{\#}\omega, \quad \eta \to f^{\#}\eta, \quad \omega\wedge\eta \to f^{\#}(\omega\wedge\eta),$$

then

$$f^{\#}(\omega\wedge\eta) = (f^{\#}\omega)\wedge(f^{\#}\eta).$$

Earlier, when discussing covariance and contravariance, we considered the pullback $\varphi^{\#}$ on 1-forms induced by a mapping φ, along with the induced transformation $\varphi_{\#}$ on vector fields. The preceding equation enables one to define a pullback on all differential forms:

THEOREM. Given $\varphi : M' \to M$, there is an induced homomorphism

$$\varphi^{\#} : \Lambda^*(M) \to \Lambda^*(M')$$

such that

$$\varphi^{\#}(\omega \wedge \eta) = (\varphi^{\#}\omega) \wedge (\varphi^{\#}\eta)$$

for all $\omega, \eta \in \Lambda^*(M)$.

The significance of this result is best appreciated in local coordinates, since it dictates a "change of variables" formula for differential forms. Consider $p \in M$ with local coordinates $(x^1, \ldots, x^n)$ and $p' \in M'$ with local coordinates $(y^1, \ldots, y^m)$ along with

$$\varphi : M' \to M$$

such that $\varphi(p') = p$. This transformation induces (via $\varphi^{\#}$) a linear transformation on $\Lambda_1^*(M)$, where

$$dx^i = \sum_{j=1}^{m} \frac{\partial x^i}{\partial y^j}\, dy^j.$$

The exterior algebra homomorphism says that given a k-form $\omega \in \Lambda_k^*(M)$, where

$$\omega = \sum_{i_1 < i_2 < \cdots < i_k}^{n} a_{i_1 i_2 \ldots i_k}(p)\, dx^{i_1} \wedge dx^{i_2} \wedge \cdots \wedge dx^{i_k},$$

we have

$$\varphi^{\#}\omega = \sum_{i_1 < \cdots < i_k}^{n} a_{i_1 i_2 \ldots i_k}(\varphi(p')) \left(\sum_{j_1=1}^{m} \frac{\partial x^{i_1}}{\partial y^{j_1}}\, dy^{j_1} \right) \wedge \cdots \wedge \left(\sum_{j_k=1}^{n} \frac{\partial x^{i_k}}{\partial y^{j_k}}\, dy^{j_k} \right).$$

This transformation is precisely the one required to leave

$$(\varphi^{\#}\omega)(X_1, \ldots, X_k) = \omega(\varphi_{\#}X_1, \ldots, \varphi_{\#}X_k).$$

Furthermore, the change of variables formula for multiple integrals takes on the following form

$$\int_{R'} \varphi^{\#}\omega = \int_{\varphi(R')} \omega.$$

This change of variable formula is most easily understood by means of a few examples.

Example 7.3 Change of variables formula in two dimensions. Suppose $R' \subset M'$ has local coordinates u, v, while $\varphi(R') \subset M$ has local coordinates s, t. Let

$$I = \int_{\varphi(R')} f(s,t)\, ds \wedge dt$$

and consider the change of variables

$$s = s(u,v), \qquad t = t(u,v).$$

Since

$$ds = \frac{\partial s}{\partial u}\, du + \frac{\partial s}{\partial v}\, dv, \quad dt = \frac{\partial t}{\partial u}\, du + \frac{\partial t}{\partial v}\, dv,$$

and

$$ds \wedge dt = \left(\frac{\partial s}{\partial u}\,du + \frac{\partial s}{\partial v}\,dv\right) \wedge \left(\frac{\partial t}{\partial u}\,du + \frac{\partial t}{\partial v}\,dv\right)$$
$$= \left(\frac{\partial s}{\partial u}\frac{\partial t}{\partial v} - \frac{\partial s}{\partial v}\frac{\partial t}{\partial u}\right) du \wedge dv = \frac{\partial(s,t)}{\partial(u,v)}\,du \wedge dv,$$

one has

$$I = \int_{R'} f\,(s(u,v), t(u,v))\,\frac{\partial(s,t)}{\partial(u,v)}\,du \wedge dv. \qquad \square$$

Example 7.4 Change of variables formula for surface integrals in three dimensions. Suppose $R' \subset M'$ has local coordinates u, v, while $\varphi(R') \subset M$ has local coordinates x, y, z. Let

$$J = \int_{\varphi(R')} B_z\,dx \wedge dy + B_y\,dz \wedge dx + B_x\,dy \wedge dz$$

where B_x, B_y, B_z, are functions of x, y, z. Consider a change of variables

$$x = x(u,v), \quad y = y(u,v), \quad z = z(u,v).$$

By the same type of calculation as in the previous example,

$$dx \wedge dy = \frac{\partial(x,y)}{\partial(u,v)}\,du \wedge dv,$$
$$dz \wedge dx = \ldots$$

one has

$$J = \int_{R'} \left(B_z \frac{\partial(x,y)}{\partial(u,v)} + B_y \frac{\partial(z,x)}{\partial(u,v)} + B_x \frac{\partial(y,z)}{\partial(u,v)}\right) du \wedge dv.$$

This is a generalization of the usual change of variables formula. $\square$

Example 7.5 Change of variables formula in three dimensions. Suppose $R' \subset M'$ has coordinates u, v, w while $\varphi(R') \subset M$ has coordinates x, y, z. Let

$$I = \int_{\varphi(R')} p\,dx \wedge dy \wedge dz.$$

Transforming coordinates,

$$x = x(u,v,w), \quad y = y(u,v,w), \quad z = z(u,v,w),$$

taking differentials, and using the triple product of Example 7.2 gives

$$I = \int_{R'} p \frac{\partial(x,y,z)}{\partial(u,v,w)}\,du \wedge dv \wedge dw. \qquad \square$$

It is now time to consider the formal definition of the exterior derivative, which will enable us to define a complex associated with the exterior algebra bundle and the corresponding cohomology in terms of differential forms.

MA-E. The Exterior Derivative

The exterior derivative will now be introduced in a formal way and illustrated in specific instances. As a preliminary to its definition it is useful to introduce certain vocabulary.

DEFINITION 2. An endomorphism l of the exterior algebra bundle $\Lambda^*(M)$ is

(1) a *derivation* if for $\omega, \eta \in \Lambda^*(M)$

$$l(\omega \wedge \eta) = l(\omega) \wedge \eta + \omega \wedge l(\eta);$$

(2) an *antiderivation* if for $\omega \in \Lambda_k^*(M), \eta \in \Lambda^*(M)$

$$l(\omega \wedge \eta) = l(\omega) \wedge \eta + (-1)^k \omega \wedge l(\eta);$$

(3) of *degree* k if $l : \Lambda_j^*(M) \to \Lambda_{j+k}^*(M)$ for all j.

The following theorem characterizes the exterior derivative as a unique mapping which satisfies certain properties.

THEOREM. There exists a unique antiderivation $d : \Lambda^*(M) \to \Lambda^*(M)$ of degree $+1$ such that

(1) $d \circ d = 0$, and
(2) for $f \in \Lambda_0^*(M)$, $df(X) = X(f)$, that is, df is the differential of f.

For a globally valid construction of the exterior derivative, the reader is referred to [War71]. Next it is advantageous to see what the exterior derivative does when a local coordinate system is introduced. To this end, an obvious corollary of the above theorem is considered in order to strip the discussion of algebraic terminology.

COROLLARY. Consider a chart about a point $p \in M$ where there is a local coordinate system with coordinates $(x^1, \ldots, x^n)$. In this chart there exist a unique mapping

$$d : \Lambda_i^*(M) \to \Lambda_{i+1}^*(M)$$

satisfying

(1) $d(d\omega) = 0$ for $\omega \in \Lambda^*(M)$,
(2) $df = \sum_{i=1}^n (\partial f / \partial x^i)\, dx^i$ for $f \in \Lambda_0^*(M)$,
(3) $d(\omega \wedge \eta) = d\omega \wedge \eta + (-1)^k \omega \wedge d\eta$, where $\omega \in \Lambda_k^*(M)$ and $\eta \in \Lambda^*(M)$.

From this corollary, it is easily verified that for a k-form

$$\omega = \sum_{i_1 < i_2 < \cdots < i_k}^{n} a_{i_1 i_2 \ldots i_k}(x^j)\, dx^{i_1} \wedge dx^{i_2} \wedge \cdots \wedge dx^{i_k}$$

the exterior derivative is given by

$$d\omega = \sum_{i_1 < \cdots < i_k}^{n} \left(\sum_{j=1}^{n} \frac{\partial a_{i_1 \ldots i_k}}{\partial x^j}(x^k)\, dx^j \right) \wedge dx^{i_1} \wedge dx^{i_2} \wedge \cdots \wedge dx^{i_k}.$$

Hence when local coordinates are introduced, the exterior derivative can be computed in a straightforward way. The properties which the exterior derivative satisfies according to the above corollary will now be examined. The property

$$d(d\omega) = 0 \quad \text{for } \omega \in \Lambda^*(M)$$

should hold if Stokes' theorem is to hold and the exterior derivative is to be considered an operator adjoint to the boundary operator (recall Chapter 1). The property

$$df = \sum_{i=1}^{n} \frac{\partial f}{\partial x^i}\, dx^i$$

is what is required to make bases of the dual space to the tangent space transform in a contravariant way. Finally the last condition of the corollary is what is required to make the following theorem true:

THEOREM. Given $\varphi : M' \to M$ and $\omega \in \Lambda^*(M)$

$$\varphi^{\#} d\omega = d(\varphi^{\#}\omega).$$

For a proof of this theorem and the previous one, see [War71], pages 65-68. The next sensible thing to do is consider a series of examples which serve the dual purpose of illustrating exterior differentiation and introducing Stokes' Theorem.

Example 7.6 Exterior differentiation in one dimension. Consider a one-dimensional manifold with local coordinate t and a 0-form.

$$\omega = f(t) \quad \text{implies} \quad d\omega = \frac{\partial f(t)}{\partial t}\, dt.$$

The fundamental theorem of calculus states that

$$\int_a^b \frac{\partial f(t)}{\partial t}\, dt = f\Big|_a^b,$$

or, rewritten in terms of differential forms,

$$\int_{[a,b]} d\omega = \int_{\partial[a,b]} \omega. \qquad \square$$

Example 7.7 Complex variables. Let f be a function of a complex variable. That is,

$$f(z) = f(x+iy) = U(x,y) + iV(x,y).$$

Consider

$$\omega = f(z)\, dz = \big(U(x,y) + iV(x,y)\big)(dx + i\, dy)$$

hence

$$\begin{aligned} d\omega &= \left(\frac{\partial U}{\partial x}\, dx + \frac{\partial U}{\partial y}\, dy + i\frac{\partial V}{\partial x}\, dx + i\frac{\partial V}{\partial y}\, dy\right) \wedge (dx + i\, dy) \\ &= \left(-\left(\frac{\partial U}{\partial y} + \frac{\partial V}{\partial x}\right) + i\left(\frac{\partial U}{\partial x} - \frac{\partial V}{\partial y}\right)\right) dx \wedge dy. \end{aligned}$$

In this case, "Green's theorem in the plane" states that

$$\int_{\partial R} \omega = \int_R d\omega$$

and the Cauchy–Riemann equations

$$\frac{\partial U}{\partial y} = -\frac{\partial V}{\partial x}, \quad \frac{\partial V}{\partial y} = \frac{\partial U}{\partial x} \text{ in } R$$

are equivalent to the statement

$$d\omega = 0 \text{ in } R.$$

Hence, if the Cauchy–Riemann equations hold in the region R

$$\int_{\partial R} f(z)\, dz = 0 = \int_{\partial R} \omega.$$

This is the Cauchy integral theorem. Furthermore, if one considers the above equation for arbitrary 1-cycles, partitions these cycles into homology classes and uses de Rham's theorem, then one obtains the residue formula of complex analysis. Also, by partitioning the above expressions into real and imaginary parts one obtains the integral formulas associated with irrotational and solenoidal flows in two dimensions. □

Example 7.8 The classical version of Stokes' theorem. Let u, v, w be local coordinates in a three-dimensional manifold and R a region in a two-dimensional submanifold. Consider the 1-form:

$$\omega = A_u(u,v,w)\, du + A_v(u,v,w)\, dv + A_w(u,v,w)\, dw.$$

Using the rules for wedge multiplication and exterior differentiation one has

$$\begin{aligned} d\omega &= dA_u \wedge du + dA_v \wedge dv + dA_w \wedge dw \\ &= \left(\frac{\partial A_v}{\partial u} - \frac{\partial A_u}{\partial v}\right) du \wedge dv + \left(\frac{\partial A_w}{\partial v} - \frac{\partial A_v}{\partial w}\right) dv \wedge dw + \left(\frac{\partial A_u}{\partial w} - \frac{\partial A_w}{\partial u}\right) dw \wedge du, \end{aligned}$$

and the classical version of Stokes' theorem becomes

$$\int_{\partial R} \omega = \int_R d\omega.$$ □

Example 7.9 The divergence theorem in three dimensions. Next consider a 2-form on a three-dimensional manifold with local coordinates u, v, w. Let

$$\omega = D_u(u,v,w)\, dv \wedge dw + D_v(u,v,w)\, dw \wedge du + D_w(u,v,w) du \wedge dv.$$

Then, using the rules,

$$\begin{aligned} d\omega &= dD_u \wedge dv \wedge dw + dD_v \wedge dw \wedge du + dD_w \wedge du \wedge dv \\ &= \left(\frac{\partial D_u}{\partial u} + \frac{\partial D_v}{\partial v} + \frac{\partial D_w}{\partial w}\right) du \wedge dv \wedge dw. \end{aligned}$$

In this case, Ostrogradskii's formula becomes

$$\int_{\partial R} \omega = \int_R d\omega.$$ □

Example 7.10 Electrodynamics. Consider the four-dimensional space-time continuum with local coordinates x, y, z, t. Let

$$\alpha = A_x\,dx + A_y\,dy + A_z\,dz - \phi\,dt$$
$$\beta = (E_x\,dx + E_y\,dy + E_z\,dz) \wedge dt + (B_x\,dy \wedge dz + B_y\,dz \wedge dx + B_z\,dx \wedge dy)$$
$$\eta = (H_x\,dx + H_y\,dy + H_z\,dz) \wedge dt - (D_x\,dy \wedge dz + D_y\,dz \wedge dx + D_z\,dx \wedge dy)$$
$$\lambda = (J_x\,dy \wedge dz + J_y\,dz \wedge dx + J_z\,dx \wedge dy) \wedge dt - \rho\,dx \wedge dy \wedge dz.$$

By a straightforward calculation it is easy to verify that Maxwell's equations can be written as

$$d\beta = 0, \qquad d\eta = \lambda.$$

If φ is a transformation of coordinates, the identity

$$\varphi^{\#} d = d\varphi^{\#}$$

is an expression of the principle of general covariance. Also, putting aside considerations of homology theory, the identity $d \circ d = 0$ enables one to write the field in terms of potentials

$$\beta = d\alpha.$$

The general covariance of Maxwell's equations is nicely expressed in [Bat10] and makes the study of electrodynamics in noninertial reference frames tractable. Following [Har08] one can rewrite Maxwell's equations in integral form by using Stokes' theorem:

$$\int_{b_2=\partial c_3'} \beta = \int_{c_3'} d\beta = 0, \qquad \int_{\partial c_3} \eta = \int_{c_3} d\eta = \int_{c_3} \lambda,$$

where b_2 is any 2-boundary and c_3, c_3' are any 3-chains. Hence, putting aside considerations of homology, we have

$$\int_{z_2} \beta = 0, \qquad \int_{\partial c_3} \eta = \int_{c_3} \lambda,$$

where z_2 is any 2-cycle. For modern uses of these equations see [Pos78, Pos84]. □

MA-F. Cohomology with Differential Forms

It is now possible to restate the ideas of Chapter 1 in a more formal way. Rewrite

$$d : \Lambda^*(M) \to \Lambda^*(M)$$

as

$$d : \bigoplus_p \Lambda_p^*(M) \to \bigoplus_p \Lambda_p^*(M),$$

and define the restriction of the exterior derivative to p-forms by

$$d^p : \Lambda_p^*(M) \to \Lambda_{p+1}^*(M).$$

As usual, one can define the set of p-cocycles (or in the language of differential forms, the space of closed p-forms) as

$$Z^p(M) = \ker\big(\Lambda_p^*(M) \xrightarrow{d^p} \Lambda_{p+1}^*(M)\big)$$

and the set of p-coboundaries (or in the language of differential forms, the space of exact p-forms) as

$$B^p(M) = \operatorname{im}\left(\Lambda^*_{p-1}(M) \xrightarrow{d^{p-1}} \Lambda^*_p(M)\right).$$

The equation

$$d_{p+1} \circ d_p = 0$$

shows that $\Lambda^*(M)$ is a cochain complex and that

$$B^p(M) \subset Z^p(M).$$

Thus one defines

$$H^P_{\mathrm{deR}}(M) = Z^p(M)/B^p(M)$$

to be the de Rham cohomology of the manifold. In order to relate the notation of this and the first chapter, define

$$C^*(M) = \Lambda^*(M),$$

where

$$C^p(M) = \Lambda^*_p(M).$$

One can define many other cohomology theories if the definition of cohomology is written out explicitly, with

$$H^p(M) = \left(Z^p(M) \cap C^p(M)\right) / \left(\left(d^{p-1}C^{p-1}(M)\right) \cap C^p(M)\right).$$

Thus if $C^p(M)$ consisted of all p-forms of compact support, or all p-forms with square integrable coefficients, one would obtain "cohomology with compact support" or "L^2 cohomology". Although these cohomology theories tend to agree on compact manifolds, they do not agree in general. The precise definition of compactly supported cohomology involves a limiting procedure which is ignored here (see [GH81, Chapter 26]). Thus for example

$$H^3_{\mathrm{deR}}(\mathbb{R}^3) \simeq 0 \not\simeq \mathbb{R} \simeq H^3_c(\mathbb{R}^3).$$

Although this result has not been proven here, it is easily deduced from [Spi79], p. 371. The regions of interest are bounded subsets of $\mathbb{R}^n$ and in this case cohomology with compact support is easily interpreted in terms of boundary conditions, an interpretation which will soon be given. As in the first chapter, the complex associated with differential forms with compact support in a region Ω will be denoted by

$$C^*_c(\Omega)$$

and the associated cocycle, coboundary, and cohomology spaces will be distinguished by the subscript c.

The de Rham cohomology vector spaces play a central role for the present purpose as does the cohomology with compact support in the context of relative cohomology. The L^2 cohomology spaces, although important in the context of finite energy constraints on variational functionals, will not be considered. There are two reasons for this. First, the properties of L^2 cohomology are harder to articulate mathematically and secondly, for bounded regions of $\mathbb{R}^3$ the de Rham cohomology and cohomology with compact support give the required insight into L^2 cohomology while unbounded domains in $\mathbb{R}^3$ can be handled by attaching a point at infinity and mapping $\mathbb{R}^3$ onto the unit sphere S^3 in $\mathbb{R}^4$ (this procedure is analogous to stereographic projections in complex variables).

MA-G. Cochain Maps Induced by Mappings Between Manifolds

Having defined the cochain complexes associated with de Rham cohomology and cohomology with compact support, we now consider how mappings between manifolds induce cochain maps between cochain complexes. Given a map

$$\varphi : M' \to M,$$

there are covariant and contravariant transformations

$$\begin{aligned} \varphi_{\#} &: T(M') \to T(M), \\ \varphi^{\#} &: \Lambda^*(M) \to \Lambda^*(M') \end{aligned}$$

on vector fields and differential forms respectively. For a given k and

$$\omega \in \Lambda_k^*(M) \text{ and } X_i \in T(M'),\ 1 \leq i \leq k,$$

one has

$$\omega\left(\varphi_{\#}(X_1), \varphi_{\#}(X_2), \ldots, \varphi_{\#}(X_k)\right) = (\varphi^{\#}\omega)(X_1, X_2, \ldots, X_k)$$

which express the invariance of the whole scheme. Having defined

$$\begin{aligned} C^*(M) &= \Lambda^*(M), \\ C^*(M') &= \Lambda^*(M'), \end{aligned}$$

the formula

$$\varphi^{\#} d_M = d_{M'} \varphi^{\#},$$

where the d on the right is the coboundary operator (exterior derivative) in the complex $C^*(M')$ while the d on the left is the exterior derivative in the complex $C^*(M)$, shows that $\varphi^{\#}$ is a cochain homomorphism. That is, if φ^p is the restriction of $\varphi^{\#}$ to p-forms, the following diagram commutes for all k:

$$\begin{array}{ccc}
\vdots & & \vdots \\
\Big\downarrow d_M^{k-1} & & \Big\downarrow d_{M'}^{k-1} \\
C^k(M) & \xrightarrow{\varphi^k} & C^k(M') \\
\Big\downarrow d_M^{k} & & \Big\downarrow d_{M'}^{k} \\
C^{k+1}(M) & \xrightarrow{\varphi^k} & C^{k+1}(M') \\
\Big\downarrow d_M^{k+1} & & \Big\downarrow d_{M'}^{k+1} \\
\vdots & & \vdots
\end{array}$$

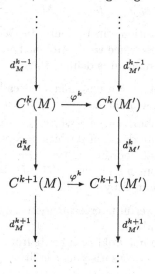

A very important special case of this construction occurs when M' is a submanifold of M and φ is the injection mapping. That is, if S is a submanifold of M and $i : S \to M$ is the injection map, the pullback

$$i^{\#} : C^*(M) \to C^*(S)$$

is a cochain homomorphism. When this happens, it is possible to construct a long exact cohomology sequence in several ways (see [Spi79, pp. 571–591] or [BT82, pp. 78–79]). This topic, however, will not be pursued here since once the de Rham theorem is established, it is easier to think in terms of cycles and the long exact homology sequence.

MA-H. Stokes' Theorem, de Rham's Theorems and Duality Theorems

As a prelude to Stokes' Theorem, the concepts of an orientation and regular domain are required. Since $\Lambda_n(M_p^*)$ is one-dimensional, it follows that $\Lambda_n(M_p^*) - \{0\}$ has two connected components. An orientation of M_p^* is a choice of connected component of $\Lambda_n(M_p^*) - \{0\}$. An n-dimensional manifold is said to be orientable if it is possible to make an unambiguous choice of orientation for M_p^* at each $p \in M$. If M is not connected, M is orientable if each of its connected components is orientable. The following proposition clears up the intuitive picture about orientation:

PROPOSITION. *If M is a differentiable manifold of dimension n, the following are equivalent:*
(1) *M is orientable;*
(2) *there is an atlas $\mathcal{A} = \{(U_i, \varphi_i)\}$ such that*

$$\left(\frac{\partial(x^1, \ldots, x^n)}{\partial(y^1, \ldots, y^n)}\right) > 0 \qquad on\ U_i \cap U_j$$

whenever $\left(U_i, (x^1, \ldots, x^n)\right), \left(U_j, (y^1, \ldots, y^n)\right) \in \mathcal{A}$;
(3) *there is a nowhere vanishing n-form on M.*

The proof of this proposition can be found in [War71, pages 138–140]. An example of a nonorientable surface is the Möbius band of Example 1.12.

The notion of a regular domain is defined thus:

DEFINITION 3. A subset D of a manifold M is called a *regular domain* if for each $p \in M$ one of the following holds:
(1) There is an open neighborhood of p which is contained in $M - D$.
(2) There is an open neighborhood of p which is contained in D.
(3) Given $\mathbb{R}^n$ with Cartesian coordinates $(r^1, \ldots, r^n)$ there is a centered coordinate system (U, φ) about p such that $\varphi(U \cap D) = \varphi(U) \cap H^n$, where H^n is the half-space of $\mathbb{R}^n$ defined by $r^n \geq 0$.

Thus if we trained a powerful microscope onto any point of D, we would see one of the cases in Figure MA-2.

A regular domain D in a manifold M inherits from M a structure of *submanifold with boundary*; points $p \in D$ satisfying condition (2) above form the *interior* of D, and points satisfying condition (3) form the *boundary* of D, denoted by ∂D. The boundary of a regular domain is itself a manifold (without boundary).

In cases where this definition is too restrictive, for example if D is a square, one can use the notion of an almost regular domain (see [LS68, pp. 424–427]).

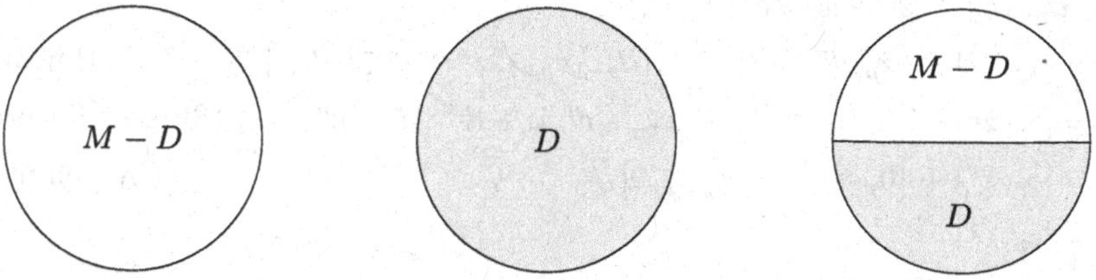

Figure MA-2.

The main result of this section is the following version of Stokes' Theorem.

THEOREM. Let D be a regular domain in an oriented n-dimensional manifold M and let ω be a smooth $(n-1)$-form of compact support. Then

$$\int_D d\omega = \int_{\partial D} i^{\#}\omega,$$

where $i : \partial D \to M$ induces the pullback

$$i^{\#} : C^*(M) \to C^*(\partial D).$$

For a nice, simple proof of this theorem, see [War71, pages 140–148].

At this point it is worthwhile interpreting integration as a bilinear pairing between differential forms and chains so that the de Rham isomorphism is easy to understand. It has been assumed all along that integration is a bilinear pairing between chains and cochains (forms). In the heuristic development of Chapter 1 this was emphasized by writing

$$\int : C_p(M) \times C^p(M) \to \mathbb{R}.$$

Furthermore, the reader was lead to believe that differential forms were linear functionals on differentiable chains. This was emphasized notationally by writing

$$\int_c \omega = [c, \omega].$$

Stokes' Theorem was then written as

$$[\partial_p c, \omega] = [c, d^{p-1}\omega] \quad \text{for all } p,$$

and hence was interpreted as saying that the exterior derivative (coboundary) operator and the boundary operator were adjoint operators. This is the setting for de Rham's theorem, for if the domain of the bilinear pairing is restricted to cocycles (closed forms) and cycles, that is,

$$\int : Z_p(M) \times Z^p(M) \to \mathbb{R},$$

then it is easy to show that the value of this bilinear pairing depends only on the homology class of the cycle and the cohomology class of the closed form (cocycle). This is easily verified by the following calculation. Take $z^p \in Z^p(M)$ where $c^{p-1} \in C^{p-1}(M)$ and $z_p \in Z_p(M)$ where $c_{p+1} \in C_{p+1}(M)$. Then

$$\begin{aligned}
&\left[z_p + \partial_{p+1}c_{p+1}, z^p + d^{p-1}c^{p-1}\right] \\
&\quad = \left[z_p, z^p\right] + \left[z_p, d^{p-1}c^{p-1}\right] + \left[\partial_{p+1}c_{p+1}, z^p + d^{p-1}c^{p-1}\right] && \text{(linearity)} \\
&\quad = \left[z_p, z^p\right] + \left[\partial_p z_p, c^{p-1}\right] + \left[c_{p+1}, d^p z_p + d^p \circ d^{p-1}c^{p-1}\right] && \text{(Stokes' Theorem)} \\
&\quad = \left[z_p, z^p\right] + \left[0, c^{p-1}\right] + \left[c_{p+1}, 0\right] && \text{(by definition)} \\
&\quad = \left[z_p, z^p\right].
\end{aligned}$$

Hence, when the domain of the bilinear pairing is restricted to chains and cochains, one obtains a bilinear pairing between homology and cohomology. The theorems of de Rham assert that this induced bilinear pairing is nondegenerate and hence there is an isomorphism

$$H_p(M) \simeq H^p_{\text{deR}}(M).$$

As noted in Chapter 1, this is not meant to be the place to prove that such an isomorphism exists since no formal way of computing homology is considered. The reader will find down-to-earth proofs of the de Rham isomorphism in de Rham [dR31] or Hodge [Hod52, Chapter 2]. A sophisticated modern proof can be found in [War71, Chapter 5], while less formal proofs can be found in the appendices of [Gol82] and [Mas80].

For the present exposition de Rham type of isomorphism is required for relative homology and cohomology groups. Though this type of isomorphism is not readily found in books (if at all) there are two methods of obtaining such an isomorphism once the usual de Rham isomorphism is established. The first approach is in Duff [Duf52]. The second approach is to reduce the problem to a purely algebraic one and use the so called five lemma. Though this second approach is straightforward, it does not appear to be in the literature.

Consider, for example, a manifold M with compact boundary ∂M. In this case there is the following long exact cohomology sequence [Spi79, Theorem 13, p. 589]:

$$\cdots \to H^k_c(M) \to H^k(\partial M) \to H^{k+1}_c(M-\partial M) \to H^{k+1}_c(M) \to H^{k+1}(\partial M) \to \cdots$$

Also there is a long exact homology sequence (see [GH81, Chapter 14] for example)

$$\cdots \leftarrow H_k(M) \leftarrow H_k(\partial M) \leftarrow H_{k+1}(M, \partial M) \leftarrow H_{k+1}(M) \leftarrow H_{k+1}(\partial M) \leftarrow \cdots$$

and the following de Rham isomorphisms are known to exist for all k:

$$\begin{aligned}
H_k(M) &\simeq H^k_c(M) \qquad (M \text{ compact}) \\
H_k(\partial M) &\simeq H^k(\partial M).
\end{aligned}$$

The above isomorphisms are induced by integration and there is also a bilinear pairing between $H^{k+1}_c(M - \partial M)$ and $H_{k+1}(M, \partial M)$ which is induced by integration. In this case the following diagram is commutative:

$$\begin{array}{ccccccccc}
H_{k+1}(\partial M) & \longrightarrow & H_{k+1}(M) & \longrightarrow & H_{k+1}(M,\partial M) & \longrightarrow & H_k(\partial M) & \longrightarrow & H_k(M)\\
\simeq\updownarrow & & \simeq\updownarrow & & \updownarrow & & \simeq\updownarrow & & \simeq\updownarrow\\
H^{k+1}(\partial M) & \longleftarrow & H_c^{k+1}(M) & \longleftarrow & H_c^{k+1}(M-\partial M) & \longleftarrow & H^k(\partial M) & \longleftarrow & H_c^k(M).
\end{array}$$

What is required is to show that the middle vertical arrow in this picture (and hence every third arrow in the long sequence of commutative squares) is an isomorphism. To do this one first considers the dual spaces $\left(H^k(\partial M)\right)^*$, $\left(H_c^k(M)\right)^*$, and $\left(H_c^k(M-\partial M)\right)^*$ and notices that, by definition there is a commutative diagram

$$\begin{array}{ccccccccc}
H^{k+1}(\partial M) & \longleftarrow & H_c^{k+1}(M) & \longleftarrow & H_c^{k+1}(M-\partial M) & \longleftarrow & H^k(\partial M) & \longleftarrow & H_c^k(M)\\
\simeq\updownarrow & & \simeq\updownarrow & & \simeq\updownarrow & & \simeq\updownarrow & & \simeq\updownarrow\\
\left(H^{k+1}(\partial M)\right)^* & \longrightarrow & H_c^{k+1}(M)^* & \longrightarrow & H_c^{k+1}(M-\partial M)^* & \longrightarrow & H^k(\partial M)^* & \longrightarrow & H_c^k(M)^*
\end{array}$$

where the vertical arrows are all isomorphisms, the two rows are exact sequences and the mappings on the bottom row are the adjoints of the mappings directly above them. Combining the above two commutative diagrams one has the following commutative diagram:

$$\begin{array}{ccccccccc}
H_{k+1}(\partial M) & \longrightarrow & H_{k+1}(M) & \longrightarrow & H_{k+1}(M,\partial M) & \longrightarrow & H_k(\partial M) & \longrightarrow & H_k(M)\\
\simeq\updownarrow & & \simeq\updownarrow & & \simeq\updownarrow & & \simeq\updownarrow & & \simeq\updownarrow\\
H^{k+1}(\partial M)^* & \longrightarrow & H_c^{k+1}(M)^* & \longrightarrow & H_c^{k+1}(M-\partial M)^* & \longrightarrow & H^k(\partial M)^* & \longrightarrow & H_c^k(M)^*
\end{array}$$

where the rows are exact sequences and one wants to know whether the middle homomorphism is an isomorphism. To see that the answer is yes consider the following lemma (see Greenberg and Harper [GH81, p. 77–78]).

FIVE LEMMA. Given a diagram of R-modules and homomorphisms with all rectangles commutative

$$\begin{array}{ccccccccc}
A_1 & \xrightarrow{a_1} & A_2 & \xrightarrow{a_2} & A_3 & \xrightarrow{a_3} & A_4 & \xrightarrow{a_4} & A_5\\
\downarrow\alpha & & \downarrow\beta & & \downarrow\gamma & & \downarrow\delta & & \downarrow\varepsilon\\
B_1 & \xrightarrow{b_1} & B_2 & \xrightarrow{b_2} & B_3 & \xrightarrow{b_3} & B_4 & \xrightarrow{b_4} & B_5
\end{array}$$

such that the rows are exact (at joints 2, 3, 4) and the four outer homomorphisms $\alpha, \beta, \delta, \varepsilon$ are isomorphisms, then γ is an isomorphism.

It is obvious that the five lemma applies in the above situation (since a vector space over $\mathbb{R}$ is an instance of an R-module) and hence

$$H_{k+1}(M, \partial M) \simeq H_c^{k+1}(M - \partial M)^*.$$

Thus

$$H_{k+1}(M, \partial M) \simeq H_c^{k+1}(M - \partial M)$$

and the relative de Rham isomorphism is proven. It is also obvious that the isomorphism would be true if ∂M were replaced by a collection of connected components of ∂M or parts of ∂M which arise from symmetry planes as in Chapter 1 all that is required is the existence of long exact (co)homology sequences and the usual de Rham isomorphism.

Having seen how the de Rham isomorphism can be understood with the help of Stokes' Theorem, a simple corollary of Stokes' Theorem will now be used to give a heuristic understanding of duality theorems. Suppose M is an oriented n-dimensional manifold, D is a regular domain in M and $\lambda \in C_c^k(M), \mu \in C^{n-k-1}(M)$. Since λ has compact support,

$$\omega \in C_c^{n-1}(M) \quad \text{if} \quad \omega = \lambda \wedge \mu.$$

Furthermore

$$d\omega = d(\lambda \wedge \mu) = (d\lambda) \wedge \mu + (-1)^k \lambda \wedge d\mu,$$

and if i is the injection of ∂D into M then, as usual, the pullback $i^{\#}$ satisfies

$$i^{\#}(\omega) = i^{\#}(\lambda \wedge \mu) = i^{\#}(\lambda) \wedge i^{\#}(\mu).$$

Substituting this expression for ω into Stokes' Theorem one obtains an important corollary:

COROLLARY (Integration by parts). If D is a regular domain in an oriented n-dimensional manifold M and

$$\lambda \in C_c^k(M), \quad \mu \in C^{n-k-1}(M)$$

then

$$(-1)^k \int_D \lambda \wedge d\mu = \int_{\partial D} (i^{\#}\lambda) \wedge (i^{\#}\mu) - \int_D d\lambda \wedge \mu$$

where

$$i : \partial D \to M$$

induces

$$i^{\#} : C^*(M) \to C^*(\partial D).$$

Just as Stokes' Theorem is often called the fundamental theorem of multivariable calculus since it generalizes the usual fundamental theorem of integral calculus, the above corollary is the multivariable version of "integration by parts". This integration by parts formula is of fundamental importance in the calculus of variations and in obtaining an interpretation of duality theorems on manifolds. These duality theorems will be considered next.

Consider first the situation of a manifold without boundary M and the Poincaré duality theorem. In this case the integration by parts formula reduces to

$$(-1)^k \int_M \lambda \wedge d\mu = -\int_M (d\lambda) \wedge \mu$$

whenever $\lambda \in C_c^k(M)$ and $\mu \in C^{n-k-1}(M)$. Also there is a bilinear pairing

$$\int_M : C_c^p(M) \times C^{n-p}(M) \to \mathbb{R},$$

where a p-form of compact support is wedge multiplied with an $(n-p)$-form to yield an n-form of compact support which is then integrated over the entire manifold. The heart of the proof of the Poincaré duality theorem involves restricting the domain of this bilinear pairing from chains to cycles and noticing that one has a bilinear pairing on dual (p and $n-p$) cohomology groups. Though the Poincaré duality theorem is not proved here, it is useful to see how this bilinear pairing on cohomology comes about. Consider

$$\int_M : Z_c^p(M) \times Z^{n-p}(M) \to \mathbb{R},$$

where for

$$z^p \in Z_c^p(M), \quad z^{n-p} \in Z^{n-p}(M)$$

one computes

$$\int_M z^p \wedge z^{n-p}.$$

To see that the value of this integral depends only on the cohomology classes of z^p and z^{n-p} one lets

$$c^{p-1} \in C_c^{p-1}(M), \quad c^{n-p-1} \in C^{n-p-1}(M)$$

and computes

$$\begin{aligned}
&\int_M (z^p + d^{p-1}c^{p-1}) \wedge (z^{n-p} + d^{n-p-1}c^{n-p-1}) \\
&\quad = \int_M z^p \wedge z^{n-p} + \int_M (d^{p-1}c^{p-1}) \wedge z^{n-p} + \int_M (z^p + d^{p-1}c^{p-1}) \wedge d^{n-p-1}c^{n-p-1} \\
&\quad = \int_M z^p \wedge z^{n-p} + (-1)^p \int_M c^{p-1} \wedge d^{n-p}z^{n-p} \\
&\qquad\qquad -(-1)^p \int_M (d^p z^p + d^p \circ d^{p-1}c^{p-1}) \wedge c^{n-p-1} \quad \text{(integrating by parts)} \\
&\quad = \int_M z^p \wedge z^{n-p} \qquad \text{(using the definition of cocycle).}
\end{aligned}$$

Thus restricting the domain of the bilinear form from cochains to cocycles induces the following bilinear pairing on cohomology:

$$\int_M : H_c^p(M) \times H_{\mathrm{deR}}^{n-p}(M) \to \mathbb{R}.$$

The Poincaré duality theorem asserts that this bilinear pairing is nondegenerate. Thus

$$H^p_c(M) \simeq H^{n-p}_{\text{deR}}(M) \quad \text{for all } p,$$

or, if M is compact,

$$H^p_{\text{deR}}(M) \simeq H^{n-p}_{\text{deR}}(M).$$

This statement of the Poincaré duality theorem is not the most general version (see [BT82, pp. 44–47] for a proof and explanation of the subtleties encountered in generalizing the above). The implicit assumption in the above argument is the finite dimensionality of the cohomology vector spaces. A nice discussion of this aspect is given in [Spi79, p. 600 and preceding pages]. As mentioned in Chapter 3, [Mas80, Chapter 9] and [GH81, Chapter 26] have proofs of the Poincaré duality theorem which do not appeal to the formalism of differential forms.

When the manifold M is not compact, the Poincaré duality theorem may be used to show the difference between de Rham cohomology and cohomology with compact support. Take for example $\mathbb{R}^n$, where, by Poincaré duality and the arguments of the last chapter, one has

$$H^{n-p}_c(\mathbb{R}^n) \simeq H^p_{\text{deR}}(\mathbb{R}^n) \simeq \begin{cases} \mathbb{R} & \text{if } p = 0, \\ 0 & \text{if } p \neq 0. \end{cases}$$

Hence, in the case of $\mathbb{R}^3$,

$$H^3_c(\mathbb{R}^3) \simeq \mathbb{R} \not\simeq 0 \simeq H^3_{\text{deR}}(\mathbb{R}^3),$$
$$H^0_{\text{deR}}(\mathbb{R}^3) \simeq \mathbb{R} \not\simeq 0 \simeq H^0_c(\mathbb{R}^3).$$

As stated in the previous chapter, the Poincaré duality theorem does not have many direct applications in boundary value problems of electromagnetics. For the present purpose, attention will be paid to compact manifolds with boundary and for these there is a variety of duality theorems. In this case, it is useful to get certain ideas established once and for all. First, if we take $i : \partial M \to M$, $i^\# : C^*(M) \to C^*(\partial M)$, then $c \in C^*_c(M - \partial M)$ if $i^\# c = 0$. Thus $z^p \in Z^p_c(M - \partial M)$ if $dz^p = 0$, $i^* z^p = 0$, and $b^p \in B^p_c(M - \partial M)$ if $b^p = dc^{p-1}$ for some $c^{p-1} \in C^{p-1}_c(M - \partial M)$.

In this case, it is customary to denote the symbol $i^\#$ by t, and avoid referring to the injection i. Thus $t\omega$ denotes the pullback of ω to ∂M and is the "tangential" part of ω. In this notation the integration by parts formula takes the form:

$$(-1)^k \int_M \lambda \wedge d\mu = \int_{\partial M} (t\lambda) \wedge (t\mu) - \int_M (d\lambda) \wedge \mu$$

where λ is a k-form, and μ is a $(n-k-1)$-form.

To see how the Lefschetz Duality Theorem comes about, consider an orientable compact n-dimensional manifold with boundary and the bilinear pairing

$$\int_M : C^p_c(M - \partial M) \times C^{n-p}(M) \to \mathbb{R}.$$

Note that if the boundary of the manifold is empty then the situation is identical to that of the Poincaré duality theorem. Restricting the domain of this bilinear pairing to cocycles (closed forms) one can easily show that there is an induced bilinear pairing on two cohomology groups. That is, considering

$$\int_M : Z^p_c(M - \partial M) \times Z^{n-p}(M) \to \mathbb{R},$$

the integration by parts formula shows that the integral $\int_M z^p \wedge z^{n-p}$ depends only on the cohomology classes of z^p and z^{n-p}, whenever $z^p \in Z^p_C(M - \partial M)$ and $z^{n-p} \in Z^{n-p}(M)$. Hence there is a bilinear pairing

$$\int_M : H^p_c(M - \partial M) \times H^{n-p}_{\text{deR}}(M) \to \mathbb{R}$$

induced by integration. The Lefschetz Duality Theorem asserts that this bilinear pairing is nondegenerate. Hence

$$H^p_c(M - \partial M) \simeq H^{n-p}_{\text{deR}}(M).$$

Again one can find the proof of this type of theorem in [Mas80, §9.7], or Greenberg and Harper [GH81, Ch. 28]. Conner [Con54] has shown that there is a generalization of the Lefschetz duality theorem. To see what this generalization is, write

$$\partial M = \bigcup_{i=1}^{m} C_i,$$

where each c_i is a connected manifold without boundary, and set

$$S_1 = \bigcup_{i=1}^{r} C_i, \quad S_2 = \bigcup_{i=r+1}^{m} C_i$$

where M is a compact, orientable n-dimensional manifold with boundary. In this case the result of Conner states that

$$H^p_c(M - S_1) \simeq H^{n-p}_c(M - S_2).$$

This result can be interpreted, as before, by saying that the bilinear pairing

$$\int_M : C^p_c(M - S_1) \times C^{n-p}_c(M - S_2)$$

descends into a nondegenerate bilinear pairing on cohomology when the domain is restricted to cocycles. This is verified by using the integration by parts formula to show that the restricted bilinear pairing does indeed depend only on cohomology classes.

In electromagnetism the situation is slightly more general in that S_1 and S_2 are not necessarily disjoint but at the intersection $S_1 \cap S_2$ a symmetry plane and a component of the boundary of some original problem meet at right angles. From the usual proofs of the Lefschetz duality theorem (which construct the double of a manifold) it is apparent that the duality theorem

$$H^p_c(M - S_1) \simeq H^{n-p}_c(M - S_2)$$

is still true. It is useful to note that the interpretation of the above duality theorems is in some sense dual to the approach taken in Chapter 3 in that the homology point of view stresses intersection numbers while the cohomology point of view stresses the bilinear pairing induced by integration. It is important to keep this interplay in mind since topological problems in electromagnetics involve the bilinear pairing in cohomology and these problems can be resolved very conveniently by thinking in terms of intersection numbers. In the case of the Alexander duality theorem there is no nice intersection number or integral interpretation. This is apparent from the proof of the Alexander Duality Theorem (see [Mas80, §9.6] or [GH81, Chapter 27]). For a different but down-to-earth exposition of intersection numbers from the point of view of differential forms and many other topics treated so far in this chapter the reader is referred to [Her77, Chapter 34, part 5].

MA-I. Existence of Cuts Via Eilenberg–MacLane Spaces

In this section we will demonstrate the existence of cuts corresponding to generators of $H^1(\Omega, \mathbb{Z})$ by showing that generators of $H_2(\Omega, \partial\Omega, \mathbb{Z})$ can be realized by orientable embedded submanifolds with boundary which are inverse images of regular points of maps from Ω into the circle. To see why this is the right thing to do, let $K(\mathbb{Z}, n)$ be the Eilenberg–MacLane space [BT82, p. 240] [Spa66, Chapter 8, §1] corresponding to the additive group $\mathbb{Z}$ and the integer n. The Eilenberg–MacLane space $K(\mathbb{Z}, n)$ is uniquely characterized, up to homotopy, by the property that its ith homotopy group is trivial except in dimension n where it is isomorphic to $\mathbb{Z}$. For example, we have

$$K(\mathbb{Z}, 1) = S^1, \tag{MA-1}$$

where S^1 is the circle. Eilenberg–MacLane spaces are of interest to us because of the isomorphism

$$H^p(A; \mathbb{Z}) \simeq [A, K(\mathbb{Z}, p)]. \tag{MA-2}$$

This means that given a space A, the kth cohomology group of A with coefficients in $\mathbb{Z}$ is isomorphic to the homotopy classes of maps from A into $K(\mathbb{Z}, k)$. In particular, if A is taken to be the region Ω where irrotational magnetic fields exist then, by (MA-1) and (MA-2), this isomorphism becomes

$$H^1(\Omega; \mathbb{Z}) \simeq [\Omega, S^1]. \tag{MA-3}$$

That is, cohomology classes of magnetic fields H with integral periods are in one to one correspondence with some homotopy class of maps from Ω to S^1. Furthermore, since S^1 can be thought of as the unit circle in the complex plane, a map from Ω to S^1 can be thought of as an assignment of an angle $\theta \in \mathbb{R}^1$ for each point of Ω such that $e^{i\theta}$ varies smoothly. Note that this prescription still allows for θ to jump by $2\pi n$ on some "cutting surface".

It is important to have an acquaintance with how the isomorphism (MA-2) comes about in general, and a clear understanding of how it works in the case of interest (MA-3). From the definition of $K(\mathbb{Z}, n)$ and the Hurewicz isomorphism [BT82, p. 225], it follows that:

$$H^p(K(\mathbb{Z}, p); \mathbb{Z}) \simeq \mathbb{Z}. \tag{MA-4}$$

In general the isomorphism (MA-2) comes about as follows. Let $[\mu]$ be the generator of the cohomology group in (MA-4) and consider a map $f : A \to K(\mathbb{Z}, p)$. The isomorphism (MA-2) is then given by the cohomology class $[f^*\mu]$ which is obtained by pulling back $[\mu]$ via f. To see this in the case of interest (MA-3) let $\mu = d\theta/2\pi$, where, by the usual abuse of notation, $d\theta$, the element of arc length on S^1 is a closed 1-form which is not exact and hence not cohomologous to zero. Also, $\mu \in H^1(S^1, \mathbb{Z})$ since its integral over S^1 is equal to one. The isomorphism (MA-3) then says that given and closed 1-form ω on Ω which has integral periods, there is a map $f : \Omega \to S^1$ such that $[\omega]$ is cohomologous to $[f^*(d\theta/2\pi)]$.

We can now talk about cuts. Consider, by means of the isomorphism (MA-3), a map $\tilde{f} : \Omega \to S^1$ which corresponds to a given integral cohomology class. If $[\tilde{f}^*\mu]$ is not cohomologous to zero, then $\tilde{f}$ is onto S^1. To see this concretely, suppose $\tilde{f}$ missed a point p on S^1 then $\tilde{f}$ misses an open neighborhood X of p. Since it is possible to construct a form μ' cohomologous to μ but with support in X, we have

$$[\tilde{f}^*\mu] \simeq [\tilde{f}^*\mu'] \simeq 0.$$

Hence we have a contradiction which shows that $\tilde{f}$ is onto if $[\tilde{f}^*\mu]$ is nonzero. Now $\tilde{f}$ can always be approximated by a smooth map f. By the Morse–Sard theorem [Hir76], the set of critical values of $f : \Omega \to S^1$ has measure zero. Thus if f represents a nontrivial cohomology class, the set of regular values, that is the set of points in S^1 such that the derivative of f is nonzero, is dense in S^1. Thus it is always possible to pick a point $p \in S^1$ and an open neighborhood $(p - \varepsilon, p + \varepsilon)$ such that every point in this neighborhood is a regular value of f.

CLAIM 1. *$f^{-1}(p)$ is an orientable 2-dimensional compact manifold with boundary embedded in Ω representing a nonzero element in $H_2(\Omega, \partial\Omega; \mathbb{Z})$ which is the Poincaré-Lefschetz dual to $[f^*\mu]$.*

In order to prove this claim, consider $f^{-1}(p)$ the inverse image of a regular point. By a generalization of the implicit function theorem for manifolds with boundary [Hir76, Chap. 1, §4], $f^{-1}(p)$ a neatly embedded 2-dimensional submanifold. That is $f^{-1}(p)$ is an embedded 2-dimensional submanifold of Ω such that $\partial f^{-1}(p) = f^{-1}(p) \bigcap \partial\Omega$ and $f^{-1}(p)$ is not tangent to $\partial\Omega$ at any point $p \in \partial f^{-1}(p)$. Also, since f is a map between compact spaces, it is a proper map and hence $f^{-1}(p)$ is a compact embedded submanifold.

It remains to be seen why $f^{-1}(p)$ is orientable and why it has $f^*(\mu')$ as a Poincaré-Lefschetz dual. Take $(p-\varepsilon, p+\varepsilon)$ to be an open interval which contains only regular points of f. We see that f maps $\mathcal{U} - f^{-1}(p)$ into $(p-\varepsilon, p+\varepsilon) - \{p\}$ and that the latter set has two connected components. Since f is continuous we see that for each connected component of $f^{-1}(p)$ there are at least two connected components of $\mathcal{U} - f^{-1}(p)$. Hence $f^{-1}(p - \varepsilon, p)$ and $f^{-1}(p, p + \varepsilon)$ lie on two globally defined "sides" of $f^{-1}(p)$ and so $f^{-1}(p)$ is orientable.

Finally, we want to verify that $f^*\mu$ is the Poincaré-Lefschetz dual $f^{-1}(p)$ [BT82, pp. 50–53]. This involves checking that

$$\text{(MA-5)} \qquad \int_\Omega f^*\mu' \wedge \gamma = \int_{f^{-1}(p)} \gamma$$

for any closed two-form γ. To do this we notice that we can construct μ' so that its support lies inside of $(p-\varepsilon', p+\varepsilon')$ with $\varepsilon > \varepsilon' > 0$. In this case, the function h defined by

$$\text{(MA-6)} \qquad h(\theta) = \int_{p-\varepsilon'}^{\theta} \mu' \quad \text{satisfies} \quad h(p-\varepsilon') = 0, \; h(p+\varepsilon') = 1.$$

Thus Equation (MA-5) is verified by the following computation:

$$\begin{aligned}
\int_\Omega f^*\mu' \wedge \gamma &= \int_{\mathcal{U}} f^*\mu' \wedge \gamma \text{ (since } \operatorname{Supp}(f^*\mu') \subset \mathcal{U}) \\
&= \int_{\mathcal{U}} f^*(dh) \wedge \gamma = \int_{\mathcal{U}} d(f^*h) \wedge \gamma \qquad \text{(since } f^* \text{ commutes with } d) \\
&= \int_{\mathcal{U}} d(f^*h \wedge \gamma) \qquad \text{(since } d\gamma = 0) \\
&= \int_{\partial\mathcal{U}} f^*h \wedge \gamma = -\int_{f^{-1}(p-\varepsilon')} f^*h \wedge \gamma + \int_{f^{-1}(p+\varepsilon')} f^*h \wedge \gamma \\
&\qquad \text{(by Stokes' theorem and the definition of } \mathcal{U}) \\
&= -0 + \int_{f^{-1}(p+\varepsilon')} \gamma \qquad \text{(by definition of } h \text{, i.e. Equation (MA-6))} \\
&= \int_{f^{-1}(p)} \gamma \qquad \text{(since } f^{-1}(p) \text{ and } f^{-1}(p+\varepsilon') \text{ are homologous).}
\end{aligned}$$

Hence the claim is proved. Summarizing the results of this section we have proven the following theorem.

THEOREM. Given $\Omega \subset \mathbb{R}^3$, a compact submanifold with boundary and a nonzero cohomology class $[\omega] \in H^1(\Omega; \mathbb{Z})$, there exists a map $f : \Omega \to S^1$ and a regular value $p \in S^1$ such that $f^{-1}(p)$ is a compact neatly embedded 2-dimensional submanifold of Ω and $[\omega]$ is the Poincaré-Lefschetz dual to $[f^{-1}(p)] \in H_2(\Omega, \partial\Omega; \mathbb{Z})$.

Given the set of generators for $H^1(\Omega, \mathbb{Z})$, this theorem, with the previous lemma yields:

COROLLARY. There are cuts S_i, $1 \le i \le \dim H^1(\Omega, \mathbb{R})$, which ensure that any irrotational vector field can be written as the gradient of some continuous single valued scalar function on $\Omega - \bigcup S_i$. The S_i are compact orientable embedded submanifolds and the "jump" in the scalar function across these surfaces depends only on the cohomology class of the vector field.

The assemblage of points on a surface is a twofold manifoldness; the assemblage of points in a tri-dimensional space is a threefold manifoldness; the values of a continuous function of n arguments is an n-fold manifoldness.

G. Chrystal, *Encyclopædia Britannica*, 1892.

MA-J. Riemannian Structures, the Hodge Star Operator and an Inner Product for Differential Forms

So far, we have considered those aspects of differential forms that are independent of the notion of distance in the manifold: the complex structure associated with the exterior algebra bundle of the manifold, the change of variable formulas for integrals, Stokes' theorem, de Rham's theorems, and duality theorems in homology and cohomology. We now turn to constructions that depend on the choice of a local inner product between vectors in the manifold's tangent bundle. Such a local inner product defines not only a metric, or distance, between points in the manifold, but also an inner product between differential forms on the manifold. It is therefore an important idea, worthy of a formal definition:

DEFINITION 4. A Riemannian structure on a differentiable manifold M is a smooth choice of a positive definite inner product $(\cdot,\cdot)$ on each tangent space M_p (recall that M_p is the tangent space to M at p).

Here *smooth* means that if the functions in the charts of an atlas for M are differentiable of order C^k and if $X, Y \in T(M)$ have components which are C^k differentiable, then the function (X, Y) is a C^k differentiable function of the coordinates of M. It is a basic fact in Riemannian geometry that any manifold admits a Riemannian structure (see for example [War71], p. 52 or [BT82], p. 42–43). A Riemannian manifold is, by definition, a differentiable manifold with a Riemannian structure, hence any differentiable manifold can be made into a Riemannian manifold.

In terms of local coordinates $(x^1, \ldots, x^n)$ about a point $p \in M$, if $X, Y \in M_p$ and

$$X = \sum_{i=1}^{n} X^i \frac{\partial}{\partial x^i}, \qquad Y = \sum_{i=1}^{n} Y^i \frac{\partial}{\partial x^i},$$

then there is a symmetric positive definite matrix (called the metric tensor) with entries

$$g_{ij} = \left(\frac{\partial}{\partial x^i}, \frac{\partial}{\partial x^j} \right)_p$$

so that

$$(X, Y)_p = \sum_{i=1}^{n} \sum_{j=1}^{n} X^i g_{ij} Y^j.$$

Since 1-forms were defined to be elements of M_p^* (the dual space to M_p), the above inner product induces one on the dual space. That is, if $\omega, \eta \in M_p^*$, where in terms of local coordinates

$$\omega = \sum_{i=1}^{n} a_i\, dx^i, \quad \eta = \sum_{j=1}^{n} b_j\, dx^j,$$

then

$$(\omega, \eta) = \sum_{i=1}^{n}\sum_{j=1}^{n} a_i g^{ij} b_j$$

where

$$g_{ij}g^{jk} = \delta_{ik} \qquad \text{(Kronecker delta)}.$$

A Riemannian structure on a differentiable manifold induces an inner product on k-forms and the immediate objective at this point is to see how this comes about. Given a Riemannian structure on the tangent bundle of a manifold it is always possible to do local computations in terms of an orthonormal basis obtained by the Gram–Schmidt procedure and patching together the results with a partition of unity. Hence, in order to define a pointwise inner product on k-forms it suffices to work in terms of local coordinates. Having made these observations, let ω_i, for $1 \le i \le n$, be an orthonormal basis for M_p^* (that is, $\Lambda_1(M_p^*)$ in some coordinate chart). This means that

$$(\omega_i, \omega_j) = \delta_{ij} \qquad \text{(Kronecker delta)}.$$

By taking all possible exterior products of these basis forms we see that $\Lambda_k(M_p^*)$ is spanned by $\binom{n}{k}$ k-forms that look like

$$\omega_{i_1} \wedge \omega_{i_2} \wedge \cdots \wedge \omega_{i_k}, \qquad 1 \le i_1 < i_2 < \cdots < i_k \le n,$$

and in particular that $\Lambda_n(M_p^*)$ is spanned by the one element

$$\omega_1 \wedge \omega_2 \wedge \cdots \wedge \omega_n.$$

This n-form is called the volume form. Next, by the symmetry of binomial coefficients,

$$\dim \Lambda_k(M_p^*) = \binom{n}{k} = \binom{n}{n-k} = \dim \Lambda_{n-k}(M_p^*).$$

Hence the two spaces are isomorphic. Consider an isomorphism (called the Hodge star operator)

$$* : \Lambda_k(M_p^*) \to \Lambda_{n-k}(M_p^*)$$

acting on the basis vectors above in the following way. Let π be a permutation of n integers and let

$$\omega_{\pi(1)} \wedge \omega_{\pi(2)} \wedge \cdots \wedge \omega_{\pi(k)}$$

be a basis vector in $\Lambda_k(M_p^*)$, so that

$$\omega_{\pi(k+1)} \wedge \omega_{\pi(k+2)} \wedge \cdots \wedge \omega_{\pi(n)}$$

becomes a basis vector in $\Lambda_{n-k}(M_p^*)$. Define

$$* \left(\omega_{\pi(1)} \wedge_{\pi(2)} \wedge \cdots \omega_{\pi(k)}\right) = \mathrm{sgn}(\pi)(\omega_{\pi(k+1)} \wedge \cdots \wedge \omega_{\pi(n)}).$$

Since the linear transformation is defined on the basis vectors of $\Lambda_k(M_p^*)$, the linear transformation is completely defined. Alternatively, one can define the operation of $*$ on basis vectors of $\Lambda_k(M_p^*)$ as follows:

$$\left(\omega_{\pi(1)} \wedge \omega_{\pi(2)} \wedge \cdots \wedge \omega_{\pi(k)}\right) \wedge * \left(\omega_{\pi(1)} \wedge \cdots \wedge \omega_{\pi(k)}\right) = \omega_1 \wedge \omega_2 \wedge \cdots \wedge \omega_n.$$

Using the usual abuse of notation, one defines the volume form

$$dV = \omega_1 \wedge \omega_2 \wedge \cdots \wedge \omega_n,$$

where it is understood that dV is not necessarily the exterior derivative of any $(n-1)$-form. If

$$\omega = \omega_{\pi(1)} \wedge \cdots \wedge \omega_{\pi(k)},$$

the rules for wedge multiplication and the definition of the star operator show that

$$dV = \omega \wedge (*\omega) = (-1)^{k(n-k)}(*\omega) \wedge \omega$$

and

$$dV = (*\omega) \wedge (*(*\omega));$$

hence

$$**\omega = (-1)^{k(n-k)}\omega, \quad 1 = *(dV).$$

By linearity, this is true for all k-forms. Furthermore if $\omega, \eta \in \Lambda_k(M_p^*)$ then

$$*(\omega \wedge *\eta) = *(\eta \wedge *\omega)$$

is a symmetric positive definite function (an inner product) on $\Lambda^*(M_p^*)$ (see Flanders [Fla89, Ch. 2] for a discussion of this result). This completes the construction of a pointwise inner product on differential forms. At this point several remarks are in order:

(1) Given an orientation on $\Lambda_n(M_p^*)$, the definition of the Hodge star operator is independent of the orthonormal basis chosen. That is, if the Hodge star operator is defined in terms of an orthonormal basis then the definition of the star operator is satisfied on any other orthonormal basis related to the first by an orthogonal matrix with positive determinant.

(2) On an orientable manifold, it is possible to choose an orientation consistently over the whole manifold and hence the star operator can be defined smoothly as a mapping

$$* : \Lambda_k^*(M) \to \Lambda_{n-k}^*(M),$$

or, equivalently,

$$* : C^k(M) \to C^{n-k}(M).$$

(3) When there is a pseudo-Riemannian structure on the manifold, that is, a Riemannian structure which is nondegenerate but not positive definite then it is still possible to define a star operator, but it does not give rise to a positive definite bilinear pairing on k-forms. Such a star operator depends on the "signature" of the metric and occurs in four-dimensional formulations of electrodynamics (see [Fla89, §2.6–2.7] and [BLG70, §3.5]).

The following examples show how the operations $(d, \wedge, *)$ are related to the operators of vector analysis.

Example 7.11 Vector analysis in 3-d orthogonal curvilinear coordinates. Suppose x^1, x^2, x^3 are orthogonal curvilinear coordinates in a subset of $\mathbb{R}^3$, that is,

$$g_{ij} = \begin{cases} h_i^2, & \text{if } i = j \\ 0 & \text{if } i \neq j, \end{cases}$$

so that $\omega^i = h_i\,dx^i$, $1 \leq i \leq 3$, is an orthonormal basis for 1-forms. In this case, if π is the permutation of three integers sending 1, 2, 3 into i, j, k, then

$$*1 = \omega_1 \wedge \omega_2 \wedge \omega_3, \qquad *(\omega_1 \wedge \omega_2 \wedge \omega_3) = 1,$$
$$*\omega_k = \operatorname{sgn}(\pi)\omega_i \wedge \omega_j, \qquad *(\omega_j \wedge \omega_k) = \operatorname{sgn}(\pi)\omega_i;$$

hence

$$\begin{aligned} *1 &= h_1 h_2 h_3\, dx^1 \wedge dx^2 \wedge dx^3, \\ *(dx^k) &= \operatorname{sgn}(\pi)\left(\frac{h_i h_j}{h_k}\right) dx^i \wedge dx^j, \\ *(dx^j \wedge dx^k) &= \operatorname{sgn}(\pi)\left(\frac{h_i}{h_j h_k}\right) dx^i, \\ *(dx^i \wedge dx^j \wedge dx^k) &= \frac{1}{h_1 h_2 h_3}\operatorname{sgn}(\pi). \end{aligned}$$

Furthermore, if $\omega = \sum_{i=1}^3 F_i\omega_i = \sum_{i=1}^3 F_i h_i\, dx^i$, $\eta = \sum_{i=1}^3 G_i\omega_i = \sum_{i=1}^3 G_i h_i\, dx^i$, and f is a function, it is a straightforward computation to show that

$$df = \sum_{i=1}^{3}\left(\frac{1}{h_i}\frac{\partial f}{\partial x^i}\right)\omega_i,$$

$$*d\omega = \begin{vmatrix} \dfrac{\omega_1}{h_2 h_3} & \dfrac{\omega_2}{h_1 h_3} & \dfrac{\omega_3}{h_1 h_2} \\ \dfrac{\partial}{\partial x^1} & \dfrac{\partial}{\partial x^2} & \dfrac{\partial}{\partial x^3} \\ h_1 F_1 & h_2 F_2 & h_3 F_3 \end{vmatrix},$$

$$*d*\omega = \frac{1}{h_1 h_2 h_3}\left(\frac{\partial}{\partial x^1}(h_2 h_3 F_1) + \frac{\partial}{\partial x^2}(h_1 h_3 F_2) + \frac{\partial}{\partial x^3}(h_1 h_2 F_3)\right),$$

$$*(\omega \wedge \eta) = \begin{vmatrix} \dfrac{\omega_1}{h_2 h_3} & \dfrac{\omega_2}{h_1 h_3} & \dfrac{\omega_3}{h_1 h_2} \\ F_1 & F_2 & F_3 \\ G_1 & G_2 & G_3 \end{vmatrix},$$

$$*(\omega \wedge *\eta) = F_1 G_1 + F_2 G_2 + F_3 G_3.$$

Thus the operations grad, curl, div, $\times$ and $\cdot$ from vector analysis are easily constructed from operations on differential forms and the correspondence is made clear by making the following identifications:

$$\begin{aligned} d^0 f &\leftrightarrow \operatorname{grad} f, \\ *d^1\omega &\leftrightarrow \operatorname{curl} \boldsymbol{F}, \\ *d^1 * \omega &\leftrightarrow \operatorname{div} \boldsymbol{F}, \\ *(\omega \wedge \eta) &\leftrightarrow \boldsymbol{F} \times \boldsymbol{G}, \\ *(\omega \wedge *\eta) &\leftrightarrow \boldsymbol{F} \cdot \boldsymbol{G}. \end{aligned}$$

In vector analysis it is customary to identify flux vector fields (arising from 2-forms) with vector fields arising from 1-forms by means of the Hodge star operator. Furthermore, one has

$$\begin{aligned} *d(df) &= *(dd)f = 0 && \text{(that is, } \operatorname{curl} \operatorname{grad} f = 0) \\ *d*(*d\omega) &= *d(**)\, d\omega = *(dd)\omega = 0 && \text{(that is, } \operatorname{div} \operatorname{curl} \boldsymbol{F} = 0), \end{aligned}$$

as well as the following identities, used when integrating by parts:

$$\begin{aligned} *d*(f\omega) &= *\,(d(f*\omega)) = *(df \wedge *\omega) + f * d * \omega, \\ *d(f\omega) &= *(df \wedge \omega) + f * d\omega, \\ *d*(*(\omega \wedge \eta)) &= *\,(d**(\omega \wedge \eta)) = *\,(d(\omega \wedge \eta)) = *\,((d\omega) \wedge \eta) - *(\omega \wedge d\eta) \\ &= *\,(*(*d\omega) \wedge \eta) - *\,(\omega \wedge *(*d\eta)), \end{aligned}$$

which translate respectively into

$$\begin{aligned} \operatorname{div}(f\boldsymbol{F}) &= (\operatorname{grad} f) \cdot \boldsymbol{F} + f \operatorname{div} \boldsymbol{F}, \\ \operatorname{curl}(f\boldsymbol{F}) &= (\operatorname{grad} f) \times \boldsymbol{F} + f \operatorname{curl} \boldsymbol{F}, \\ \operatorname{div}(\boldsymbol{F} \times \boldsymbol{G}) &= (\operatorname{curl} \boldsymbol{F}) \cdot \boldsymbol{G} - \boldsymbol{F} \cdot (\operatorname{curl} \boldsymbol{G}). \end{aligned}$$

Thus, once the algebraic rules for manipulating differential forms are understood, commonly used vector identities can be derived systematically. □

Example 7.12 Vector analysis in 2-d orthogonal curvilinear coordinates. Suppose x^1, x^2 are orthogonal curvilinear coordinates in a 2-dimensional manifold, that is

$$g_{ij} = \begin{cases} h_i^2 & i = j, \\ 0 & i \neq j, \end{cases}$$

so that $\omega^i = h_i\, dx^i$, $1 \leq i \leq 2$, is an orthonormal basis for 1-forms. In this case

$$*1 = \omega_1 \wedge \omega_2, \quad *\omega_1 = \omega_2, \quad *\omega_2 = -\omega_1, \quad *(\omega_1 \wedge \omega_2) = 1.$$

Hence

$$*(dx^1 \wedge dx^2) = \frac{1}{h_1 h_2}, \qquad *1 = h_1 h_2\, dx^1 \wedge dx^2,$$

$$*(dx^1) = \frac{h_2}{h_1}\, dx^2, \qquad *dx^2 = -\frac{h_1}{h_2}\, dx^1.$$

Furthermore, if $\omega = \sum_{i=1}^{2} F_i\omega_i$, $\eta = \sum_{i=1}^{2} G_i\omega_i$, and f is any function, it is a straightforward computation to show that

$$df = \frac{1}{h_1}\frac{\partial f}{\partial x^1}\omega_1 + \frac{1}{h_2}\frac{\partial f}{\partial x^2}\omega_2,$$

$$*df = -\frac{1}{h_2}\frac{\partial f}{\partial x^2}\omega^1 + \frac{1}{h_1}\frac{\partial f}{\partial x^1}\omega_2,$$

$$*d\omega = \frac{1}{h_1 h_2}\Big(\frac{\partial}{\partial x^1}(h_2 F_2) - \frac{\partial}{\partial x^2}(h_1 F_1)\Big),$$

$$*d*\omega = \frac{1}{h_1 h_2}\Big(\frac{\partial}{\partial x^1}(h_2 F_1) + \frac{\partial}{\partial x^2}(h_1 F_2)\Big),$$

$$*(\omega \wedge *\eta) = F_1 G_1 + F_2 G_2.$$

Thus the operators grad, $\overline{\text{curl}}$, curl, div and $\cdot$ are easily constructed from operations on differential forms and the correspondence is made explicit by making the identifications

$$\begin{aligned} d^0 f &\leftrightarrow \operatorname{grad} f, \\ *d^0 f &\leftrightarrow \overline{\operatorname{curl}}\, f, \\ *d^1\omega &\leftrightarrow \operatorname{curl} \boldsymbol{F}, \\ *d^1 * \omega &\leftrightarrow \operatorname{div} \boldsymbol{F}, \\ *(\omega \wedge *\eta) &\leftrightarrow \boldsymbol{F}\cdot\boldsymbol{G}. \end{aligned}$$

In addition one sees that

$$\begin{aligned} *d(df) &= *(ddf) = 0 && \text{(that is, } \operatorname{curl}\operatorname{grad} f = 0), \\ *d*(*df) &= *d(**)\,df = - * ddf = 0 && \text{(that is, } \operatorname{div}\overline{\operatorname{curl}}\, f = 0), \\ *d*(df) &= *d(*df) && \text{(that is, } \operatorname{div}\operatorname{grad} f = -\triangle f = \operatorname{curl}\overline{\operatorname{curl}}\, f), \end{aligned}$$

and the following identities, commonly used when integrating by parts:

$$\begin{aligned} *d*(f\omega) &= *d(f*\omega) = *(df \wedge *\omega) + f*d*\omega, \\ *d(f\omega) &= *(d(f\omega)) = *(df\wedge\omega) + f*d\omega \\ &= -*\big((*(*df))\wedge\omega\big) + f*d\omega, \end{aligned}$$

which translate into

$$\begin{aligned} \operatorname{div}(f\boldsymbol{F}) &= (\operatorname{grad} f)\cdot\boldsymbol{F} + f\operatorname{div}\boldsymbol{F}, \\ \operatorname{curl}(f\omega) &= -\big(\overline{\operatorname{curl}}\, f\big)\cdot\boldsymbol{F} + f\operatorname{curl}\boldsymbol{F}. \end{aligned}$$

These are the formulas used in [Ned78]. Once again with the use of the formalism of differential forms commonly used vector identities can be derived systematically. □

Hopefully the reader has realized that the formalism of differential forms encompasses the types of computations encountered in vector analysis and more general computations in n-dimensional manifolds. For simple calculations involving Maxwell's equations in four dimensions, see Flanders [Fla89, §2.7, §4.6] and [BLG70, Chapter 4]. Returning to the topic of inner products, recall that for an orientable Riemannian manifold, the expression

$$*(\omega \wedge *\eta), \qquad \text{with } \omega, \eta \in C^k(M),$$

can be used to define a smooth symmetric positive definite bilinear form on $\Lambda_k(M_p^*)$ for all $p \in M$. Hence let

$$\langle \omega, \eta \rangle_k = \int_M (*(\omega \wedge *\eta))\, dV$$

be an inner product on $C^k(M)$. This inner product will be of fundamental importance in deriving orthogonal decompositions. Before moving on, we recall three fundamental properties of the star operator:

$$\begin{aligned} **\omega &= (-1)^{k(n-k)}\omega && \text{with } \omega \in C^k(M), \\ \omega \wedge *\eta &= \eta \wedge *\omega && \text{with } \eta, \omega \in C^k(M), \\ *dV &= 1 && \text{with } dV \text{ the volume } n\text{-form.} \end{aligned}$$

These expressions enable one to express the inner product above in four different ways. Note that

$$(*(\omega \wedge *\eta))\, dV = **((*(\omega \wedge *\eta))\, dV) = *(*(\omega \wedge *\eta)) = (\omega \wedge *\eta).$$

This expression and the symmetry of the inner product give

$$\begin{aligned} \langle \omega, \eta \rangle_k &= \int_M (*(\omega \wedge *\eta))\, dV = \int_M \omega \wedge *\eta \\ &= \int_M (*(\eta \wedge *\omega))\, dV = \int_M \eta \wedge *\omega. \end{aligned}$$

For simplicity, assume that M is compact. The above inner product makes $C^k(M)$ into a Hilbert space. This is the first step toward obtaining useful orthogonal decompositions.

MA-K. The Operator Adjoint to the Exterior Derivative

Having an inner product on the exterior k-bundle of an orientable Riemannian manifold M (which will henceforth be assumed compact) and an operator

$$d^p : C^p(M) \to C^{p+1}(M),$$

one wants to know the form of the Hilbert space adjoint

$$\delta_{p+1} : C^{p+1}(M) \to C^p(M)$$

satisfying

$$\langle d^p\omega, \eta \rangle_{p+1} = \langle \omega, \delta_{p+1}\eta \rangle_p + \text{ boundary terms} \qquad \text{for } \omega \in C^p(M),\ \eta \in C^{p+1}(M).$$

This type of formula will now be deduced from the integration by parts formula which was developed as a corollary to Stokes' Theorem. Let

$$\omega \in C^p(M), \quad \mu \in C^{n-p-1}(M).$$

Then

$$\int_M d\omega \wedge \mu = \int_{\partial M} t\omega \wedge t\mu - (-1)^p \int_M \omega \wedge d\mu.$$

Next let $\mu = *\eta$ for some $\eta \in C^{p+1}(M)$, so that

$$\langle d\omega, \eta\rangle_{p+1} = \int_M d\omega \wedge *\eta = \int_{\partial M} t\omega \wedge t(*\eta) - (-1)^p \int_M \omega \wedge d(*\eta).$$

However, using that $(-1)^{p(n-p)} **\gamma = \gamma$ for $\gamma \in C^p(M)$ and that

$$-(-1)^p(-1)^{p(n-p)} = (-1)^{np+1+p(1-p)} = (-1)^{np+1},$$

one has

$$\begin{aligned}\langle d\omega, \eta\rangle_{p+1} &= -(-1)^p \int_M \omega \wedge d(*\eta) + \int_{\partial M} t\omega \wedge t(*\eta)\\ &= -(-1)^p(-1)^{p(n-p)} \int_M \omega \wedge *(*d*\eta) + \int_{\partial M} t\omega \wedge t(*\eta)\\ &= \int_M \omega \wedge *((-1)^{np+1} * d * \eta) + \int_{\partial M} t\omega \wedge t(*\eta).\end{aligned}$$

Hence

$$\langle d^p\omega, \eta\rangle_{p+1} = \langle \omega, \delta_{p+1}\eta\rangle_p + \int_{\partial M} t\omega \wedge t(*\eta),$$

where

$$\delta_{p+1} = (-1)^{np+1} * d^{n-p-1} * \qquad \text{on } (p+1)\text{–forms}.$$

To gain an intuitive understanding of what is happening on the boundary, we rework the boundary term. Up to now the operator t which gives the tangential components of a differential form was considered to be the pullback on differential forms induced by the map

$$i : \partial M \to M.$$

That is, $t = i^{\#}$. Given a Riemannian metric on M and if ∂M is smooth then given a point $p \in \partial M$ one can find a set of orthogonal curvilinear coordinates such that p has coordinates $(0,\dots,0)$, ∂M has local coordinates $(u^1,\dots,u^{n-2},u^{n-1},0)$, and $(u^1,u^2,\dots,u^n)$, $u^n \leq 0$, are a set of local coordinates in M. In terms of these local coordinates a k-form looks like

$$\omega = \sum_{1\leq i_1<i_2<\cdots<i_k\leq n} a_{i_1 i_2\dots i_k}\, du^{i_1} \wedge du^{i_2} \wedge \cdots \wedge du^{i_k}.$$

On ∂M, the component of this form tangent to ∂M is given by replacing one inequality in the summation by a strict inequality:

$$t\omega = \sum_{1\leq i_1<i_2<\cdots<i_k< n} a_{i_1 i_2\dots i_k}\, du^{i_1} \wedge du^{i_2} \wedge \cdots \wedge du^{i_k},$$

while the normal component is given by

$$n\omega = \omega - t\omega.$$

Clearly, each term in $n\omega$ involves dx^n. This definition of the normal component of a differential form seems to be due to Duff [Duf52] and is heavily used in subsequent literature (see for instance the papers by Duff, Spencer, Morrey, and Conner in the bibliography). By considering the k-form ω written as

$$\omega = t\omega + n\omega$$

in the orthogonal coordinate system above, it is apparent that $*\omega$ can be decomposed in two ways:

$$*(\omega) = *(t\omega) + *(n\omega), \qquad (*\omega) = t(*\omega) + n(*\omega).$$

Thus subtracting the above two equations, one deduces that

$$*t\omega - n * \omega = *n\omega - t * \omega.$$

Noticing that each term in the right-hand side of this equation involves dx^n and that no term in the left-hand side involves dx^n, we deduce that

$$t * \omega = *n\omega, \qquad n * \omega = *t\omega.$$

Furthermore, since exterior differentiation commutes with pullbacks, one has

$$dt\omega = td\omega \qquad dt * \omega = t\, d * \omega.$$

Starring this last formula and using the earlier formulas relating normal and tangential components, we conclude successfully that

$$\begin{aligned} *dt *\omega &= *t\, d*\omega, \\ *d*n\omega &= n *d*\omega \\ \delta n\omega &= n\delta\omega. \end{aligned}$$

Thus, in summary,

$$\begin{aligned} n\omega &= \omega - t\omega, \\ n * \omega &= *t\omega, \\ *n\omega &= t * \omega, \\ dt\omega &= td\omega, \\ \delta n\omega &= n\delta\omega. \end{aligned}$$

Finally, these identities can be used to rewrite the integration by parts formula. From the identities involving the normal components of a differential form, one has

$$\langle d^*\omega, \eta\rangle = \langle \omega, \delta_{k+1}\eta\rangle + \int_{\partial M} t\omega \wedge *n\eta.$$

Next, suppose $\partial M = S_1 \cup S_2$, where $S_1 \cap S_2$ is $(n-2)$-dimensional and where S_1 and S_2 are collections of connected components of ∂M or parts of M where symmetry planes exist. In this latter case S_1 and S_2 may not be disconnected but meet at right angles. The above integration by parts formula can then be reworked into the following form, which will be essential in the derivation of orthogonal decompositions:

$$\langle d^k\omega, \eta\rangle_{k+1} - \int_{S_1} t\omega \wedge *n\eta = \langle \omega, \delta_{k+1}\eta\rangle_k + \int_{S_2} t\omega \wedge *n\eta.$$

MA-L. The Hodge Decomposition and Ellipticity

On a compact orientable Riemannian manifold, an inner product structure on $C^*(M)$ and an operator adjoint to the exterior derivative enables one to define the Laplace–Beltrami operator

$$\Delta_p = d^{p-1}\delta_p + \delta_{p+1}\, d^p$$

(an elliptic operator on p-forms) and harmonic forms (solutions of the equation $\Delta\omega = 0$). Furthermore, when the manifold has no boundary, one has the Hodge decomposition theorem which generalizes the Helmholtz Theorem of vector analysis. For compact orientable manifolds without boundary, the Hodge decomposition theorem asserts that

$$C^p(M) = \mathrm{im}(d^{p-1}) \oplus \mathrm{im}(\delta_{p+1}) \oplus \mathcal{H}^p(M),$$

where $\mathcal{H}^p(M)$ is the space of harmonic p-forms. Using the tools of elliptic operator theory and the de Rham isomorphism, one can show that

$$\dim \mathcal{H}^p(M) = \beta_p(M) < \infty$$

and that the basis vectors for the de Rham cohomology vector spaces may be represented by harmonic forms. A self contained proof of the Hodge decomposition as well as an explanation of the relevant machinery from elliptic operator theory can be found in [War71, Chapter 6]. Alternatively, a short and sweet account of the Hodge decomposition theorem along the lines of this appendix is given in Flanders [Fla89, §6.4] while a succinct proof of the theorem in the case of 2-dimensional surfaces is usually given in any decent book on Riemann surfaces (see for example Springer [Spr57] or [SS54]).

For orthogonal decompositions of p-forms on orientable Riemannian manifolds with boundary, the tools of elliptic operator theory are less successful in obtaining a nice orthogonal decomposition which relates harmonic forms to the relative cohomology groups of the manifold. The history of this problem starts with the papers of Kodaira [Kod49], Duff and Spencer [DS52] and ends with the work of Friedrichs [Fri55], Morrey [Mor66], and Conner [Con56]. A general reference for this problem is [Mor66, Chapter 7]. The basic problem encountered in the case of a manifold with boundary is that the space of harmonic p-forms is generally infinite-dimensional and the questions of regularity at the boundary are quite thorny. An excellent recent reference to the theory and applications of Hodge theory on manifolds with boundary is Schwarz [Sch95]. There is a way of getting an orthogonal decomposition for p-forms on manifolds with boundary which completely avoids elliptic operator theory by defining harmonic p-fields (p-forms which satisfy $d^p\omega = 0, \delta_p\omega = 0$). Such a decomposition is called a Kodaira decomposition after [Kod49] introduced the notion of a harmonic field and the associated decompositions of p-forms. It turns out that for compact orientable Riemannian manifolds without boundary the proof of the Hodge decomposition theorem shows that harmonic fields and harmonic forms are equivalent concepts. In some sense the setting of the Kodaira decomposition generalizes that of the Hodge decomposition, but the conclusions are much weaker.

MA-M. Orthogonal Decompositions of p-Forms and Duality Theorems

The immediate objective is to show that the structure of a complex with an inner product enables one to derive useful orthogonal decompositions of p-forms. As usual, let M be a compact orientable n-dimensional Riemannian manifold with boundary, where

$$\partial M = S_1 \cup S_2$$

and $S_1 \cap S_2$ is an $(n-2)$-dimensional manifold where a symmetry plane meets the boundary of some original problem at right angles. Consider the cochain complexes $C^*(M)$, $C^*_c(M - S_1)$ and recall that

$$\begin{aligned}
C^p_c(M - S_1) &= \{\omega \mid \omega \in C^p(M), t\omega = 0 \text{ on } S_1\},\\
Z^p_c(M - S_1) &= \{\omega \mid \omega \in C^p_c(M - S_1), d\omega = 0\},\\
B^p_c(M - S_1) &= \{\omega \mid \omega = d\nu, \nu \in C^{p-1}_c(M - S_1)\},\\
H^p_c(M - S_1) &= Z^p_c(M - S_1)/B^p_c(M - S_1).
\end{aligned}$$

Next define the complex $\widetilde{C}_*(M, S_2)$, where

$$\widetilde{C}_*(M, S_2) = \{\omega \mid \omega \in C^p(M), n\omega = 0 \text{ on } S_2\}$$

and the "boundary operator" in this complex is the Hilbert space formal adjoint δ of the exterior derivative d. Note that $\widetilde{C}_*(M, S_2)$ is actually a complex, since if η is a $(p+1)$-form in this complex the condition $n\eta = 0$ on S_2 implies

$$\delta_{p+1} n\eta = n(\delta_{p+1}\eta) = 0 \text{ on } S_2$$

and

$$(-1)^n \delta_p \delta_{p+1}\eta = *d^{n-p} **d^{n-p-1} * \eta = (-1)^{p(n-p)} * d^{n-p}\, d^{n-p-1} * \eta = 0.$$

Hence define

$$\begin{aligned}
\widetilde{Z}_p(M, S_2) &= \{\eta \mid \eta \in \widetilde{C}_p(M, S_2), \delta_p\eta = 0\},\\
\widetilde{B}_p(M, S_2) &= \{\eta \mid \eta = \delta_{p+1}\gamma, \gamma \in \widetilde{C}_{p+1}(M, S_2)\},\\
\widetilde{H}_p(M, S_2) &= \widetilde{Z}(M, S_2)/\widetilde{B}_p(M, S_2).
\end{aligned}$$

The "cycles" of this complex are called coclosed differential forms while the "boundaries" are called coexact differential forms. The first step in deriving an orthogonal decomposition of p-forms on $C^p(M)$ is to recall the inner product version of the integration by parts formula:

$$\langle d^k\omega, \eta\rangle_{k+1} - \int_{S_1} t\omega \wedge *n\eta = \langle \omega, \delta_{k+1}\eta\rangle_k + \int_{S_2} t\omega \wedge *n\eta.$$

If $k = p$ and $\omega \in Z^p_c(M - S_1)$, the left side of this formula vanishes and we see that closed p-forms are orthogonal to coexact p-forms in $\widetilde{C}_*(M, S_2)$. Alternatively, if

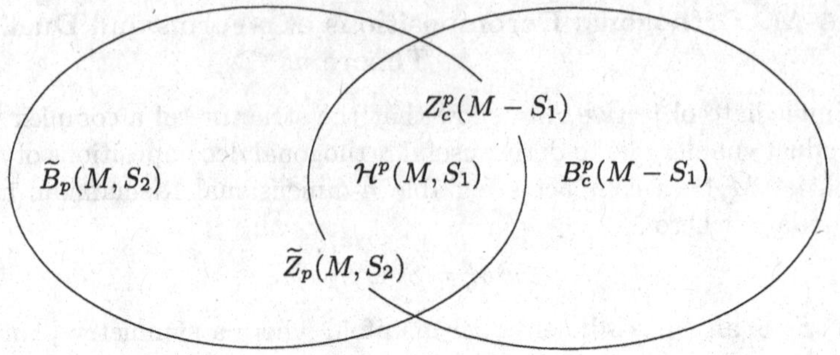

Figure MA-3.

$k+1=p$ and $\eta \in \widetilde{Z}_p(M,S_2)$, it is easily seen that coclosed p-forms are orthogonal to the exact p-forms in $C^*_c(M-S_1)$. Actually,

$$\text{(MA-7)} \qquad \begin{aligned} C^p(M) &= Z^p_c(M-S_1) \oplus \widetilde{B}_p(M,S_2), \\ C^p(M) &= \widetilde{Z}_p(M,S_2) \oplus B^p_c(M, M-S_1), \end{aligned}$$

since these identities express the fact that if A is an operator between Hilbert spaces, then

$$(\operatorname{im} A)^\perp = \ker A^{\mathrm{adj}}.$$

Next, since $C^p_c(M-S_1)$ and $\widetilde{C}_p(M,S_2)$ are complexes, one has

$$\text{(MA-8)} \qquad \begin{aligned} \widetilde{B}_p(M,S_2) &\subset \widetilde{Z}_p(M,S_2), \\ B^p_c(M-S_1) &\subset Z^p_c(M-S_1). \end{aligned}$$

Finally, defining the relative harmonic p-fields as

$$\mathcal{H}^p(M,S_1) = \widetilde{Z}_p(M,S_2) \cap Z^p(M-S_1),$$

the orthogonal decomposition is immediate once Equations (MA-7) and (MA-8) above are expressed in terms of a Venn diagram of orthogonal spaces as shown in Figure MA-3. Thus

$$\begin{aligned} \widetilde{Z}_p(M,S_2) &= \widetilde{B}_p(M,S_2) \oplus \mathcal{H}^p(M,S_1), \\ Z^p_c(M-S_1) &= B^p_c(M-S_1) \oplus \mathcal{H}^p(M,S_1), \\ C^p(M) &= \widetilde{B}_p(M,S_2) \oplus \mathcal{H}^p(M,S_1) \oplus B^p_c(M-S_1), \end{aligned}$$

where the direct summands are mutually orthogonal with respect to the inner product $\langle \cdot, \cdot \rangle_p$.

To relate this orthogonal decomposition to the relative cohomology of the pair (M,S_1), consider the identities

$$\begin{aligned} Z^p_c(M-S_1) &= B^p_c(M-S_1) \oplus \mathcal{H}^p(M,S_1), \\ H^p_c(M-S_1) &= Z^p_c(M-S_1)/B^p_c(M-S_1). \end{aligned}$$

This gives

$$H^p_c(M-S_1) \simeq \mathcal{H}^p(M,S_1),$$

that is, in each de Rham cohomology class there is exactly one harmonic field. A more concrete way of seeing this is to write the above orthogonal decomposition explicitly in terms of differential forms and use the de Rham isomorphism. That is, any $\omega \in C^p(M)$ can be decomposed into three unique, mutually orthogonal factors, say

$$\omega = d\nu + \delta\gamma + \chi,$$

where $\nu \in C_c^{p-1}(M - S_1)$, $\gamma \in \widetilde{C}_{p+1}(M, S_2)$, and $\chi \in \mathcal{H}^p(M, S_2)$. Furthermore,

$$\begin{aligned} \omega \in Z_c^p(M - S_1) &\implies \omega = d\nu + \chi, \\ \omega \in \widetilde{Z}_p(M, S_1) &\implies \omega = \delta\gamma + \chi. \end{aligned}$$

Thus if $\omega \in Z_c^p(M - S_1)$ and $z_p \in Z_p(M, S_1)$, then

$$\int_{z_p} \omega = \int_{z_p} d\nu + \int_{z_p} \chi = \int_{z_p} \chi, \qquad \text{since } \int_{z_p} d\nu = 0.$$

Hence there is at least one harmonic field in each de Rham cohomology class. However it is easy to show that there cannot be more than one distinct harmonic field in each de Rham cohomology class. Suppose that

$$\omega_1 - \omega_2 = d\beta,$$

with $\beta \in C_c^{p-1}(M - S_1)$, $\omega_1 = d\nu_1 + h_1$, and $\omega_2 = d\nu_2 + h_2$. Then

$$h_1 - h_2 = d(\beta - \nu_1 + \nu_2) \in B_c^p(M - S_1);$$

but by the orthogonal decomposition we have

$$h_1 - h_2 \in \mathcal{H}^p(M, S_1), \qquad \mathcal{H}^p(M, S_1) \perp B_c^p(M - S_1),$$

so $h_1 = h_2$ and there is necessarily exactly one harmonic field in each de Rham cohomology class. Thus, explicitly, it has been shown that

$$H_c^p(M - S_1) \simeq \mathcal{H}^p(M, S_1);$$

hence

$$\infty > \beta_p(M, S_1) = \beta_c^p(M - S_1) = \dim \mathcal{H}^p(M, S_1),$$

where the first equality results from the de Rham theorem. The preceding isomorphism shows that the projection of ω on $\mathcal{H}^p(M, S_1)$ is deduced from the periods of ω on a basis of $H_p(M, S_1)$.

At this point we summarize the preceding discussion with a theorem.

THEOREM (Orthogonal Decomposition of p-forms). Given M, S_1, S_2 as usual and

$$\mathcal{H}^p(M, S_1) = Z_c^p(M - S_1) \cap \widetilde{Z}_p(M, S_2),$$

one has

(1) direct sum decompositions

$$\begin{aligned} C^p(M) &= B_c^p(M - S_1) \oplus \mathcal{H}^p(M, S_1) \oplus \widetilde{B}_p(M, S_2), \\ Z_c^p(M - S_1) &= B_c^p(M - S_1) \oplus \mathcal{H}^p(M, S_1), \\ \widetilde{Z}_p(M, S_2) &= \widetilde{B}_p(M, S_2) \oplus \mathcal{H}^p(M, S_1), \end{aligned}$$

where the direct summands are mutually orthogonal with respect to the inner product $\langle \cdot , \cdot \rangle_p$; and

(2) a unique harmonic field in each de Rham cohomology class, that is, an isomorphism

$$\mathcal{H}^p(M, S_1) \simeq H^p_c(M - S_1),$$

so that

$$\infty > \beta_p(M, S_1) = \beta^p_c(M - S_1) = \dim \mathcal{H}^p(M, S_1).$$

Having established the orthogonal decomposition theorem, it is useful to see how duality theorems come about as a result of the Hodge star operator. As a preliminary, several formulas must be derived. Recall that if $\omega \in C^p(M)$ then

$$**\omega = (-1)^{p(n-p)}\omega, \quad \delta_p\omega = (-1)^{n(p+1)+1} * d * \omega.$$

Thus

$$*\delta_p = (-1)^{n(p+1)+1} **d^{n-p}*= (-1)^{n(p+1)+1+(p+1)(n-p-1)}\, d^{n-p}*,$$
$$*d^{n-p} = (-1)^{p(n-p)} *d^{n-p} ** = (-1)^{p(n-p)+n(p+1)+1}\delta_p * .$$

Cleaning up the exponents with modulo 2 arithmetic gives

$$*\delta_p = (-1)^p d^{n-p}*, \quad *d^{n-p} = -(-1)^{n-p}\delta_p * ,$$

and hence

$$*\delta_{n-p} = (-1)^{n-p}\, d_p * , \quad *d^p = -(-1)^p \delta_{n-p} * .$$

Next recall the identities

$$*t = n*, \quad *n = t* .$$

These six formulas will now be used to make some useful observations. Let $\lambda \in C^k(M)$ and $\mu \in C^{n-k}(M)$, where $*\lambda = \mu$. In this case

$$\delta_k\lambda = 0 \iff 0 = *\delta_k\lambda = (-1)^k d^{n-k} * \lambda = (-1)^k d^{n-k}\mu,$$
$$d^k\lambda = 0 \iff 0 = *d^k\lambda = -(-1)^k\delta_{n-k} * \lambda = -(-1)^k\delta_{n-p}\mu,$$
$$t\lambda = 0 \iff 0 = *t\lambda = n * \lambda = n\mu,$$
$$n\lambda = 0 \iff 0 = *n\lambda = t * \lambda = t\mu.$$

In other words, if $\lambda, \mu \in C^*(M)$ and $*\lambda = \mu$, we have the equivalences

$$\delta\lambda = 0 \iff d\mu = 0, \qquad d\lambda = 0 \iff \delta\mu = 0,$$
$$t\lambda = 0 \iff n\mu = 0, \qquad n\lambda = 0 \iff t\mu = 0.$$

With a little reflection, the above four equivalences show that the Hodge star operator induces the following isomorphisms, where $1 \le i \le 2, 1 \le k \le n$:

$$C^p(M) \simeq C^{n-p}(M),$$
$$\widetilde{C}_k(M, S_i) \simeq C^{n-k}_c(M - S_i),$$
$$\widetilde{Z}_k(M, S_i) \simeq Z^{n-k}_c(M - S_i),$$
$$\widetilde{B}_k(M, S_i) \simeq B^{n-k}_c(M - S_i),$$
$$\widetilde{H}_k(M, S_i) \simeq H^{n-k}_c(M - S_i).$$

What is particularly interesting is the following computation, where $1 \leq i, j \leq 2$, $i \neq j$, $1 \leq l \leq n$:

$$\begin{aligned} *\mathcal{H}^l(M, S_i) &= *\big(Z_c^l(M - S_i) \cap \widetilde{Z}_l(M, S_j)\big) \\ &= *\big(Z_c^l(M - S_i)\big) \cap *\big(\widetilde{Z}_l(M, S_j)\big) \\ &= \big(\widetilde{Z}_{n-l}(M, S_i)\big) \cap \big(Z_c^{n-l}(M - S_j)\big) \\ &= \mathcal{H}^{n-l}(M, S_j). \end{aligned}$$

To interpret this result, notice that the derivation of the orthogonal decomposition is still valid if S_1 and S_2 are interchanged everywhere. Hence, juxtaposing the two orthogonal decompositions

$$\begin{gathered} C^p(M) = B_c^p(M - S_1) \oplus \mathcal{H}^p(M, S_1) \oplus \widetilde{B}_p(M, S_2), \\ C^{n-p}(M) = \widetilde{B}_{n-p}(M, S_1) \oplus \mathcal{H}^{n-p}(M, S_2) \oplus B_c^{n-p}(M - S_2)\,, \end{gathered}$$

one sees that each term in these decompositions is related to the one directly above or below it by the Hodge star operator. Also, the star operation performed twice maps $\mathcal{H}^p(M, S_1)$ and $\mathcal{H}^{n-p}(M, S_2)$ isomorphically back onto themselves, since in this case

$$(-1)^{p(n-p)} ** = \text{Identity}.$$

Hence

$$\mathcal{H}^p(M, S_1) \simeq \mathcal{H}^{n-p}(M, S_2).$$

At this point it is useful to summarize the isomorphisms in (co)homology derived in this chapter where coefficients are taken in $\mathbb{R}$ (of course) and M is an orientable, compact, n-dimensional Riemannian manifold with boundary where $\partial M = S_1 \cup S_2$ in the usual way.

THEOREM.

$$\begin{array}{ccccc} \mathcal{H}^k(M, S_1) & \simeq & H_c^k(M - S_1) & \simeq & H_k(M, S_1) \\ \updownarrow \simeq & & & & \\ \mathcal{H}^{n-k}(M, S_2) & \simeq & H_c^{n-k}(M - S_2) & \simeq & H_{n-k}(M, S_2) \end{array}$$

This theorem expresses the relative de Rham isomorphism (on the right), the representability of relative de Rham cohomology classes by harmonic fields (in the center), and the duality isomorphism induced on harmonic fields by the Hodge star (on the left). For the inspiration behind this theorem see [Con54].

In order to let the orthogonal decomposition sink in, it is useful to rewrite it in terms of differential forms and then consider several concrete examples. Thus consider the following theorem, which restates the orthogonal decomposition theorem in a more palatable way:

THEOREM. If M, S_1, and S_2 have their usual meaning and $\omega \in C^p(M)$, then ω has the unique representation

$$\omega = d\nu + \delta\gamma + \chi,$$

where

$$\begin{array}{lll} t\nu = 0, & t\chi = 0 & \text{on } S_1, \\ n\gamma = 0, & n\chi = 0 & \text{on } S_2, \\ d\chi = 0, & \delta\chi = 0 & \text{in } M \text{ and on } \partial M. \end{array}$$

Furthermore:

(1) The three factors are mutually orthogonal with respect to the inner product $\langle \cdot, \cdot \rangle_p$.
(2) If $d\omega = 0$ in M and $t\omega = 0$ on S_1 one can take $\gamma = 0$.
(3) If $\delta\omega = 0$ in M and $n\omega = 0$ on S_2 then one can take $\nu = 0$.

We illustrate the theorem with a couple of examples.

Example 7.13 The Hodge decomposition and 3-d vector analysis: $n = 3$, $p = 1$. In vector analysis it is customary to identify flux vector fields (arising from 2-forms) with vector fields arising from 1-forms by means of the Hodge star operator. Keeping this in mind, the identifications established in Example 7.11 show that in the case of $n = 3$, $p = 1$ the theorem above can be rewritten as follows. If M is an orientable, compact three-dimensional manifold with boundary embedded in $\mathbb{R}^3$, then any vector field $\boldsymbol{V}$ can be uniquely expressed as

$$\boldsymbol{V} = \operatorname{grad}\varphi + \operatorname{curl}\boldsymbol{F} + \boldsymbol{G},$$

where

$$\begin{array}{rll} \varphi = 0, & \boldsymbol{n} \times \boldsymbol{G} = 0 & \text{on } S_1, \\ \boldsymbol{F} \cdot \boldsymbol{n} = 0, & \boldsymbol{n} \cdot \boldsymbol{G} = 0 & \text{on } S_2, \\ \operatorname{curl}\boldsymbol{G} = 0, & \operatorname{div}\boldsymbol{G} = 0 & \text{in } M \text{ and on } \partial M. \end{array}$$

Furthermore:

(1) The three factors are mutually orthogonal with respect to the inner product

$$\langle \boldsymbol{U}, \boldsymbol{V} \rangle_1 = \int_M \boldsymbol{U} \cdot \boldsymbol{V}\, dV.$$

(2) If $\operatorname{curl}\boldsymbol{V} = 0$ in M and $\boldsymbol{n} \times \boldsymbol{V} = 0$ on S_1, we may set $\boldsymbol{F}$ to zero;
(3) If $\operatorname{div}\boldsymbol{V} = 0$ in M and $\boldsymbol{V} \cdot \boldsymbol{n} = 0$ on S_2, we may set φ to zero.

In practical problems it is customary to describe $\boldsymbol{G}$ through a (possibly multivalued) scalar potential. The dimension of the space of harmonic vector fields that satisfy the conditions imposed on $\boldsymbol{G}$ is $\beta_1(M - S_1) = \beta_2(M - S_2)$. □

Example 7.14 The Hodge decomposition and 2-d vector analysis: $n = 2$, $p = 1$. Let M be an orientable compact 2-dimensional Riemannian manifold with boundary where $\partial M = S_1 \cup S_2$ in the usual way. Using the identifications established in Example 7.12 one can rephrase the orthogonal decomposition theorem as follows. Any vector field $\boldsymbol{V}$ on M can be written as

$$\boldsymbol{V} = \operatorname{grad}\phi + \overline{\operatorname{curl}}\,\psi + \boldsymbol{G},$$

where

$$\begin{aligned} \phi &= 0, & \boldsymbol{n} \times \boldsymbol{G} &= 0 && \text{on } S_1, \\ \psi &= 0, & \boldsymbol{G} \cdot \boldsymbol{n} &= 0 && \text{on } S_2, \\ \operatorname{curl} \boldsymbol{G} &= 0, & \operatorname{div} \boldsymbol{G} &= 0 && \text{in } M \text{ and on } \partial M. \end{aligned}$$

Furthermore:

(1) The three factors are mutually orthogonal with respect to the inner product given by the metric tensor in the usual way:

$$\langle \boldsymbol{U}, \boldsymbol{V} \rangle_1 = \int_M \boldsymbol{U} \cdot \boldsymbol{V} dV.$$

(2) if $\operatorname{curl} \boldsymbol{V} = 0$ in M and $\boldsymbol{n} \times \boldsymbol{V} = 0$ on S_1 we may set ψ to zero.

(3) if $\operatorname{div} \boldsymbol{V} = 0$ in M and $\boldsymbol{V} \cdot \boldsymbol{n} = 0$ on S_2 we may set ϕ to zero.

In practical problems $\boldsymbol{G}$ is invariably described in terms of a (possibly multivalued) scalar potential or stream function. The dimension of the space of harmonic vector fields which satisfy the conditions imposed on $\boldsymbol{G}$ is $\beta_1(-S_1) = \beta_1(M - S_2)$. □

One final remark is appropriate. In the case of electrodynamics there is no positive definite inner product on p-forms since the metric tensor is not positive definite. One can, however, define all of the spaces found in the orthogonal decomposition and obtain a direct sum decomposition of $C^p(M)$ even though there is no positive definite inner product.

The committee which was set up in Rome for the unification of vector notation did not have the slightest success, which was only to have been expected.

F. Klein, *Elementary Mathematics from an Advanced Standpoint*, 1925.

Bibliography

[AK98] V. I. Arnold and B. A. Khesin, *Topological methods in hydrodynamics*, Springer-Verlag, New York, 1998.

[AL65] Holt Ashley and Marten Landahl, *Aerodynamics of wings and bodies*, Addison-Wesley, Reading, MA, 1965, Section 2-7.

[Alt55] M. Altman, *A generalization of Newton's method*, Bulletin de l'academie Polonaise des sciences **III** (1955), no. 4, 189–193, Cl.III.

[Arm83] M.A. Armstrong, *Basic topology*, Springer-Verlag, New York, 1983.

[Bat10] H. Bateman, *The transformation of the electrodynamical equations*, Proc. Lond. Math. Soc., II, vol. 8, 1910, pp. 223–264.

[BB69] N. Balabanian and T.A. Bickart, *Electrical network theory*, John Wiley, New York, 1969.

[BLG70] N. N. Balasubramanian, J. W. Lynn, and D. P. Sen Gupta, *Differential forms on electromagnetic networks*, Butterworths, London, 1970.

[Bos81] A. Bossavit, *On the numerical analysis of eddy-current problems*, Computer Methods in Applied Mechanics and Engineering **27** (1981), 303–318.

[Bos82] A. Bossavit, *On finite elements for the electricity equation*, The Mathematics of Finite Elements and Applications IV (MAFELAP 81) (J.R. Whiteman, ed.), Academic Press, 1982, pp. 85–91.

[Bos98] ———, *Computational electromagnetism: Variational formulations, complementarity, edge elements*, Academic Press, San Diego, 1998.

[Bra66] F.H. Branin, *The algebraic-topological basis for network analogies and the vector calculus*, Proc. Symp. Generalised Networks, Microwave Research, Institute Symposium Series, vol. 16, Polytechnic Institute of Brooklyn, April 1966, pp. 453–491.

[Bra77] ———, *The network concept as a unifying principle in engineering and physical sciences*, Problem Analysis in Science and Engineering (K. Husseyin F.H. Branin Jr., ed.), Academic Press, New York, 1977.

[Bri93] Erik Brisson, *Representing geometric structures in d dimensions: Topology and order*, Discrete and Computational Geometry **9** (1993), 387–426.

[Bro84] M. L. Brown, *Scalar potentials in multiply connected regions*, Int. J. Numer. Meth. Eng. **20** (1984), 665–680.

[BS90] Paul Bamberg and Shlomo Sternberg, *A course in mathematics for students of physics: 2*, Cambridge U. Press, NY, 1990, Ch. 12.

[BT82] Raoul Bott and Loring W. Tu, *Differential forms in algebraic topology*, Springer-Verlag, New York, 1982, pp. 40–42, 51, 234, 258, 240.

[BV82] A. Bossavit and J.C. Verite, *A mixed FEM-BIEM method to solve 3-d eddy-current problems*, IEEE Trans. Mag. **MAG-18** (1982), no. 2, 431–435.

[BV83] ———, *The "TRIFOU" code: Solving the 3-d eddy-currents problem by using H as state variable*, IEEE Trans. Mag. **MAG-19** (1983), no. 6, 2465–2470.

[Cai61] S.S. Cairns, *Introductory topology*, The Ronald Press Company, 1961.

[Cam83] G. Cambrell, *Upper and lower bounds, standard and complementary variational principles of primal and dual kind, tonti diagrams for linear self-adjoint problems: Parts i and ii*, Tech. report, Dept. of Elec. Eng. CADLab Seminars, McGill University, 1983.

[CEG86] Thomas F. Coleman, Anders Edenbrandt, and John R. Gilbert, *Predicting fill for sparse orthogonal factorization*, Journal of the Association for Computing Machinery **33** (1986), 517–532.

[Chi68] G.E Ching, *Topological concepts in networks; an application of homology theory to network analysis*, Proc. 11th. Midwest Conference on Circuit Theory, University of Notre Dame, 1968, pp. 165–175.

[CK96] Paula Chammas and P. R. Kotiuga, *Sparsity vis a vis Lanczos methods for discrete helicity functionals*, Proceedings of the Third International Workshop on Electric and Magnetic Fields (A. Nicolet and R. Belmans, eds.), 1996.

[Coh93] Henri Cohen, *A course in computational algebraic number theory*, Springer-Verlag, New York, 1993.

[Con54] P. E. Conner, *The Green's and Neumann's problems for differential forms on Riemannian manifolds*, Proc. Nat. Acad. Sci. U.S.A. **40** (1954), 1151–1155.

[Con56] ———, *The Neumann's problem for differential forms on Riemannian manifolds*, Mem. Amer. Math. Soc. **1956** (1956), no. 20, 56.

[Cou36] R. Courant, *Differential and integral calculus*, vol. II, Wiley-Interscience, New York, 1936.

[Cro78] F. H. Croom, *Basic concepts of algebraic topology*, Springer-Verlag, New York, 1978, Chaps. 2, 7.3, 4.5.

[Des81] Georges A. Deschamps, *Electromagnetics and differential forms*, IEEE Proc. **69** (1981), no. 6, 676–696.

[Dod74] Jozef Dodziuk, *Combinatorial and continuous Hodge theories*, Bull. Amer. Math. Soc. **80** (1974), no. 5, 1014–1016.

[Dod76] ———, *Finite-difference approach to the Hodge theory of harmonic forms*, Amer. J. Math. **98** (1976), no. 1, 79–104.

[dR31] Georges de Rham, *Sur l'Analysis situs des variétés à n dimensions*, Journ. de Math. **X** (1931), no. II, 115–200.

[dR73] ———, *Variétés différentiables. Formes, courants, formes harmoniques*, troisième édition revue et augmentée, publications de l'institut de mathématique de l'université de nancago, iii, actualités scientifiques et industrielles, no. 1222b ed., Hermann, Paris, 1973, First edition 1955. English edition published by Springer, 1984, is translation of 1973 French edition.

[DR91] E. Domínguez and J. Rubio, *Computers in algebraic topology*, The Mathematical Heritage of C.F. Gauss (George M. Rassias, ed.), World Scientific Publ. Co., Singapore, 1991, pp. 179–194.

[DS52] G. F. D. Duff and D. C. Spencer, *Harmonic tensors on Riemannian manifolds with boundary*, Ann. of Math. (2) **56** (1952), 128–156.

[Duf52] G. F. D. Duff, *Differential forms in manifolds with boundary*, Ann. of Math. (2) **56** (1952), 115–127.

[Dys72] F.J. Dyson, *Missed opportunities*, Bull. Amer. Math. Soc. **78** (1972), no. 5, 635–652.

[ES52] Samuel Eilenberg and Norman Steenrod, *Foundations of algebraic topology*, Princeton University Press, Princeton, New Jersey, 1952.

[Fla89] H. Flanders, *Differential forms with applications to the physical sciences*, Dover, New York, 1989, pp. 79–81.

[Fri55] K. O. Friedrichs, *Differential forms on Riemannian manifolds*, Comm. Pure Appl. Math. **8** (1955), 551–590.

[Gau77] C. F. Gauss, *Zur mathematischen Theorie der electodynamischen Wirkungen*, Werke V, Teubner, 1877, p. 605.

[GH81] Marvin J. Greenberg and John R. Harper, *Algebraic topology*, Benjamin/Cummings, Reading, MA, 1981, P. 235, 63–66.

[Gib81] P.J. Giblin, *Graphs, surfaces and homology: An introduction to algebraic topology*, second ed., Chapman and Hall, London, 1981.

[GK95] P. W. Gross and P. R. Kotiuga, *A challenge for magnetic scalar potential formulations of 3-d eddy current problems: Multiply connected cuts in multiply connected regions which necessarily leave the cut complement multiply connected*, Electric and

Magnetic Fields: From Numerical Models to Industrial Applications (New York) (A. Nicolet and R. Belmans, eds.), Plenum, 1995, Proceedings of the Second International Workshop on Electric and Magnetic Fields, pp. 1–20.

[GK01a] ———, *Data structures for geometric and topological aspects of finite element algorithms*, Progress In Electromagnetics Research (F. L. Teixeira, ed.), vol. PIER 32, EMW Publishing, 2001, pp. 151–169.

[GK01b] ———, *Topological constraints and computational complexity of finite element-based algorithms to make cuts for magnetic scalar potentials*, Progress In Electromagnetics Research (F. L. Teixeira, ed.), vol. PIER 32, EMW Publishing, 2001, pp. 207–245.

[Gol82] S.I. Goldberg, *Curvature and homology*, Dover Publications, New York, 1982.

[GP74] Victor Guillemin and Alan Pollack, *Differential topology*, Prentice-Hall, Englewood Cliffs, New Jersey, 1974, Page 21.

[GR65] Robert C. Gunning and Hugo Rossi, *Analytic functions of several complex variables*, Prentice-Hall, Englewood Cliffs, N.J., 1965.

[Har08] R. Hargraves, *Integral forms and their connection with physical equations*, Transactions of the Cambridge Philosophical Society **XXI** (1908), 107–122.

[Her77] R. Hermann, *Differential geometry and the calculus of variations*, 2nd ed., Math. Sci. Press, Brookline Mass., 1977.

[Hip02] R. Hiptmair, *Finite elements in computational electromagnetism*, Acta Numerica (2002), 237–339.

[Hir76] Morris Hirsch, *Differential topology*, GTM #33, Springer-Verlag, New York, 1976.

[Hod52] W. V. D. Hodge, *The theory and applications of harmonic integrals*, 2d ed ed., Cambridge, at the University Press, 1952, The first edition, published in 1941 and reissued in 1989, is subject to the criticism raised in [Wey43]. The second edition benefited from the existence of a rigourous proof of the Hodge theorem.

[HS85] C. S. Harrold and J. Simkin, *Cutting multiply connected domains*, IEEE Trans. Mag. **21** (1985), no. 6, 2495–2498.

[HY61] J.G. Hocking and G.S. Young, *Topology*, Addison-Wesley, 1961.

[Jac74] N. Jacobson, *Basic Algebra I*, W.H. Freeman and Company, San Francisco, 1974.

[Jac80] ———, *Basic Algebra II*, W.H. Freeman and Company, San Francisco, 1980.

[KG90] P. R. Kotiuga and R. Giles, *A topological invariant for the accessibility problem of micromagnetics*, J. Appl. Phys. **67** (1990), no. 9, 5347–5349.

[KI59] K. Kondo and M. Iri, *On the theory of trees, cotrees, multitrees, and multicotrees*, RAAG Memoirs **2** (1959), 220–261.

[Kle63] F. Klein, *On Riemann's theory of algebraic functions and their integrals*, Dover, New York, 1963.

[Kod49] K. Kodaira, *Harmonic fields in Riemannian manifolds (generalised potential theory)*, Annals of Mathematics **50** (1949), 587–665.

[Koh72] J. J. Kohn, *Differential complexes*, Les Presses de l'Université Montréal, 1972.

[Kot82] P. R. Kotiuga, *Well-posed three dimensional magnetostatics formulations*, Master's thesis, McGill University, Montreal, August 1982.

[Kot84] ———, *Hodge decompositions and computational electromagnetics*, Ph.D. thesis, McGill University, Montreal, 1984.

[Kot87] ———, *On making cuts for magnetic scalar potentials in multiply connected regions*, J. Appl. Phys. **61** (1987), no. 8, 3916–3918.

[Kot88] ———, *Toward an algorithm to make cuts for magnetic scalar potentials in finite element meshes*, J. Appl. Phys. **64** (1988), no. 8, 3357–3359, Erratum: 64, (8), 4257, (1988).

[Kot89a] ———, *An algorithm to make cuts for scalar potentials in tetrahedral meshes based on the finite element method*, IEEE Trans. Mag. **25** (1989), no. 5, 4129–4131.

[Kot89b] ———, *Helicity functionals and metric invariance in three dimensions*, IEEE Trans. Mag. **25** (1989), no. 4, 2813–2815.

[Kot89c] ———, *Topological considerations in coupling scalar potentials to stream functions describing surface currents*, IEEE Trans. Mag. **25** (1989), no. 4, 2925–2927.

[Kot91] ———, *Essential arithmetic for evaluating three dimensional vector finite element interpolation schemes*, IEEE Trans. Mag. **MAG-27** (1991), no. 6, 5208–5210.

[Kro59] G. Kron, *Basic concepts of multidimensional space filters*, AIEE Transactions **78** (1959), no. Part I, 554–561.

[Lam32] Sir Horace Lamb, *Hydrodynamics*, Dover, New York, 1932, (1879).

[Lan70] Cornelius Lanczos, *The variational principles of mechanics*, University of Toronto Press, Toronto, 1970.

[LS68] L.H. Loomis and S. Sternberg, *Advanced calculus*, Addison-Wesley, Reading, Massachusetts, 1968.

[Lue69] D. G. Luenberger, *Optimization by vector space methods*, John Wiley, 1969.

[Mac70] A.G.J. MacFarlane, *Dynamical system models*, George Harrap, London, 1970.

[Mas67] W. S. Massey, *Algebraic topology: An introduction*, GTM, vol. 56, Springer Verlag, 1967.

[Mas80] ———, *Singular homology theory*, GTM #70, Springer-Verlag, New York, 1980.

[Max91] James Clerk Maxwell, *A treatise on electricity and magnetism*, third ed., Oxford University Press, Clarendon, England, 1891, Republished Dover Publications, Inc., New York, 1954.

[MIK59] Y. Mizoo, M. Iri, and K. Kondo, *On the torsion characteristics and the duality of electric, magnetic, and dielectric networks*, RAAG Memoirs **2** (1959), 262–295.

[MN82] A. Milani and A. Negro, *On the quasi-stationary Maxwell equations with monotone characteristics in a multiply connected domain*, Journal of Mathematical Analysis and Applications **88** (1982), 216–230.

[Mor66] C.B. Morrey, *Multiple integrals in the calculus of variations*, Springer-Verlag, Berlin, 1966.

[Mül78] Werner Müller, *Analytic torsion and R-torsion of Riemannian manifolds*, Advances in Mathematics **28** (1978), 233–305.

[Mun84] James R. Munkres, *Elements of algebraic topology*, Addison-Wesley, Reading, MA, 1984, p. 377–380.

[Ned78] J. C. Nedelec, *Computation of eddy currents on a surface in* $\mathbb{R}^3$ *by finite element methods*, SIAM J. Numer. Anal. **15** (1978), no. 3, 580–594.

[Neu79] Lee Neuwirth, *The theory of knots*, Sci. Am. **240** (1979), no. 6, 110–124.

[PF84] J. Penman and J.R. Fraser, *Unified approach to problems in electromagnetism*, IEE Proc. A **131** (1984), no. 1, 55–61.

[PF90] Alex Pothen and Chin-Ju Fan, *Computing the block triangular form of a sparse matrix*, ACM Transactions on Mathematical Software **16** (1990), 303–324.

[Pin66] Tad Pinkerton, *An algorithm for the automatic computation of integral homology groups.*, Math. Algorithms **1** (1966), 27–44.

[Pos78] E. J. Post, *The gaussian interpretation of ampere's law*, Journal of Mathematical Physics **19** (1978), no. 1, 347.

[Pos84] ———, *The metric dependence of four-dimensional formulations of electromagnetism*, J. Math. Phys. **25** (1984), no. 3, 612–613.

[PP62] W. K. H. Panofsky and M. Phillips, *Classical electricity and magnetism*, Addison-Wesley, Reading, MA, 1962, pp. 8–10, 20–23, 125–127.

[Rot55a] J. P. Roth, *An application of algebraic topology to numerical analysis: On the existence of a solution to the network problem*, Proc. Nat. Acad. Sci. U. S. A. **41** (1955), 518–521.

[Rot55b] ———, *The validity of Kron's method of tearing*, Proc. Nat. Acad. Sci. U. S. A. **41** (1955), 599–600.

[Rot59] ———, *An application of algebraic topology: Kron's method of tearing*, Quart. App. Math. **XVII** (1959), no. 1, 1–24.

[Rot71] ———, *Existence and uniqueness of solutions to electrical network problems via homology sequences*, Mathematical Aspects of Electrical Network Theory, SIAM-AMS Proceedings III, 1971, pp. 113–118.

[Rot88] Joseph J. Rotman, *An introduction to algebraic topology*, Springer-Verlag, NY, 1988.

[Sai94] I. Saitoh, *Perturbed H-method without the Lagrange multiplier for three dimensional nonlinear magnetostatic problems*, IEEE Trans. Mag. **30** (1994), no. 6, 4302–4304.

[Sch95] G. Schwarz, *Hodge decomposition: A method for solving boundary value problems*, Lecture Notes in Mathematics, 1607, Springer-Verlag, 1995.

[SF73] Gilbert Strang and George Fix, *An analysis of the finite element method*, Wellesley-Cambridge Press, Wellesley, MA, 1973.

[SF90] P.P. Silvester and R.L. Ferrari, *Finite elements for electrical engineers*, Cambridge U. Press, NY, 1990, 2nd Ed.

[Sle68] P. Slepian, *Mathematical foundations of network analysis*, Springer-Verlag, Berlin, 1968.

[Sma72] S. Smale, *On the mathematical foundations of electrical network theory*, J. Differential Geom. **7** (1972), 193–210.

[Spa66] E. Spanier, *Algebraic topology*, Springer-Verlag, New York, 1966.

[Spi65] Michael Spivak, *Calculus on manifolds. A modern approach to classical theorems of advanced calculus*, W. A. Benjamin, Inc., New York-Amsterdam, 1965.

[Spi79] ———, *A comprehensive introduction to differential geometry. Vol. I*, second ed., Publish or Perish Inc., Wilmington, Del., 1979.

[Spr57] George Springer, *Introduction to Riemann surfaces*, Addison-Wesley, Reading, MA, 1957, Section 5-8.

[SS54] Menahem Schiffer and Donald C. Spencer, *Functionals of finite Riemann surfaces*, Princeton University Press, Princeton, N. J., 1954. MR 16,461g

[SS70] L. M. Sibner and R. J. Sibner, *A non-linear Hodge-de-Rham theorem*, Acta Math. **125** (1970), 57–73.

[SS79] ———, *Nonlinear Hodge theory: Applications*, Adv. in Math. **31** (1979), no. 1, 1–15.

[SS81] ———, *A sub-elliptic estimate for a class of invariantly defined elliptic systems*, Pacific J. Math. **94** (1981), no. 2, 417–421.

[ST80] H. Seifert and W. Threlfall, *A textbook on topology*, Academic Press, 1980, Translated by W. Heil from original 1934 German edition.

[Ste54] A. F. Stevenson, *Note on the existence and determination of a vector potential*, Quart. Appl. Math. **12** (1954), 194–198. MR 16,36a

[Sti93] John Stillwell, *Classical topology and combinatorial group theory, second edition*, Springer-Verlag, NY, 1993, Ch. 3,4.

[Str41] J. A. Stratton, *Electromagnetic theory*, McGraw-Hill, New York, 1941, pp. 227–228.

[Tar02] Timo Tarhasaari, *Mathematical structures and computational electromagnetics*, Ph.D. thesis, Tampere University of Technology, Tampere, Finland, 2002.

[Tei01] F. L. Teixeira (ed.), *Geometric methods for computational electromagnetics*, Progress In Electromagnetis Research, no. 32, Cambridge, MA, EMW Publishing, 2001.

[Tho69] W. Thompson, *On vortex motion*, Transactions of the Royal Society of Edinburgh **XXV** (1869), 217–280.

[Thu97] William P. Thurston, *Three-dimensional geometry and topology*, Princeton University Press, Princeton, New Jersey, 1997.

[Ton68] Enzo Tonti, *Gauge transformations and conservation laws*, Atti Accad. Naz. Lincei Rend. Cl. Sci. Fis. Mat. Natur. (8) **XLV** (1968), 293–300.

[Ton69] ———, *Variational formulation of nonlinear differential equations. II*, Acad. Roy. Belg. Bull. Cl. Sci. (5) **55** (1969), 262–278.

[Ton72a] ———, *A mathematical model for physical theories. I, II*, Atti Accad. Naz. Lincei Rend. Cl. Sci. Fis. Mat. Natur. (8) **52** (1972), 175–181; ibid. (8) **52 (1972), 350–356.**

[Ton72b] ———, *On the mathematical structure of a large class of physical theories*, Atti Accad. Naz. Lincei Rend. Cl. Sci. Fis. Mat. Natur. (8) **52** (1972), 48–56.

[Ton77] ———, *The reason of the analogies in physics*, Problem Analysis in Science and Engineering (F. H. Branin Jr. and K. Husseyin, eds.), Academic Press, New York, 1977.

[Vai64] M. M. Vainberg, *Variational methods for the study of nonlinear operators*, Holden-Day, San Francisco, 1964.

[VB89] A. Vourdas and K. J. Binns, *Magnetostatics with scalar potentials in multiply connected regions*, IEE Proc. A **136** (1989), no. 2, 49–54.

[Vic94] James W. Vick, *Homology theory*, second ed., Springer-Verlag, New York, 1994.

[Wal57] A.H. Wallace, *An introduction to algebraic topology*, vol. 1, Pergamon Press, New York, 1957.

[War71] F.W. Warner, *Foundations of differentiable manifolds and lie groups*, Scott, Foresman and Company, Glenview, Illinois, 1971.

[Wei52] André Weil, *Sur les théorèmes de de Rham*, Comment. Math. Helv. **26** (1952), 119–145.

[Wey23] Hermann Weyl, *Repartición de corriente en una red conductora*, Rev. Mat. Hisp. Amer. **5** (1923), 153–164.

[Wey43] ———, *On Hodge's theory of harmonic integrals*, Ann. of Math. **44** (1943), 1–6.

[Whi37] Hassler Whitney, *On matrices of integers and combinatorial topology*, Duke Math. Jnl. **3** (1937), no. 1, 35–45.

[Whi50] ———, *r-Dimensional integration in n-space*, Proceedings of the International Congress of Mathematicians, vol. 1, 1950.

[Whi57] ———, *Geometric integration theory*, Princeton University Press, Princeton, New Jersey, 1957.

[Yan70] Kentaro Yano, *Integral formulas in Riemannian geometry*, Marcel Dekker Inc., New York, 1970, Pure and Applied Mathematics, No. 1.

Symbols in the list are sometimes also used temporarily for other purposes...

G. H. Hardy and E. M. Wright, *An Introduction to the Theory of Numbers*, 5th ed.

Summary of Notation

Because the material in the book draws on a variety of fields, there are some resulting conflicts or ambiguities in the notation. In general, these ambiguities can be cleared up by context, and the authors have attempted to avoid situations where like notation overlaps in the same context. Some examples are:

(1) The symbol $[\cdot,\cdot]$ can have three meanings: bilinear form, commutator, and homotopy classes of maps.

(2) χ can have three meanings: Euler characteristic, stream function for a surface current, or a gauge function.

(3) π can be a permutation map or the ratio of circumference to diameter of a circle. In addition, π_k signifies the kth homotopy group, while π_1^k signifies the kth term in the lower central series of the fundamental group.

(4) R can be a resistance matrix, the de Rham map, or a region in $\mathbb{R}^3$.

(5) Pullbacks and pushforwards of many varieties can be induced from a single map. For example, an inclusion map i can induce maps denoted by i^*, i_*, $i^\#$, $i_\flat$, $\tilde{\imath}$, etc.

(6) The symbols ξ, λ, α, β, η, ρ, ν, and θ have meanings particular to Chapter 7 (see Figure 7.4, page 211).

Other multiple uses of notation are noted below.

$\beta_p(R)$	pth Betti number = Rank $H_p(R)$
δ^{ij}	Kronecker delta; 1 if $i = j$, 0 otherwise
δ	Inner product space adjoint to the exterior derivative
δ	Connecting homomorphism in a long exact sequence
∂	Boundary operator
∂^T	Coboundary operator.
$\breve{\partial}$	Boundary operator on dual mesh (related to ∂^T)
ε	Dielectric permittivity
ζ_j^i	jth 1-cocycle on dual mesh, indexed on 1-cells of DK: $1 \le i \le \breve{m}_1$

η	Wave impedance
θ	Normalized angle of $f : R \longrightarrow S^1$
θ^i_j	θ discretized on nodes of unassembled mesh
λ	Wavelength
λ_i	Barycentric coordinates, $1 \leq i \leq 4$
μ	Magnetic permeability
π	Ratio of circumference to diameter of a circle
π	Permutation map
π_i	ith homotopy group (but π_0 distinguishes path components and is not a group)
ρ	Volume electrical charge density
σ	Electrical conductivity
σ_s	Surface electrical charge density
$\sigma_{p,i}$	ith p-simplex in a triangulation of R
τ_e	Dielectric relaxation time, $\tau_e = \varepsilon/\sigma$
Φ_i	ith magnetic flux
ϕ	Electric scalar potential
χ	Euler characteristic
χ	Stream function for surface current distribution
χ	Gauge function
χ_e, χ_m	Electric and magnetic susceptibilities
ψ	Magnetic scalar potential
ψ^+ (ψ^-)	Value of ψ on plus (minus) side of a cut
ω	Radian frequency
Ω	Subset of $\mathbb{R}^n$
$\boldsymbol{A}$	Magnetic vector potential
$\boldsymbol{B}$	Magnetic flux density vector
$B^p(K;R)$	p-coboundary group of K with coefficients in module R
$B_p(K;R)$	p-boundary group or K with coefficients in module R
$B^p(K,S;R)$	Relative p-coboundary group of K (relative to S)
$B_p(K,S;R)$	Relative p-boundary group of K (relative to S)
$B^p_c(M-S)$	Relative exact form defined via compact supports; $S \subset \partial M$
$\tilde{B}_p(M,S_1)$	Coexact p-forms in $\tilde{C}_p(M,S_1)$
c	Speed of light in a vacuum, $(\varepsilon_0\mu_0)^{-1/2}$
c	Curve (or contour of integration)
c_p	p-chain
c^p	p-cochain
curl	Curl operator
$\overline{\text{curl}}$	Adjoint to the curl operator in two dimensions
C	Capacitance matrix
C	Constitutive law (see Figure 7.4)

C^i_{jk}	Connection matrix, $1 \le i \le m_3$, $1 \le j \le 4$, $1 \le k \le m_0$
$C^i_{p,jk}$	Connection matrix of p-dimensional mesh
$C^p(K;R)$	p-cochain group of K with coefficients in module R
$C_p(K;R)$	p-chain group or K with coefficients in module R
$C^p(K,S;R)$	Relative p-cochain group of K (relative to S)
$C_p(K,S;R)$	Relative p-chain group or K (relative to S)
$C^p_c(M-S)$	Differential forms with compact support on $M-S$; $S \subset \partial M$
$\tilde{C}_p(M,S_1)$	p-forms in the complex defined by δ, the formal adjoint of d in $C^*_c(M-S_2)$
d	Coboundary operator; exterior derivative
d	Thickness of current-carrying sheet
div	Divergence operator
div_S	Divergence operator on a surface
D	Differential operator
D, δ	Skin depth
$\boldsymbol{D}$	Electric displacement field
DK	Dual cell complex of simplicial complex K
$\boldsymbol{E}$	Electric field intensity
E_M	Magnetic energy
f_p	"Forcing function" associated with the pth cut (a vector with entries f_{pi})
f	Frequency
f	Generic function
$f^*(\mu)$	Pullback of μ by f
F	Rayleigh dissipation function
F_p	Free subgroup of pth homology group
F^p	Free subgroup of pth cohomology group
F	Primary functional
$F^\perp$	Secondary functional needed for convexity
F^s_0	Number of FLOPs per CG iteration for node-based interpolation of scalar Laplace equation
F_0	Number of FLOPs per CG iteration for node-based vector interpolation
F_1	Number of FLOPs per CG iteration for edge-based vector interpolation
$\mathcal{F}$, $\mathcal{G}$	Spaces of vector fields with elements F and G, respectively
grad	Gradient operator
G	Convex functional
$\boldsymbol{H}$	Magnetic field intensity
$H^p(R;\mathbb{Z})$	pth cohomology group of R with coefficients in $\mathbb{Z}$
$H_p(R;\mathbb{Z})$	pth homology group of R, coefficients in $\mathbb{Z}$

$H^p(R, \partial R; \mathbb{Z})$	pth cohomology group of R relative to ∂R, coefficients in $\mathbb{Z}$
$H_p(R, \partial R; \mathbb{Z})$	pth homology group of R relative to ∂R, coefficients in $\mathbb{Z}$
$H^p_c(M - S)$	$Z^p_c(M - S)/B^p_c(M - S)$; harmonic forms
$\mathcal{H}^p(M, S_1)$	$\tilde{Z}^p_c(M, S_2) \cap Z^p(M - S_1)$; harmonic fields
i	inclusion map
im	Image of map
I	Electrical current
I_i	ith current
I_f,I_p	Free and prescribed lumped-parameter currents
Int $(\cdot, \cdot)$	Oriented intersection number
$\mathcal{I}$	Intersection number matrix
$\mathcal{I}_p(m, l)$	Indicator function, $1 \leq p \leq \beta_1(R)$, $1 \leq m \leq 4$, $1 \leq l \leq m_3$
j	Map inducing a third map in a long exact sequence
$\boldsymbol{J}$	(Volumetric) current density vector
$\boldsymbol{J}_{\text{av}}$	Average current density in effective depth of current sheet
$\mathcal{J}^i_j \in \mathbb{Z}$	Nodal jumps on each element, $1 \leq i \leq m_3$, $1 \leq j \leq 4$
ker	Kernel of map
$\boldsymbol{K}$	Surface current density vector
K	Simplicial complex
$\mathcal{K}^k_{mn}$	Stiffness matrix for kth element in mesh
$\mathcal{K}$	Global finite element stiffness matrix
l_{max}	"Characteristic length" of electromagnetic system
L	Inductance matrix
L	Lagrangian
$L^2\Lambda^q(X)$	Space of square-integrable differential q-forms on manifold X
Link$(\cdot, \cdot)$	Linking number of two curves
m_p	Number of p-simplexes in a triangulation of R
$\breve{m}_p$	Number of p-cells in dual complex
$\boldsymbol{M}$	Magnetization
M	Manifold
n_c, n_v	Number of prescribed currents and number of prescribed voltages
n_p	Number of p-simplexes in a triangulation of ∂R
$\boldsymbol{n}$	Normal vector to a codimension 1 surface
$\boldsymbol{n}'$	Normal to a two-dimensional manifold with boundary embedded in $\mathbb{R}^3$
nz(A)	Number of nonzero entries of a matrix A
$\mathcal{O}(n^\alpha)$	Order n^α
$\boldsymbol{P}$	Polarization density
$\boldsymbol{P}$	Poynting vector
P	Period matrix

P_J	Eddy current power dissipation
Q_i	ith charge
R	Resistance matrix
R	de Rham map, $R : L^2\Lambda^q(X) \rightarrow C^q(K)$ (K a triangulation of X)
R	Region in $\mathbb{R}^3$, free of conduction currents
$\tilde{R}$	Three-dimensional manifold with boundary, subset of $\mathbb{R}^3$
R_S	Surface resistivity
S	Surface
S', S'_{ck}	Current-carrying surface after cuts for stream function have been removed, and the kth connected component of S'
S_q	qth cut
S^1	Unit circle, $S^1 = \{p \in \mathbb{C} \mid \ \lvert p \rvert = 1\}$
T	Kinetic energy
$\boldsymbol{T}$	Vector potential for volumetric current distributions
T_p	Torsion subgroup of pth homology group
T^p	Torsion subgroup of pth cohomology group
T^*	Cotangent space
u_k	Nodal potential, $1 \leq k \leq m_0$
v	vertex
V	Voltage
V_j	Prescribed voltage, $1 \leq i \leq n_v$
V	Potential energy
w_e	Electric field energy density
w_m	Magnetic field energy density
W	Whitney map $W : C^q(K) \rightarrow L^2\Lambda^q(X)$
W_e	Electric field energy
W_m	Magnetic field energy
X	Riemannian manifold
X_0^s	# nonzero entries in stiffness matrix for node-based scalar interpolation
X_0	# nonzero entries in stiffness matrix for node-based vector interpolation
X_1	# nonzero entries in stiffness matrix for edge-based vector interpolation
$\bar{z}$	Complex conjugate of z
z^p	p-cocycle
z_p	p-cycle
$Z^p(K; R)$	p-cocycle group of K with coefficients in module R
$Z_p(K; R)$	p-cycle group or K with coefficients in module R
$Z^p(K, S; R)$	Relative p-cocycle group of K (relative to S)

$Z_p(K, S; R)$	Relative p-cycle group or K (relative to S)
$Z^p_c(M - S)$	Relative closed form defined via compact supports; $S \subset \partial M$
$\tilde{Z}_p(M, S_1)$	Coclosed p-forms in $\tilde{C}_p(M, S_1)$
$S^{+(-)}$	Positive (negative) side of an orientable surface with respect to a normal defined on the surface
A^c	Set-theoretic complement of A
$[A, B]$	Homotopy classes of maps $f : A \to B$, i.e. $\pi_0(\text{Map}(A, B))$
$[\cdot, \cdot]$	Homotopy classes of maps
$[\cdot, \cdot]$	Commutator
$[\cdot, \cdot]$	Bilinear pairing
$[\cdot]$	Equivalence class of element $\cdot$
$\wedge$	Exterior multiplication
$*$	Hodge star
$\cap$	Set-theoretic intersection
$\cup$	Set-theoretic union
$\cup$	Cup product

Examples and Tables

Index